Making the Palace Machine Work

Making the Palace Machine Work

Mobilizing People, Objects, and Nature in the Qing Empire

Edited by
Martina Siebert,
Kai Jun Chen, and
Dorothy Ko

Routledge
Taylor & Francis Group
LONDON AND NEW YORK

First published in 2021 by Amsterdam University Press Ltd.

Published 2025 by Routledge
4 Park Square, Milton Park, Abingdon, Oxon OX14 4RN
605 Third Avenue, New York, NY 10158

Routledge is an imprint of the Taylor & Francis Group, an informa business

ISBN: 9789463720359 (hbk)
ISBN: 9781041182351 (pbk)
ISBN: 9781003699187 (ebk)
NUR 692

Cover illustration: Artful adaptation of a section of the 1750 *Complete Map of Beijing of the Qianlong Era* (Qianlong Beijing quantu 乾隆北京全圖) showing the Imperial Household Department by Martina Siebert based on the digital copy from the Digital Silk Road project (http://dsr.nii.ac.jp/toyobunko/II-11-D-802, vol. 8, leaf 7)

Cover design: Coördesign, Leiden

DOI: 10.5117/9789463720359

For Product Safety Concerns and Information please contact our EU representative:
GPSR@taylorandfrancis.com
Taylor & Francis Verlag GmbH, Kaufingerstraße 24, 80331 München, Germany

Table of Contents

List of Figures, Tables, Charts, and Maps

Figures

Maps

Map 1a Imperial City

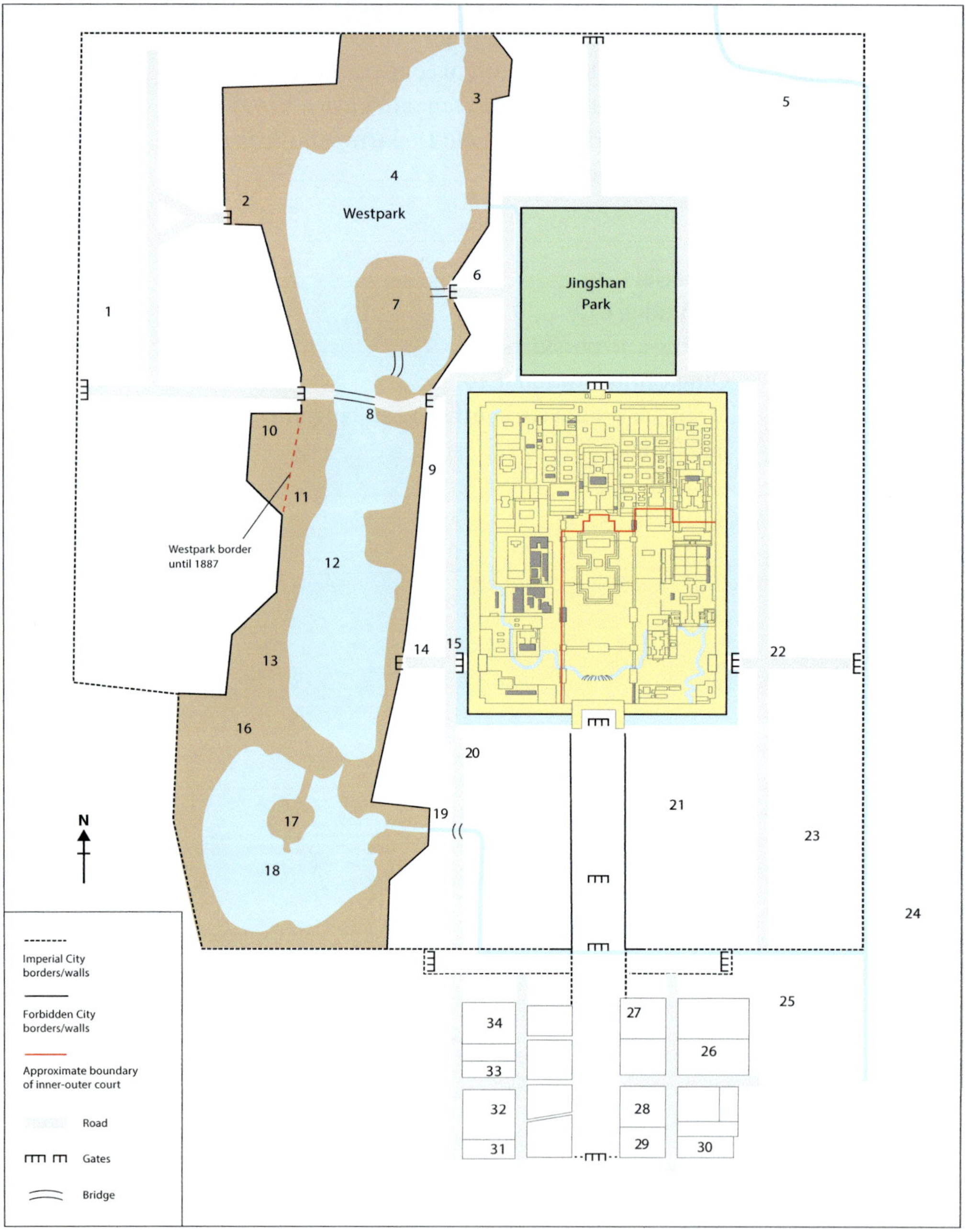

Map by Mary Yang

Map 1b Forbidden City

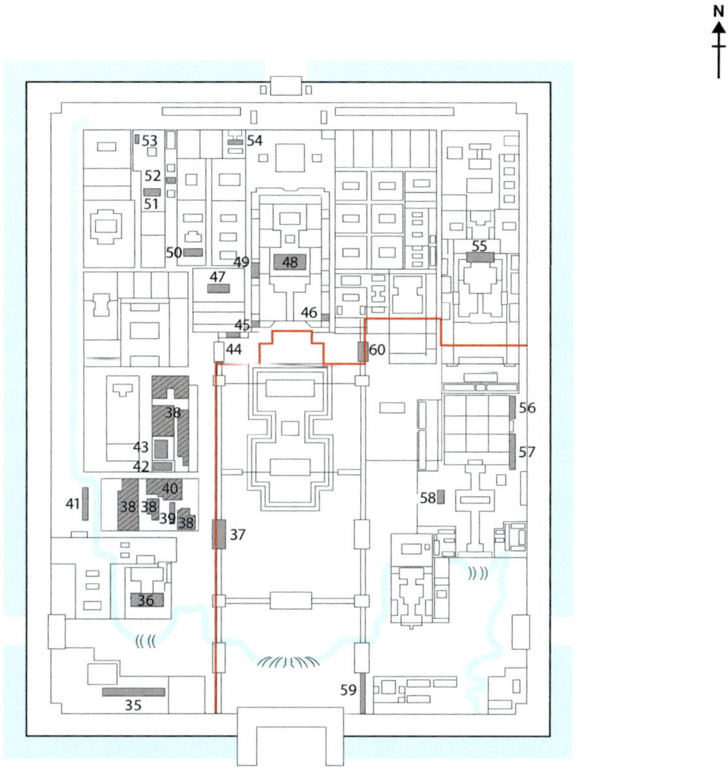

1 Catholic church at Xishiku
2 Sandalwood Temple
3 Temple of Sericulture
4 North Lake
5 Imperial Bureau of Weaving and Dying
6 Office of Inspection and Censors for the Affairs at the Imperial Household Department
7 White Stupa
8 Gold-Turtle Jade-Rainbow Bridge
9 Office of Control and Punishment
10 Catholic church at Canchikou
11 Pavilion of Purple Light
12 Middle Lake
13 Yiluan Palace
14 Bureau of Imperial Gardens and Parks
15 Inner Department of Works
16 Garden of Abundant Beneficence
17 Yingtai
18 South Lake
19 Office of Insignia
20 Court of Treasury of the Imperial Workshops
21 Imperial Ancestral Temple
22 Imperial Armory
23 Textile Vault (of Board of Revenue)
24 Ministry for Ruling Outer Provinces
25 Imperial Chariot Garage
26 Board of Work
27 Court of Imperial Clan
28 Board of Revenue
29 Board of Rites
30 Court of Imperial Academy of Medicine
31 Board of Punishment
32 Court of Censors
33 Court of Imperial Sacrifice
34 Imperial Procession Guard
35 Ceramic Vaults
36 Wuying Palace
37 Hongyi Pavilon
38 Courtyards of the Imperial Workshops
39 Silver Vault
40 Courtyard of the Imperial Household Department
41* Grand Storage Office
42 Treasury of the Imperial Workshops
43 Audit Bureau and Document House
44 Grand Council
45 Southern Study
46 Imperial Dispensary
47 Yangxin Palace
48 Qianqing Palace
49 Maoqin Hall
50 Qixiang Palace
51 Zhongzheng Hall
52 Jianfu Palace
53 Bilin Lodge
54 Hall of Zhengyi mingdao
55 Ningshou Palace
56 Medicinal Warehouse
57 Imperial Academy of Medicine duty office
58 Imperial Stud
59 Archives of the Grand Secretariat
60 Jingyun Gate

* The map above shows the position of the Grand Storage Office according to the Qianlong map. An internal map of the Beijing Palace Museum locates the office within the Courtyard of the Imperial Household Department (no. 40).

Map 2 Beijing and outskirts

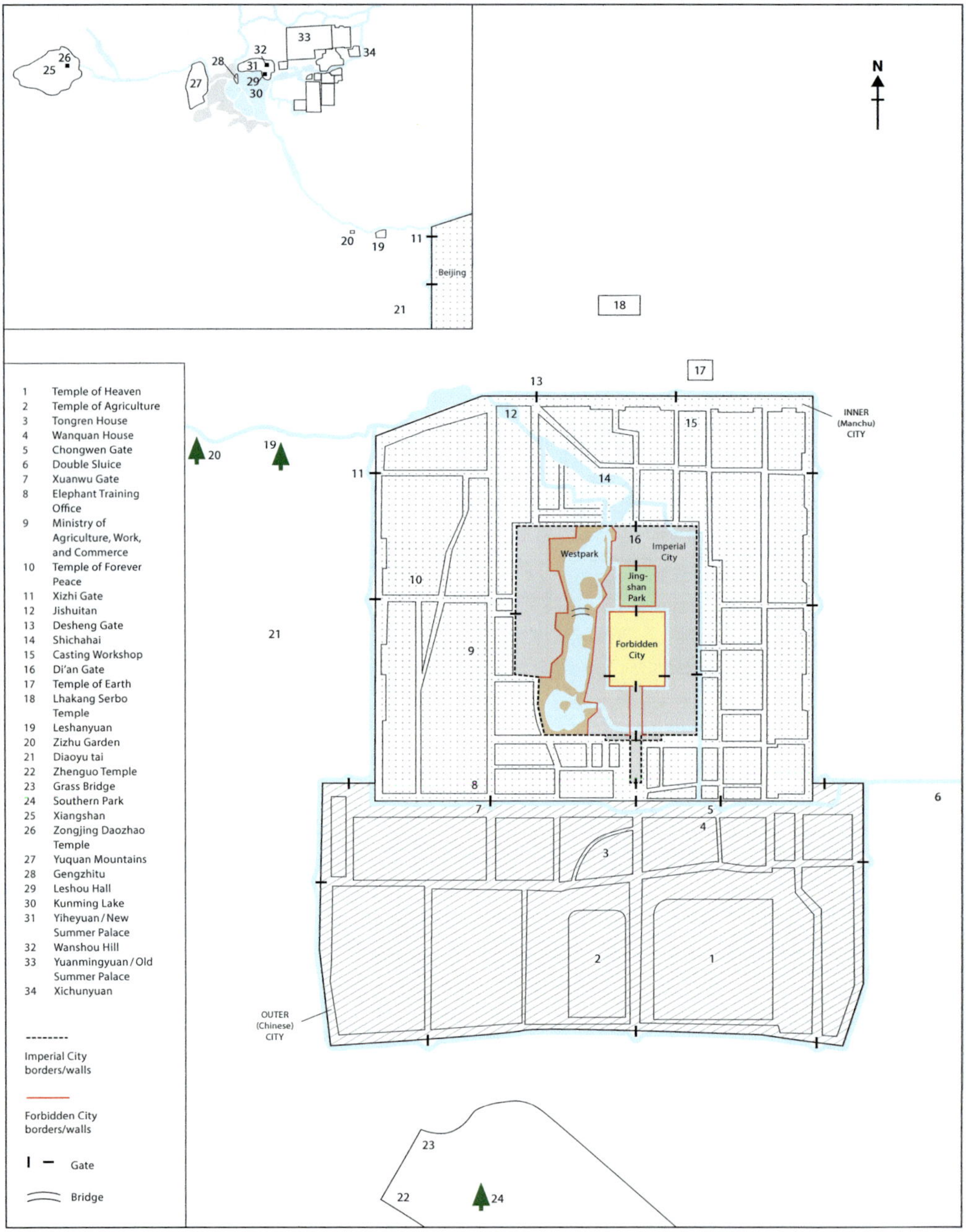

Map 3 Qing China around 1840

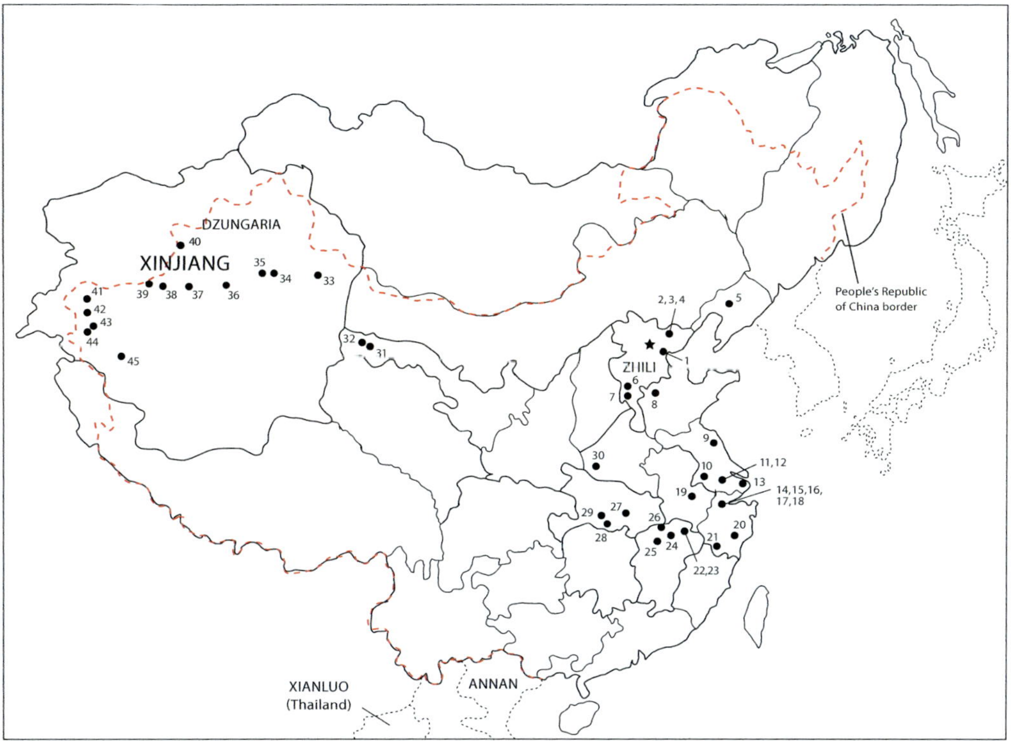

Map by Mary Yang

- 1 Tanggu (near Tianjin)
- 2 Putuo Zongcheng Temple (Chengde)
- 3 Rehe (present Chengde)
- 4 Xumi Foshou Temple (Chengde)
- 5 Shenyang / Mukden
- 6 Neiqui county (Zhili)
- 7 Cizhou (Zhili)
- 8 Dong'e county (Shandong)
- 9 Huai'an Customs House
- 10 Jiangning (present Nanjing)
- 11 Imperial Textile Manufactory (Suzhou)
- 12 Suzhou (Jiangsu)
- 13 Yangshupu Garden (in Shanghai)
- 14 Tiger Grotto kilns
- 15 Turtle Hill kilns
- 16 Jiaotanxia kilns
- 17 Lin'an (present Hangzhou)
- 18 Renhe County (Zhejiang)
- 19 Wuhu (Anhui)
- 20 Chuzhou (Zhejiang)
- 21 Longquan kilns
- 22 Porcelain Bureau in Fuliang County
- 23 Imperial Ceramic Manufactory (Jingdezhen)
- 24 Raozhou prefecture (Jiangxi)
- 25 Nanchang Government Repository
- 26 Jiujiang Customs House
- 27 Wuhan (Hubei)
- 28 Jiangling (Hubei)
- 29 Jingzhou (Hubei)
- 30 Junzhou (present Henan)
- 31 Suzhou (Gansu)
- 32 Jiayu gate (Gansu)
- 33 Hami (Xinjiang)
- 34 Pizhan (Xinjiang)
- 35 Turfan (Xinjiang)
- 36 Karashar (Xinjiang)
- 37 Kucha (Xinjiang)
- 38 Aksu (Xinjiang)
- 39 Ush (Xinjiang)
- 40 Yili (Xinjiang)
- 41 Kashgar (Xinjiang)
- 42 Ying-a-za-er (Xinjiang)
- 43 Yarkand (Xinjiang)
- 44 Mi'erdai mountains (Xinjiang)
- 45 Hotan (Khotan)
- ★ Beijing

Acknowledgments

The editors would like to thank the contributors for their patience and perseverance through several rounds of editing during the long process of preparing this volume. Most contributors have been with us since 2015 and took part in one or both workshops organized by Kai Jun Chen and Martina Siebert during their time as post-doc and senior researcher respectively at the Max Planck Institute for the History of Science in Berlin. Although Professor Xiaodong Xu's paper is not included in this volume, her contributions at the workshops were invaluable, as were those of the other workshop participants. We also want to thank the three anonymous reviewers who invested time and energy identifying quirks and flaws, helping us to turn it into a coherent and readable volume. The support of the Weatherhead Institute of East Asian Studies at Columbia University makes possible the availability of the volume in open access; the Dean of Faculty and the Department of East Asian Studies of Brown University provided funds towards translation and preparation of the manuscript. Saskia Gieling of AUP guided us professionally through the various stages of the publication process.

Conventions for the Notation of Time, Weights, and Measures[1]

Chronology of Dynasties and Periods

Xia dynasty, ca. 2070-1600 BCE
Shang dynasty, 1600-ca. 1945 BCE
Zhou dynasty, ca. 1045-256 BCE
    Western Zhou, ca. 1045-771 BCE
    Eastern Zhou, 771-256 BCE
Spring and Autumn period, 770-476 BCE
Warring States period, 476-221 BCE
Qin dynasty, 221-206 BCE
Han dynasty, 206 BCE-220 CE
    Western Han, 206 BCE-8 CE
    Eastern Han, 25-220
Three Kingdoms period, 220-280
Six Dynasties period, 265-589
    Western Jin, 265-316
    Eastern Jin, 317-420
    Southern and Northern dynasties, 420-589
Sui dynasty, 581-618
Tang dynasty, 618-907
Five Dynasties and Ten Kingdoms period, 907-979
Song dynasty, 960-1279
    Northern Song, 960-1127
    Southern Song, 1127-1279
Yuan (Mongol) dynasty, 1271-1368
Ming dynasty, 1368-1644

1 The information here is intended as a quick guide for the convenience of the reader. The beginning and end dates of dynasties are controversial, tied as they are to claims of legitimacy. The values of units of measure also varied in time and space. The conversions provided here are approximations and may differ from those used by the authors of individual chapters. The dynasty dates and modern conversions of measurements are in part adopted from Christine Moll-Murata, *State and Crafts in the Qing Dynasty, 1644-1911* (Amsterdam University Press, 2018), 11-13 and Endymion Wilkinson, *Chinese History: A New Manual* (Cambridge, Ma.: Harvard University Asia Center, 2013), 551-562.

Qing dynasty (1644-1911) reign names[2]

Shunzhi, 1644-1661
Kangxi, 1662-1722
Yongzheng, 1723-1735
Qianlong, 1736-1795
Jiaqing, 1796-1820
Daoguang, 1821-1850
Xianfeng, 1851-1861
Tongzhi, 1862-1874
Guangxu, 1875-1908
Xuantong, 1909-1911

Manchukuo (*Manzhouguo*), 1932-1945
Republic of China, 1912-1949
People's Republic of China, 1949-
Republic of China (Taiwan), 1949-

Lengths
chi 尺: foot: a unit of length; about 32 cm (12.6 inches) in the Ming-Qing period = 10 *cun* 寸 (inches)
zhang 丈: a unit of length = 10 *chi*

Land Measures
mu 畝: a land measure; official value in the Ming = 608 sq. m., or 6.5 *mu* per acre. Official value in the Qing = 614 sq. m., or 6.5 *mu* per acre

Weights and Volume
The present book uses the common English translations of *liang* 兩 (tael) and *jin* 斤 (catty) in lieu of the Chinese. All other units of measures appear in Chinese:
jin 斤 (catty in the volume) = 16 *liang* 兩 = 160 *qian* 錢; about 596.8 g (21.2 oz) in the Ming-Qing period
liang 兩 (tael in the volume) = 10 *qian* 錢; approximately 37.3 kg
qian 錢 (sometimes rendered as 'mace'); approximately 3.73 g
Shi 石 can be a measure of weight or volume.

2 Authors in this volume refer to a specific year of a reign period in three ways: (1) spelling out fully; for example, 17th year of the Qianlong reign (1752), (2) abbreviating to Qianlong 17 (1752) or, (3) in the footnotes, QL17 (1752). Following Chinese conventions, the full date of an archival source is given in this order: reign year/lunar month/day. For example, Qianlong (or QL) 26/7/4; an 'r' before the month indicates an intercalary month (*runyue* 閏月). Western equivalents are given in the order of day, month, year. For example, 3 August (or Aug.) 1761.

Shi as a measure of weight:
1 *shi* = 120 catties; approximately 71.6 kg (159 lbs)
Shi as a measure of volume:
1 *shi* of rice = 138.75 catties; approximately 82.8 kg

Currencies
liang 兩 (tael) of unminted silver; 100% purity (unless otherwise stated) = 10 *qian* 錢 = 100 *fen* 分 = 1,000 *li* 釐.
Kuping liang 庫平兩 (Treasury ounce): imperial standard set by the Board of Revenue, used for official accounts = 37.31 g. Many other regional *liang* weights existed.
wen 文 (cash), brass coin of copper-lead-zinc-tin alloy, with a nominal standard exchange rate of 1,000 cash to one tael of silver. Many local variances of exchange rates existed.
diao 吊 (string) = 1,000 *wen* (cash). The cash coins were bound with strings in ten bundles of a hundred each, which was sometimes accounted as strings (*diao*) of 1,000.

Note on Translation

Following East Asian conventions, the names of Chinese and Manchu people are given by their family name first (Tang Ying). The only exceptions are the names of authors of the present volume (Yijun Wang).

Well-known Manchu and non-Han individuals are referred to by their common English translated names (Gao-pu; Fuk'anggan), whereas others are transliterations in *pinyin* of their names in Chinese, as is given in the archival documents, without short dashes in between syllables (Yiling'a).

Chinese proper names, office titles, and technical terms in each chapter are given in *pinyin* romanization followed by Chinese characters at first appearance. All frequently used terms and titles are also translated into English (*Neiwufu* 內務府, Imperial Household Department). Henceforth, only the English is used. The English translations of office titles and sources have been harmonized throughout the volume and are those of the editors.

The titles of non-English books and articles in the footnotes of each chapter are given without English translations. The only exceptions are the chapter or entry titles from anthologies or collectanea that are not individually listed in the bibliography. All other titles of non-English books and articles are translated into English in the bibliography and have been harmonized as far as possible.

Note on the Frontmatter Maps and Cover Image

The four maps at the front of the volume present the locations of the main palaces, offices of the Imperial Household Department and Qing administration, as well as geographical features mentioned in the chapters. All four maps are abstractions aiming to give readers an idea of the whereabouts of locations and are by nature selective. The two most pertinent sources for Maps 1a-b and Map 2, of the Imperial and Forbidden City and Qing Beijing, are the geo-referenced *Complete Map of Beijing of the Qianlong Era* (Qianlong Beijing quantu 乾隆北京全圖) of 1750 made available by the Digital Silk Road project of the National Institute of Informatics in Tokyo[1] and an annotated map of the Forbidden City compiled for internal use by the Department of Architectural Heritage (*Gujian bu* 古建部) of the Palace Museum in Beijing in 2002. In addition, two Prussian maps of Beijing produced in 1907 and 1914[2] as well as a Qing manuscript map of Beijing[3] and a number of maps in scholarly publications[4] have been consulted. The research for Map 3, of Qing China around 1840, relied largely on the Harvard China Historical GIS project.[5]

Readers should be aware that the locations of Qing offices could move, and that the warehouses mentioned in the archival sources are difficult to

1 http://dsr.nii.ac.jp/toyobunko/II-11-D-802/ presenting the Toyo Bunko facsimile of the original map made in the 1940ies. The cover image is an artful adaptation by Martina Siebert of a detail of leaf 7 in volume 8 of this map.

2 *Beijing quantu* 北京全圖 = *Peking*, compiled by Deutsch Ostasiatisches Expeditions Korps. Feldtopograph, Prussia (Germany), 1914, Library of Congress: http://hdl.loc.gov/loc.gmd/g7824b.ct001942 (accessed October 2020) and *Peking und Umgebung*, Berlin: Reichsamt für Landesaufnahme, 1907, Library of Congress: http://hdl.loc.gov/loc.gmd/g7824b.ct001944 (accessed October 2020)

3 *Jingshi jiucheng quantu* 京师九城全圖 (Complete map of the nine inner cities of Beijing), Library of Congress: http://hdl.loc.gov/loc.gmd/g7824b.ct001960 (accessed October 2020).

4 'Diagram of the Imperial Palace', in Preston Torbert, *The Ch'ing Imperial Household Department: A Study of Its Organisation and Principal Functions, 1662-1796* (Cambridge, MA: Harvard University Press, 1977), 31; 'The Peking Palace inner and outer courts in the seventeenth and eighteenth centuries', in Beatrice S. Bartlett, *Monarchs and Ministers: The Grand Council in Mid-Ch'ing China, 1723-1820* (Berkeley: University of California Press, 1991), 14; 'Etablissements Catholiques de Pékin, 1950', in Alan Richard Sweeten, *China's Old Churches The History, Architecture, and Legacy of Catholic Sacred Structures in Beijing, Tianjin, and Hebei Province* (Leiden: Brill, 2019), 70.

5 The China Historical Geographical Information System, http://chgis.fas.harvard.edu/ (accessed October 2020).

locate as they sometimes existed in multiples on different administrative levels; their physical locations could also change over time. In addition, the Qianlong 1750 map and the 2002 map are both specific to their times, i.e. of Beijing under early- to mid-Qianlong rule and of the Forbidden City in a more recent academic reconstruction of its late-Qing layout.

The cover image of this volume, a detail of the Qianlong map showing the Imperial Household Department and the Imperial Workshops, draws attention to this mutability of places and their function by highlighting the damages and cracks that the map has suffered in the 250 years since its production. That some of the pertinent information on the map is erased by time serves as a useful reminder that our knowledge of the Qing palace machine is not – and may never be – complete. But it also suggests that with the materials we do have at hand, it is still possible to piece together the lively picture of the Qing palace presented in this volume.

We thank graphic designer Mary Yang of Open Rehearsal for contributing her expertise and making the frontmatter maps.

Introduction

Martina Siebert, Kai Jun Chen, and Dorothy Ko

The Qing, the last dynasty of the Chinese imperium, ruled for over 260 years (1644-1911). At the end of the 19th century it occupied a territory of roughly 13 million square kilometres and claimed sovereignty over more than 400 million people. One of the questions this book examines is how – on a sheer logistical level – was a complex empire of this size governed before the age of telegrams, telephones, and internet? Instead of looking to the Qing emperor, often perceived as an autocratic Son of Heaven who exercised absolute power over his subjects, our inquiry begins with the palace compound itself, in the heart of the capital city, Beijing.

Behind the deliberate staging of splendour and order, imperial palaces were complex and mobile structures with mundane functions that required diverse strands of management: from long-term fiscal planning to chores like sweeping the floors, from the choreography of state rituals to the provisioning of daily meals to princesses and servants. The minutiae of these tasks had to be coordinated on multiple levels so that the palace could function effectively as a unit. The administrative organization of the imperial palace of the Qing dynasty provides a telling example of how representational, religious, diplomatic, and day-to-day activities were planned and executed. An abundance of archival and material sources, coupled with eyewitness reports, has facilitated our investigation into the inner workings of the Qing palace. The chapters of this volume take readers inside the public halls, private chambers, and treasure vaults of the Qing palace, to explore the secrets of its operation.

The 'palace' means many things, both tangible and intangible, to the editors and authors of this volume: the buildings, gardens, and bodies of water in and near the Forbidden City; the rules and regulations that defined the duties of personnel and workflows; the paper trails tracking the circulation of materials and monies; and – last but not least – the people, plants, and animals whose lives were entangled with the life of the court, such as high officials and hereditary bannermen, lowly workers and maids, elephants,

Siebert, Martina, Kai Jun Chen, and Dorothy Ko (eds), *Making the Palace Machine Work: Mobilizing People, Objects, and Nature in the Qing Empire*. Taylor & Francis Group, 2021
DOI: 10.5117/9789463720359_INTRO

medicinal herbs, and lotus plants. Thus construed, the Qing palace was not just an architectural compound, nor can it be reduced to its bureaucratic or social dimensions, as is often assumed. This volume proposes that the Qing palace was a shifting material and social assemblage that was made and unmade by daily routines and ad hoc adjustments. Building on Latour's concept of 'networks of actants' and Bennett's concept of 'assemblage', this book focuses on an early modern, non-Western case, to refine our understanding of the power hierarchy and dynamic operation in this network or assemblage of human and non-human agents.[1]

The Palace as Machine

Our interest in mapping the dynamic operations and ad hoc everyday decisions in the palace network or assemblage is, by definition, a post-structuralist one. This interest in moving parts, rather than a static structure, is expressed in our choice of the metaphor of 'machine' in describing the Qing palace in the book's title. Structure *does* matter in a machine. After all, routine decisions had to be made within the parameters of established imperial rites and decrees as well as the financial resources available, which were all steered by the empire's ideological and political agenda. But structures are revealed, if not forged, only through routine operations. Over time, some expedient measures would become the new established practice. Thus, the dynasty administration was like a machine wrought of component parts that move, like rotating gears. In using the model of a 'machine' to explore the workings of the Qing palace, and by extension of the Qing empire, this book highlights the dynamic and material nature of politics, the power of logistics, and the contributions from lowly workers and non-human actors that made the Qing empire work.

Historians of engineering generally talk about two kinds of machines: the simple Archimedes machines, and the complex industrial and post-industrial machines. The simple Archimedean machines provide a 'mechanical advantage' by using a lever, wheel and axle, pulley, inclined plane, wedge, or screw to multiply force.[2] In contrast, complex machines use what the 19th-century

1 Bruno Latour, *Reassembling the Social: An Introduction to Actor-Network-Theory* (Oxford; New York: Oxford University Press, 2005); Jane Bennett, *Vibrant Matter: A Political Ecology of Things* (Durham, NC: Duke University Press, 2010). For a global comparison of imperial palaces, see Jeroen Duinda, et al., eds., *Royal Courts in Dynastic States and Empires: A Global Perspective* (Leiden; Boston: Brill, 2011).

2 Isaac Asimov, *Understanding Physics* (New York: Barnes and Noble, 1966), 88.

engineer Franz Reuleaux (1829-1905) conceived of as 'kinetic chains'. He asserted that a 'machine is a combination of *resistant bodies* so arranged that by their means the mechanical forces of nature can be compelled to do work accompanied by certain *determinate motions*' (author's emphasis).[3] These 'resistant bodies' and 'determinate motions' are contradictory forces that work in tension and in tandem to accomplish tasks. Other thinkers use 'the machine' as a metaphor to highlight the open-ended quality of all operations.[4] All of these meanings are applicable to the Qing palace machine, as this anthology will reveal.

Using the model of a palace-as-machine affords several new insights. First, it calls attention to the enormous physical presence and scale of the Qing palace compound, in the same way that turbine machines can occupy monumental buildings. Anthropologist Michael Meeker highlights the architecture of the Ottoman Palace as an 'instrument of edification as well as of government ... designed to confound and enthral, if not intimidate and terrify'. He sees a building itself as an instrument of power that 'mediates between the eye of its resident [the ruler] and the imperium beyond its walls'. The layout of windows, balconies and towers constitutes a 'device of omnipresent inspection and regulation' and provides an object lesson in 'internalizing a discipline' for imperial subjects.[5] The long, meandering corridors of the Forbidden City in Beijing and wide empty grounds in front of the main buildings might have the same intimidating effect on officials and visitors, as viewers of Bertolucci's film 'The Last Emperor' may recall. In that film, the palace buildings as brick-and-mortar extensions of Puyi's body, the source of his political persuasion, also marked the depth of his psychic isolation and physical confinement. This volume shows how the Qing palace machine was both material and fluid, at once constraining and enabling for its operators – including the emperor himself.

A second insight afforded by the metaphor of the Qing palace as a machine concerns the nature of its infrastructure and power. A machine is comprised of functional parts designed to be combined into more complex

3 Franz Reuleaux, *Kinematics of Machinery: Outline of A Theory of Machines*, translated and edited by Alexander B.W. Kennedy (New York: Dover Publications, 1963), 35.

4 One notable thinker in this regard is Gilbert Simondon. See the collection of reflections on Simondon's philosophy in *Gilbert Simondon: Being and Technology*, eds. Arne De Boever, Alex Murray, Jon Roffe and Ashley Woodward (Edinburgh: Edinburgh University Press, 2012); see especially the explanations on the pages 208 and 216-217 in the 'Glossary of key terms' on 'Automaton / Open machine' and 'Machine'.

5 Michael E. Meeker, *A Nation of Empire: The Ottoman Legacy of Turkish Modernity* (Berkeley: University of California Press, 2001). Quotations cited are from pp. 118, 120, 131.

wholes which, once combined, enable the performance of high-level tasks. Conceiving the Qing palace as such a complex constellation of interrelated parts allows scholars to alternate their analytic lens between the micro- and macro-levels with greater ease than that afforded by traditional approaches to institutional history. This is especially the case for one key institution examined in this volume, the Imperial Household Department, which was both the 'brain' that planned the myriad daily tasks and the 'hands' that executed them. By focusing on this key institution, the volume seeks to illuminate the design principles of the palace as a whole. Exploring the logic, logistics, and interaction of these parts – among themselves and as part of the whole – provides a novel glimpse inside the operational structure and sequence of the Qing palace machine.

The findings are striking when placed in a comparative frame between early modern China and Europe. Readers will find that the Qing palace machine, and especially the Imperial Household Department, was in part 'Western' and 'modern' by design. The Department not only exhibited the 'mechanical advantage' principle of Archimedean simple machines, but also rested on the premise of 'mechanical objectivity', by planning and treating all issues in a regulated and predictable way. This created the same kind of 'freedom from will' that is often ascribed to computers today.[6] For instance, the meticulousness with which the Imperial Household Department drew up its regulations and workflows projected a planner's faith that, once the perfect machine had been designed and built, all matters would be assessed accordingly and no ad hoc tinkering would be necessary. Furthermore, the planners believed that the existence of this perfect machine, being immune from human errors, would deter illicit activities. But, as with all 'real machines, [when] in operation ... [they] are open rather than closed systems'.[7]

The distance between design principles and the realities of practice became even more glaring as time went on. In the 18th century, a series of

6 See Stephen Toulmin on the changing use – metaphorical and beyond – of a 'machine' in physics which, in the mid-19th century, gave way to the concept of 'fields', and Lorraine Daston and Peter Galison on the Renaissance ideal of a 'mechanical objectivity' that offered a 'freedom from will' in order to overcome any 'wilful intervention' by which a scientist might influence scientific results or theories (Stephen Toulmin, 'From Clocks to Chaos: Humanising the Mechanistic World-View', in *The Machine as Metaphor and Tool*, ed. Hermann Haken et al. (Berlin: Springer, 1993), 139-153; Lorraine Daston and Peter Galison, *Objectivity* (Brooklyn: Zone Books, 2008), 115-190, 123.

7 This quote by anthropologist Tim Ingold takes its inspiration from philosopher Gilbert Simondon, *On the Mode of Existence of Technical Objects* (Minneapolis, MN: Univocal Publishing, 2017), 61-62. See Ingold, *Being Alive: Essays on Movement, Knowledge, and Description* (London; New York: Routledge, 2011), 18.

military campaigns in Central Asia created logistical challenges of enormous proportions; when the empire was won with the annexation of Xinjiang 新疆 (New Territories) in the 1760s, the complexity and magnitude of the tasks faced by the Qing court were compounded. The palace machine's transformation was also necessitated by the weakening of the imperial household's financial strength and a changing global order from the late-18th century on. The officials of the Imperial Household Department and the emperor both understood that it was impossible to enhance the capacities of the palace machine by merely 'mechanical' means, that is to say, by applying impetus from outside the machine, or transferring movements from one part of the machine to another. They strived to perfect the machine at the overall design level by improving workflows and incorporating new features into the regulations. To allow for the adjustment of project expenditures with wide fluctuations in market prices, for example, they introduced a mechanism of feedback control; in requiring sub-departments to raise the funds needed for projects instead of awaiting central dispensation, they adopted what was, in fact, a modern management principle of self-organization. Contrary to the narrative of decline in the standard history books, the Qing palace machine proved to be resilient and adoptable to new realities in the 18th and 19th centuries, be it expansion or precarity.

The Machinery of the Imperial Household Department

The Imperial Household Department (Chinese: *Neiwufu* 内務府; Manchu: *Dorgi baita be uheri kadalara yamum*; literally, the 'Bureau in Charge of All Internal Affairs') was a unique Manchu institution that was the 'ghost in the machine'. Designed in part as a mirror image of the bureaucracy of the public state, this emperors' private institution was staffed by hereditary members from the upper three banners of the Manchu banner system.[8] The Department had been founded in 1638, when the Manchu court was still in Shengjing 盛京 (present-day Shenyang 瀋陽), Manchuria.[9] After

8 For an introduction to the Eight Banners system as a socio-military organization, see Mark Elliott, *The Manchu Way: The Eight Banners and Ethnic Identity in Late Imperial China* (Stanford: Stanford University Press, 2001). Ideas and projects developed within the Imperial Household Department probably influenced how the state dealt with similar issues. The interconnections between the 'public state' and the emperor's 'private' bureaucracy are discussed in a number of chapters, but are not the focus of this anthology.

9 Evelyn S. Rawski, *The Last Emperors: A Social History of Qing Institutions* (Berkeley: University of California Press, 2001), 179, 355-56, note 73.

the conquest of Beijing it became an instrument in a power struggle waged by the early Qing monarchs against the Thirteen Yamen (*Shisan yamen* 十三衙門). Dominated by eunuchs, the latter had served the Ming royal house for similar purposes of filling the imperial coffers and running their imperial residences.[10] In 1667 the Kangxi emperor (r. 1662-1722) succeeded in formally abolishing the eunuch power structure and replaced it with the Ten Offices – or Seven Offices and Three Courts (*Qisi Sanyuan* 七司三院) – which comprised the basic hierarchical structure of the Imperial Household Department. The Department reported directly to the emperor.[11]

The Department's jurisdiction entailed managing the vast system of Imperial Workshops (*Zaoban huojichu* 造辦活計處 or just *Zaobanchu*) that was established in 1680 to provision the entire material needs of the court, ranging from ceremonial banners, cannons, to everyday articles and objects of art.[12] The chapters in this volume examine the operation and products of the Imperial Workshops, as well as the knowledge cultures thus engendered. Other offices analysed in this anthology, including the Grand Storage Office (*Guangchusi* 廣儲司) which managed the finances and vaults, the Imperial Stud (*Shangsiyuan* 上駟院), the Garden Bureau (*Fengchenyuan* 奉宸苑), and the Imperial Dispensary (*Yuyaofang* 御藥房), were crucial to the functioning of the palace machine and have not been scrutinized in previous research. A number of chapters also reach beyond the Imperial Household Department to show how the emperors' household bureaucracy and finances were intertwined with those of the public.[13] Moreover, the Board of Work (*Gongbu* 工部) and the Ministry for Ruling

10 For eunuchs' extensive function in the Ming dynasty, see Shih-shan Henry Tsai, *The Eunuchs in the Ming Dynasty* (Albany, NY: State University of New York Press, 1996).

11 For an initial exploration of the bureaucratic structure of the Imperial Household Department up to the Qianlong reign, see Preston M. Torbert, *The Ch'ing Imperial Household Department: A Study of Its Organisation and Principal Functions, 1662-1796* (Cambridge, MA: Harvard University Press, 1977); and Qi Meiqin 祁美琴, *Qingdai neiwufu* 清代內務府 (The Qing Imperial Household Department) (Beijing: Zhongguo renmin daxue chubanshe, 1998).

12 In the past thirty years, art historians and cultural historians have published extensively on many specific art workshops in the Imperial Workshops (*Zaobanchu*), but the structural operation of the institution can be found in Wu Zhaoqing 吳兆清, 'Qingdai zaobanchu de jigou he jiangyi 清代造辦處的機構和匠役 (The Institution and Personnel of the Imperial Workshops of the Qing Dynasty)', *Historical Archive* 歷史檔案 04 (1991): 79-89; and in Chi Jo-hsin 嵇若昕 'Cong "huojidang" kan yongqian liangchao de neiting qiwu yishu guwen 從"活計檔"看雍乾兩朝的內廷器物藝術顧問 (The Advisors for Decorative Art to Emperor Yung-chen and Emperor Ch'ien-lung: Based on the Artisans of the Imperial Workshops)', *Soochow University Journal of History* 東吳歷史學報 16 (2005): 53-105.

13 See especially chapters Five and Ten on the intertwining of private industries (porcelain and pharmacy) and the court.

Outer Provinces (*Lifanyuan* 理藩院) are two public agencies addressed in the volume that also tended to the palace machine's needs and demands.[14]

Establishing the Imperial Household Department was an ingenious move, satisfying the multiple agendas of the Qing emperor in one stroke: keeping the Manchu nobility who had helped put the Aisin Gioro family on the throne gainfully occupied, eliminating dangerous residual power structures, managing the enormous economic interests and stakes of the royal house, as well as creating a propaganda, money-making, and 'ruling machine' for the Qing emperors.[15] The Yongzheng (r. 1723-1735) and Qianlong (r. 1736-1796) emperors completed this work by developing the institution into a hierarchical structure with several levels and numerous sub-units that continued to run the palace and the emperor's personal coffers until the very end of the Qing rule.[16] While some of these sub-units were veritable bureaucracies in themselves, and over time acquired a certain de facto autonomy in decision-making, each and every sub-unit had to report to the centre – the Central Administration of the Imperial Household Department (*Neiwufu zongguan* 內務府總管) – and were held ultimately accountable to the emperor. By design and in practice, the Imperial Household Department was an elaborate and, as will be seen, generally efficient machine.

Viewing the Qing palace as a machine highlights two main themes of this anthology: first, the material and tangible nature of the palace's operation, especially its enormous appetite for the money, labour, and materials that served as its 'fuel'. Second, it reveals the dynamic nature of the interwoven parts and coordinated workflows, as continuously unfolding processes within a set frame of rules and regulations. Machines are as much about

14 The Ministry for Ruling the Outer Provinces was a unique Qing bureau that managed relations with Mongolia, Tibet, and the new territories in the northwest. See Dittmar Schorkowitz and Ning Chia, eds, *Managing Frontiers in Qing China: The Lifanyuan and Libu Revisited* (Leiden: Brill, 2017).

15 See Chang Te-Ch'ang, 'The Economic Role of the Imperial Household in the Ch'ing Dynasty', *The Journal of Asian Studies*, vol. 31, no. 2 (Feb. 1972): 243-273, and Torbert, *The Ch'ing Imperial Household Department*, 1-26, on the evolution and strategic background of establishing the Imperial Household Department. For a book-length exploration of the Imperial Household Department as a whole see also Qi, *Qingdai neiwufu*.

16 Each sub-unit in addition had different levels of officials. Torbert counts four levels in the Imperial Workshops and six in the Grand Storage Office. Torbert, *The Ch'ing Imperial Household Department*, 34. In 1796 the Imperial Household Department employed 1,623 officials. In the late-19th century it controlled up to 56 sub-departments (Torbert, ibid., 28-29). For the evolution of departments in the Kangxi period, see the table in Qi, *Qingdai neiwufu*, 61-21. For descriptions of the duties of each office, see ibid., 64-84.

stillness and regularity as they are about motion and flow; they embody abstract design principles as much as they occupy concrete physical space.

Methods and Key Themes of this Anthology

In focusing on the Imperial Household Department, a key institution that has so far not received the academic attention it deserves, this volume seeks to demonstrate a new approach to Qing history in the longue durée. This approach builds on, yet departs from, previous scholarship under the rubric of institutional history, which has tended to focus on the structural and static aspects of bureaucratic organization and its relationship with monarchical power.[17] The chapters in this book show that much more is at stake. Extensive archival and artefactual source materials that have become available in the recent decades render the Imperial Household Department an ideal case to explore the technological, organizational, and knowledge-making processes that made the complex system work. It was through these processes, this book argues, that such mundane attributes as technical expertise and managerial skills became coextensive with political power.

The dynamic perspective adopted in this book reveals not only the tensions between action or motion and resistance in the Qing machine but also the entanglements between the inside and outside of the palace. Like all mechanical devices, the palace machine needed constant maintenance and repair, a supply of new or re-configured old parts, as well as a steady injection of fuel and the disposal or recycling of waste. The palace machine was neither a self-sufficient nor self-perpetuating entity. Taking a long-term view of this process from the beginning to the end of the Qing and beyond, the chapters of this book reveal a history wrought of different rhythms than those afforded by conventional political or social history. As time (and the machine) ground on, systematic reforms became necessary in the 19th century. Growing infrastructural failures such as broken dikes in hydraulic projects resulted in catastrophic floods in the northern provinces; a civil war raged in the southern heartland.[18] The increasingly threatening presence of

17 Torbert's 1977 study of the Imperial Household Department is an example of this. See also Etienne Balazs, *Chinese Civilization and Bureaucracy* (New Haven: Yale University Press, 1966) and others on bureaucratic versus monarchical power.

18 See the study of both bannermen and Han Chinese officials' work on water conservancy in Randall Dodgen, *Controlling the Dragon: Confucian Engineers and the Yellow River in the Late Imperial China* (Honolulu: University of Hawai'i Press, 2001).

Western powers on East Asian territories posed an additional danger. The palace machine was able to innovate in the face of these challenges, thus adapting to drastically altered realities in the empire and the global order. In the mid- to late-18th century and again at the end of the 19th century, periods commonly construed as times of imperial decline, the Imperial Household Department underwent major overhauls that cut across various subdivisions, although smaller and less remarkable changes were always part of its routine operation.[19]

While the temporal coverage of this book encompasses the entire Qing dynasty, its spatial reach also extends far beyond the palace compound. The Imperial Household Department machine appeared to be omnipresent in the empire. To complete large projects, it routinely joined forces with other administrative agents in the formal, 'public' bureaucracy, especially the Board of Revenue and the Board of Work. The Department also collaborated with regional garrisons and provincial governors to exert control over borderlands and isolated local counties. Most important of all, the Department was in charge of several enormous manufacturing and financial apparatus scattered in the provinces: the Ceramic and Textile Manufactories,[20] Custom Houses,[21] and the bureaus overseeing salt production and trade.[22] The Department used two ways to exert control over these remote offices: first, by appointing supervisors from the court and sending them to those places, where they also served as the emperor's 'spies', and second, by claiming sovereign ownership of their funding, materials, and products.

The temporally dynamic and spatially dispersed frame of this book reveals a simple but profound argument: that the Imperial Household Department was present beyond the capital city and its activities, far exceeding the

19 Some of the salient late-Qing institutional innovations include: the creation, modification and merging of key sub-units, the transformation of the functions of some sub-units, and new working relations between the Imperial Household Department and the Board of Works (*Gongbu*).

20 Jonathan Spence, *Ts'ao Yin and the K'ang-hsi Emperor: Bondservant and Master* (New Haven: Yale University Press, 1966).

21 Chen Kuo-Tung 陳國棟, 'Qingdai zhongye yihou zhongyao shuichai zhuanyou neiwufu baoyi danren de jidian jieshi 清代中葉以後重要稅差專由內務府包衣擔任的幾點解釋 (A Few Explanations on the Booi's Exclusive Appointment from the Imperial Household Department to Custom Houses since the mid Qing)', in Hsu Cho-Yun 許倬雲 et al., eds., *Di'erjie zhongguo shehui jingjishi yantaohun lunwenji* 第二屆中國社會經濟史研討會論文集 (Taipei: Hanxue Yanjiu ziliao ji fuwu zhongxin 漢學研究資料及服務中心, 1983), 173-204.

22 Wang Shih-Ming 王士銘, 'Ji shi guanyuan yeshi nucai: qianlong chao changlu yanzheng 既是官員也是奴才: 乾隆朝長蘆鹽政 = Either a government official or a private representative of the Emperor – A study of the supervisory officers to the Chang-lu Salt Administration'. MA thesis, National Chi Nan University, 2007.

provisioning of the imperial family, encompassed those of the bureaucratic state and the commercial market. As such, the Department affords a unique window into the manoeuvrings of the Qing court in its efforts to extend its control over multiethnic human subjects, animals, plants, technology, natural resources, material cultures and art forms[23] – all subjects of Qing statecraft. Indeed, understanding how the Qing palace machine worked means understanding how the Qing empire worked.

Organization and Chapters

Investigating the design, working principles, and practice of such a complex and extensive machine requires the collaboration of scholars from multiple disciplines and knowledge cultures. Eight of the nine chapters of this volume were first developed as part of an international project led by Martina Siebert and Kai Jun Chen at the Max Planck Institute for the History of Science in Berlin. The project brought together a team of researchers from the academy and museums in China, Taiwan, Germany, and the U.S., representing fields as diverse as economic history, history of technology, art history, labour history, textile studies, and material engineering. These chapters are grouped into three parts, each focusing on one design aspect of the Qing palace machine. Each part opens with a short introduction and a vignette essay that places one object or document at centre stage.

Part One explores the basic operating principles of the palace machine. Two salient principles are revealed in all three chapters of Part One: the conformity to hierarchy and genealogy in the allocation of roles, and the reliance on paper-tools to regulate and document every step of a procedure. In Chapter One Moll-Murata explains how 'Regulations and Precedents' (*zeli* 則例) and personal accounts allows historians to examine the personnel driving the palace machine, from lowly maids to imperial family members. The career trajectories of a unique group of experts, the Manchu bannermen, are examined across several generations in Chapter Two, by Chen. This kinship-based perspective reveals shifts in family expertise in accordance with the Qing court's changing technical demands. Chapter Three, by

23 The Imperial Household Department appears to be understudied, especially when compared with two Manchu institutional innovations in charge of military and diplomatic affairs, namely the Grand Council (*Junjichu* 軍機處) and the Court for Ruling Outer Affairs (*Lifanyuan*). See Beatrice S. Bartlett, *Monarchs and Ministers: The Grand Council in Mid-Ch'ing China, 1723-1820* (Berkeley: University of California Press, 1991); Schorkowitz and Chia, *Managing Frontiers in Qing China*.

Y. Wang and Bae, turns from the movement of people to the circulation of material, by following the flow of *kupiao* 庫票 (acquisition tickets) in and out of the palace workshops and storages. This chapter also elucidates the logic of the four-pillar accounting system used in the palace workshops and the commercial houses as a sophisticated modern device.

Part Two examines three categories of material artefacts that contributed to the visual and cultural splendour of the Qing court: porcelain, jade, and gilded roofs and statues. The Qing exerted firm fiscal control over the production of imperial porcelain, a time-honoured material embodiment of Chinese technological and cultural sophistication. The careful management and disposal of misfired pieces studied by G. Wang in Chapter Four shows the symbolic and material value that china had to the court. The same is true for jade. In Chapter Six Wu describes a key transitional moment when the territory of Xinjiang was added to the Qing empire, thus giving the court access to quality jade boulders. The logistical challenges involved in transporting these boulders over long distances and the dangers of embezzlement, however, reveal the shaky nature of the dominion claimed by the emperor. Chapter Five by Su and Lai investigates the procurement procedures and technology needed for gilding artefacts using such precious metals as gold, silver, and copper in the 18th century. The gilded roof tiles and Buddhist statues of the newly-built religious complex in the imperial summer residence of Chengde are veritable expressions of Qing prowess. The perfection of an advanced fire-gilding technique from Tibet and the emperor's desire to attract Mongol pilgrims, like acquiring jade boulders from Xinjiang, signify the cosmopolitan and multiethnic nature of the high Qing empire. With the availability of such new human, material, and symbolic resources, the palace machine that fuelled the expanded Qing mandate had to adapt accordingly.

Part Three probes the Qing palace machine's handling of plants and animals, exploring how these less predictable components were incorporated into its workings. Each chapter focuses on a different ecology of integration and usage of organic matter. In Chapter Seven Siebert explores the history of the imperial lotus ponds in and around the palace grounds in Beijing and the sale of lotus roots grown in the lakes of the imperial Westpark. She ponders the disjuncture between the miniscule income and the comprehensive set of regulations devised by the Imperial Household Department. Guan's study of the provisioning of medicinal herbs and minerals for the emperor and his family in Chapter Eight reveals a different logic and value system. To secure the best ingredients from every locale in the country, the palace machine utilized established tribute networks and commandeered experts from commercial medicine shops. The economic world outside the palace walls was thus incorporated into the body politic. The transport of elephants, a tribute animal from Vietnam, to the capital studied by Yu in Chapter Nine reveals the precarity of preserving life and the challenges this posed to the palace machine.

Taken together, the handling of lotus plants, medicinal herbs and minerals, and elephants reveals a sensitivity toward the finitude of time and the precariousness of living matter which had to be integrated into economic rationality as an underlying principle driving the Qing palace machine. Thus, while Part One reveals the commitment to hierarchical ordering and techniques of accounting that made this economic rationality practicable, and Part Two recognizes both the economic and symbolic values of material things as the political rationale behind this economic rationality, Part Three exposes the challenges posed by the preservation and integration of unpredictable living organisms into this rationality.

In summary, this book seeks to illuminate the hidden design, working principles, and values that guided the operation of the Qing palace machine, by examining them as a set of dynamic and complex material processes. It does not aim to provide an exhaustive survey of the Imperial Household Department or any of its subdivisions. In the final analysis, the palace machine was at once animate and inanimate, organic and inorganic, hierarchical and dispersed. It was a machine wrought from sub-machines, from materials and objects, rules and routines, as well as from human and non-human beings. These entities are rendered visible in this book as agents in co-producing the myriad life-worlds within and beyond the palace that constituted the Qing empire.

About the Authors

Martina Siebert works as subject librarian at the East Asia department of the Staatsbibliothek zu Berlin and as independent scholar. She researches and publishes on the history of Chinese exploration into nature and technology with a focus on the styles and agendas of presenting and organizing that knowledge in writing.

Kai Jun Chen is Assistant Professor at Brown University. He specialises in the history of imperial institution, handicraft technology, and material culture studies. His monograph in progress examines the technocratic culture in the Imperial Ceramic Industry of the Qing dynasty. His other scholarly publications investigate the production of technological knowledge in the global circulation, collection, and replication of luxury artifacts, especially porcelains.

A native of Hong Kong, DOROTHY KO is Professor of History at Barnard College, Columbia University. She is a cultural historian of early modern China whose research focuses on gender, technology, and material culture. Her recent book, *The Social Life of Inkstones: Artisans and Scholars in Early Qing China* (Washington, 2017) is a finalist of the Morey Prize of the College Art Association. An earlier book, *Cinderella's Sisters: A Revisionist History of Footbinding* (California, 2005) won the Joan Kelly Memorial Prize of the American Historical Association.

Part I

Operating the Machine: Personnel and Paper Trails

The Chinese word *ji* 機 is commonly taken to mean machines or mechanical devices such as the weaving loom. Yet the basic meaning given to *ji* in the earliest Chinese character dictionary, the *Shuowen jiezi* 說文解字 compiled in the 1st century C.E., was 'trigger' (*zhufa* 主發, literally 'controlling release'), which referred to the trigger of a crossbow or a certain triggering force (*jiguan* 機關) within interlocking parts of machinery. This semantic connection between machine and trigger highlights the necessity of both structure and motion to complete an action. Part One of this book examines these dynamics between structure and action in the Qing palace machine.

An inherent difficulty in studying the Qing palace machine is that many aspects of its operation were shroud in secrecy. Indeed, a second meaning of *ji* is confidentiality (*jimi* 機密).[1] The three chapters in Part One seek to unlock the hidden values and logics of this machine by tracking the visible products it produced through sequential steps. Like all good planners, the administrators of the Qing machine established a linked chain of tasks and commands. If all went well, an order from above would initiate action from subordinates, then the completion of one task would trigger the next. Focusing on the people, administrative units, and paper trails thus generated in the Qing palace, these three chapters delineate the overall design and structure of the machine in action. In so doing, they reveal an operational flexibility that belied the rigidity of stated rules. Evidence collected from running memorials, genealogies, and account books shows that, although the formal bureaucratic scheme controlling personnel and finance promulgated in the *Collected Statues of the Great Qing* (Da Qing huidian 大清會典) remained firmly in place, the actual operation often diverged from it.

Chapters One and Two focus on the human operators in the Imperial Household Department hierarchy, from the lowly servant-workers to the emperor (Moll-Murata) and back to the technical experts and managers (Chen). Regardless of their status, the training of these personnel can be seen as a form of ideological control stemming from the Qing mandate of imperial command over knowledge. The emperor and his advisers determined which facets of technical knowledge were crucial to imperial interests and needed to be safeguarded or shared. In the course of defining and re-defining who

1 This connotation of secrecy or concealment of *ji* found a pronounced expression in the naming of the Grand Council (*Junjichu* 軍機處, lit. 'Office of Military Secrets'), an important political agency established in the mid-Qing under emperors' direct control which aimed to achieve greater efficiency and secrecy in military decisions and related intelligence. See Beatrice S. Bartlett, *Monarchs and Ministers: The Grand Council in Mid-Ch'ing China, 1723-1820* (Berkeley: University of California Press, 1990).

was entitled to have access to specific material, funding, information, designs and products, the machine constructed a hierarchy of skills and expertise along its chain of command that sometimes diverged from the official status rankings.

Chapter Three (Y. Wang and Bae) analyses the paper-tools in the book-keeping system that registered the flow of personnel and materials through the imperial workshops, which constituted a key part of the palace machine. More than merely inert records, these accounting ledgers show the workings of triggering mechanisms in the numerous steps required to fulfil a production order from the emperor. The ledgers also track the movement of each minute item in the palace workshop as it was being conveyed between warehouses, workshops, and its destination, or undergoing quality control in between these locales. The balances of future and current debit and credit were precisely logged to several decimal points in the ledgers. The movements and pauses of materials, artefacts, monies, and personnel make visible the functioning of the machine.

Taken together, the three chapters in Part One show a considerable gap between intent and results, or between theory and practice, in the Qing palace machine. To the machine operators and historians in posterity, the meticulous accounting and voluminous paper trail created an illusion of total control on the parts of workshop supervisors and, ultimately, the emperor. But the mere fact that the same people charged with such precise reckoning – the officials, warehouse managers, and craftsmen – were also held responsible for missing fractions of a gram of silver or a delay in delivery suggests that there was a rift between the ideal of frictionless operation and the reality of dealing with unexpected occurrences, human error, or sub-standard outcomes. Guilt and punishment constituted an inevitable part of the working of the machine.

In conclusion, the Qing palace machine thrived on the dynamic balance of two principles: the principle of a stable structure and transparent rules, and that of speed and motion. Part One opens with a vignette essay by Shuxian Zhang that illustrates the principle of motion in the Qing palace machine. Focusing on a decorative wooden frame installed at Empress Dowager Cixi's (1835-1908) new bedroom in the wake of her 60th birthday, the vignette traces the surprising degree of mobility of not only people but also of seemingly immobile structure through palatial spaces and hierarchies. Interior architectural elements which appear to be fixed to a certain space at first glance are revealed to be no less mobile than things of commodity value or beings with legs or wings.

Vignette Essay I

Moving Pieces

On the Reuse of Interior Decoration Components in Qing Palaces

Shuxian Zhang

Abstract
Interior decorations were architectural components used to divide indoor space. As the palaces of the Qing court went through constant renovation, these components were replaced, refurbished, or moved. By examining archival documents together with extant architectural examples, this essay shows how the Qing court reused, sold, and recycled raw materials from old components of interior decoration in order to reduce cost.

Keywords: Qing dynasty, palace, interior decoration, recycling

In contrast to their seemingly fossilized appearance today, Qing imperial palaces, as the most concrete core space of the palace machine, were used and lived-in spaces that were repeatedly renovated and changed, either according to a ruling emperor's taste or following practical necessities. Furniture and objets d'art were the most mobile or exchangeable, as is amply proved by the 'inventories of things on display' (*chenshe dang* 陳設檔). This vignette sheds some light on the repurposing of wooden decorative components that were intended to be permanently installed features of a specific space, such as frames or plaques, but became moveable and adaptable in the process of refurbishing.

From the archival documents we learn that there were various possible 'career' paths for these objects and materials. Some of them were sent back to the Imperial Workshops (*Zaobanchu* 造辦處) that in most cases had originally produced them, to serve as raw materials in future projects. This is what happened, for example, to a room divider made from eight lacquered

Siebert, Martina, Kai Jun Chen, and Dorothy Ko (eds), *Making the Palace Machine Work: Mobilizing People, Objects, and Nature in the Qing Empire*. Taylor & Francis Group, 2021
DOI: 10.5117/9789463720359_ZHANGS

cedar wood panels that was dismounted in the Qianqing Palace (*Qianqing gong* 乾清宮), one of the central inner audience halls under Qing rule in the Forbidden City.[1] Some room accessories were turned into second-hand goods and sold for cash to outside the palace walls. When the four jade characters '*bu wei wu xian* 不為物先' (Do not get ahead of the natural rhythm of things; do not place things first) in Qianlong's handwriting were partly damaged in the process of being replaced with an identical set made of brass, they were sold or 'exchanged into their value in money' (*bianjia* 變價) at the Chongwen Gate. The characters had been originally mounted on a plaque furnishing the Jianfu Palace (*Jianfu gong* 建福宮) – one of Qianlong's recreational spaces in the palace grounds.[2] A third option was direct relocation to another palace. This kind of movement was either triggered by a request from the emperor – or the empress dowager – demanding that a specific item be moved to another place, or was proposed by officials in charge of refurbishing a space, who would suggest a new location for the item. For example, in 1749 three interior screens and a matching decorative frame made from the wood of *Phoebe nanmu* (*nanmu* 楠木) that used to decorate the Hall of Zhengyi Mingdao (*Zhengyi mingdao tang* 正宜明道堂) in the Forbidden City were handed over to the official Sanhe 三和 (1698-1773, Manchu: Nala Sanhe) to be reused at Wanshou Hill (*Wanshou shan* 萬壽山) near the new imperial summer palace, Yiheyuan, northwest of Beijing.[3] In many cases these re-installed elements needed to be adjusted in size and form to fit into their new place. The Qianlong emperor sometimes even involved himself in the process of these refurbishments. In the 17th year of the Qianlong reign (1752), in the course of renovating the scenic copy of the Hangzhou 'Autumn Moon over a Calm Lake' (*Pinghu Quiyue* 平湖秋月) at the Old Summer Palace Yuanmingyuan north of Beijing, certain parts of the interior decoration of the Bilin Lodge (*Bilin guan* 碧琳館) – one hall within the aforementioned

1 Qianlong 14/10/3 (11 Dec. 1749), in *Qing gong Neiwufu Zaobanchu dang'an zonghui* 清宮內務府造辦處檔案總匯 (Archives from the Imperial Workshops of the Imperial Household Department), ed. Chinese University of Hong Kong and The First Historical Archives of China (Beijing: Renmin chubanshe, 2005), vol. 16, 649.

2 Qianlong 21/2/28 (28 March 1756), *Qing gong Neiwufu Zaobanchu dang'an zonghui*, vol. 21, 293. For the Neiwufu's economical activities at the Chongwen gate see for example Wu Meifeng 吳美鳳, '"Chongwen men bianjia" – Qingdai Neiwufu bianjia wujian zhi duoshao "交崇文門變價" 清代內務府變價物件知多少', *Zijincheng* 紫禁城, no. 12 (2017), 106-123.

3 Qianlong 14/4/23 (7 June 1749), *Qing gong Neiwufu Zaobanchu dang'an zonghui*, vol. 17, 97. The Manchu official San-he served in numerous high-ranking official posts of the Qing administration under Qianlong. In 1741 he was chief supervisor of the Imperial Household Department (*Zongguan Neiwufu dachen* 總管內務府大臣), and by 1749, the time of the transfer, he had already served as a minister at the Board of Works (*Gongbu shangshu* 工部尚書). Sanhe died in 1773.

Forbidden City's Jianfu Palace – were dismounted to be reused for this new purpose. To evaluate the new positioning and necessary adjustments of the inner decorations at Pinghu Qiuyue, the Imperial Workshops were instructed to 'make a cardboard model (*hepai yang* 合牌樣) of the reworked bed with drawers'. When the eunuch Hu Shijie 胡世杰 (fl. 1739-1776)[4] showed this model to the emperor, Qianlong responded as follows: 'Production according to the model is granted. Repurpose the bed currently installed at the Bilin Lodge. If there need to be any changes, make the changes whenever possible by modifying the old parts; if the old parts cannot be altered to fit, make them anew according to the model'.[5]

Several instances of this kind of refurbishing and refitting can be traced in the paper trails of the Imperial Workshops' archival resources. It is very rare to find any physical proof of these material relocations made during the Qianlong era in extant palace buildings today because renovations moved on and tastes continued to change. However, there is still a great deal of physical evidence of palace renovation projects from the late Qing period. In particular, preparations for the Empress Dowager Cixi's 60th birthday in the 18th year of the Guangxu reign (1892) led to large-scale renovations of the Ningshou Palace district (*Ningshou gong qu* 寧壽宮區) on the eastern side of the Forbidden City. One such modification was transforming the western room of one of the Ningshou Palace's halls, the Leshou Hall (*Leshou tang* 樂壽堂), into Cixi's bedroom, containing two beds – one installed at the back and the other at the front of the room.

The instruction for doing this reads as follows: 'The western heating chamber together with the adjacent inner room is to be transformed into a bedroom by removing the separating wall between them and installing instead for each of the beds a decorative frame in front and along both sides that touches the floor (*luodi zhao* 落地罩). Install the beds as beds facing each other (*duimian chuang* 對面床)'.[6]

Figure I.1, a photograph taken in 2018, shows the state in which this room has survived into our present time: in the front part of the room a

4 The eunuch Hu Shijie stood in close personal contact with the Qianlong emperor, giving him suggestions on political and aesthetic matters. His exact birth and death dates are not known but his name appears regularly in archival documents between 1739 and 1776. There is no mention of him from 1777 on.

5 Qianlong 17/10/19 (24 Nov. 1752), *Qing gong Neiwufu Zaobanchu dang'an zonghui*, vol. 18, 805.

6 Guangxu 18/7/18 (8 Sept. 1892), see Fang Yuqin 方裕瑾, 'Guangxu shiba nian zhi ershi nian Ningshou gong nei gaijian gongcheng shulüe 光緒十八年至二十年寧壽宮內改建工程述略', in *Zhongguo zijincheng xuehui lunwen ji* 中國紫禁城學會論文集, 2nd collection (Beijing: Zijincheng chubanshe 1997), 261.

Figure I.1 Decoration and condition of the front bed in Leshou Hall in 2018

Photograph by the author

Figure I.2 Lower part frame of Leshou Hall bed

Photograph by the author

floor-touching frame is installed to separate the bed from the room. This frame consists of two components. Firstly, on the top and upper sides, hollowed-out rhombus-structured frames and secondly, on the lower sides and bottom part, frames inset with jade carvings of bamboo leaves on a ground of bamboo slivers in tortoise-shell design (Figure I.2 shows this detail).

This two-component bed frame must have been installed later, on top of the original bed, as both are visibly too small for the bed they are currently installed on. To adjust the frame to the size of this specific bed and room, an additional structure was added between the decorative frame boards and the jambs that fix the structure to the walls and ceiling. The bed support (*chuang bang* 床幫) at the very bottom of the construction, on the other hand, exactly matches the size of the bed. Moreover, the motifs and craftwork of that bed support evinces clear differences from the other parts by displaying a *kui*-dragon (*kuilong* 夔龍) motif carved in wood. Most probably an original lower bed was made into a higher bed with the help of these additional decorative frames, which were dismounted from a high bed and a different-sized room.

The Ningshou Palace district was constructed in the 37th year of Qianlong reign (1772) and one of the major design technologies used throughout was intarsia made from various materials. So the parts of the bed frame shown in Figure I.2 above are representative of the style Qianlong had chosen for the district. The bed support with its *kui*-style motif also probably originated from the Qianlong era and might as well have been installed in the Ningshou Palace district. But, obviously, it had not been part of one set with the rest of the decorative frame. When this bed had been installed in the building more than one hundred years later in 1893, the decorative frames from various other buildings of the Ningshou Palace district were repurposed and reworked to fit this new necessity – a process seemingly standard enough to not leave any prominent traces in the extant archival evidence.

Today, most of the Leshou Hall in the Ningshou Palace district is open to the public, but the western room in which the bed is installed is still dozing in a Snow White-like slumber. Part of the interior decoration of the room and the bed have thus been preserved unchanged (and unrenovated) for over a century, constituting rare evidence of how material (and human) puzzle pieces of the Qing palace were moved around, redefined, and adjusted to fit new spatial contexts and purposes.

Translated from Chinese by Martina Siebert in close consultation with the author

About the Author

SHUXIAN ZHANG 張淑嫻 is Researcher at the Palace Museum in Beijing. She specialises in the architecture and interior decoration of Ming Qing palaces. Among her publications are the monographs *Ming Qing wenren yuanlin yishu* 明清文人園林藝術 (The Art of Literati Garden during the Ming and Qing Dynasties) and recently *Jinchuang xiuhu: qingdai huanggong neiyan zhuangxiu yanjiu* 金窗繡戶——清代皇宮內檐裝修研究 (Gilded Windows and Carved Doors: Study of Interior Decoration of the Qing Palaces).

1 Working the Qing Palace Machine

The Servants' Perspective

Christine Moll-Murata

Abstract

This chapter asks about the personnel working at and for the Qing court. It explores their numbers, working conditions, labour relations, and social positions with a temporal focus on the mid and late Qing. Labour relations, in accordance with the definitions of the Global Collaboratory on the History of Labour Relations, include the non-working, reciprocal, tributary, and commodified types. All of these types were represented at the Qing courts in various constellations. The paper outlines work incentives and sanctions based on *Palace Regulations and Precedents* (Qinding gongzhong xianxing zeli) and personal accounts of a palace maid and a eunuch in the early twentieth century and gives insights into the interaction of humans with the institutional mechanisms of the palace machine.

Keywords: labour relations, Qing palaces, service personnel, maids, eunuchs

Who, if not humans, made the palace machine run, and were at the same time moved by it? Within this volume, looking into how human and non-human elements involved in the multiple interactions and mechanisms of the Qing palace machine, this chapter asks about the actual men and women working in the machine. Who were they, and how many of them were there? What were their working conditions and labour relations? Why do their daily myriad activities warrant the use of the concept of 'machine'?

In order to address these questions, this chapter first offers a quantitative approach to the palace personnel. It then explores labour relations and organization based on the late-Qing era *Imperially Commissioned Current*

Siebert, Martina, Kai Jun Chen, and Dorothy Ko (eds), *Making the Palace Machine Work: Mobilizing People, Objects, and Nature in the Qing Empire.* Taylor & Francis Group, 2021
DOI: 10.5117/9789463720359_CH01

Regulations and Precedents of the Court (Qinding Gongzhong xianxing zeli 欽定宮中現行則例), investigates the self-perceptions of palace workers that are available for the very late-Qing period and early Republic of China, and reconsiders, on the basis of those data and insights, the machine as a thinking model.

Humans in the Palace Machine: Numbers and Types of Positions

Quantifying the number of people working in the palace is not quite as straightforward an exercise as might be expected, despite the voluminous compendia of palace administration contained in 'Regulations and Precedents' (*zeli* 則例) and the *Precedent Cases of Statutes of the Great Qing* (Da Qing huidian shili 大清會典事例; from here on *Precedent Cases*).[1] Authoritative research on the institution and activities of the Imperial Household Department[2] cite figures that have been taken from a wide array of archival materials as well as from *zeli* and other contemporary sources. Yet no single, neat and comprehensive account of the entire palace personnel in Qing China exists. Moreover, a view of the number of people at work in the Qing palaces must be established by considering all the different status groups employed there. Naturally, the numbers of personnel waxed and waned over subsequent reign periods, and normative quotas and real-life assignments could be at variance at all periods. This tension between the nominal data on paper and actual numbers in practice was one of the general characteristics of the palace machine, particularly because the authorities attempted – probably beyond institutional capacity – to systemize data regarding labour, materials, funding, and styles.

The most accessible data are the nominal numbers of officials (*guan* 官) in the service of the palace and the Imperial Household Department. Table 1.1 shows the figures stated in the *Precedent Cases* in the introductory chapters to the section on the Imperial Household Department.

1 The latest version of the *Precedent Cases* was finished in 1899 under the editorship of Kungang 崑岡 et al., *Da Qing huidian shili* (Guangxu ban) 大清會典事例 (光緒版) (reprinted Taipei: Chongwen shuju, 1963).

2 Preston M. Torbert, *The Ch'ing Imperial Household Department: A Study of Its Organization and Principal Functions, 1662-1796* (Cambridge, MA: Harvard University Press, 1977); Qi Meiqin 祁美琴, *Qingdai neiwufu* 清代內務府 (The Qing Imperial Household Department) (Beijing: Zhongguo renmin daxue chubanshe, 1998); Evelyn S. Rawski, *The Last Emperors: A Social History of Qing Imperial Institutions* (Berkeley and Los Angeles: University of California Press, 1998).

Table 1.1 Nominal number of officials in the Imperial Household Department

Date	Numbers of official posts (*que* 缺)*
Early Kangxi (1662)	402
Late Kangxi (1722)	939
Yongzheng	1,285
Jiaqing 1 (1796)	1,623

* Not including officials with the highest supervising function, i.e. the *Zongguan Neiwufu dachen*, who varied in number from two to nine, but generally remained between four and six.
Source: Torbert, pp. 28-9, 33, quoting from the Guangxu edition of 1899 of the *Precedent Cases*

The numbers of dependent and relatively restricted workers are not included in these figures of officials, but are mentioned elsewhere in separate statements. That concerns the numbers of bondservants, i.e. *baoyi* 包衣 and *aha* 阿哈 who, according to Rawski, mainly worked in domestic service and in agriculture respectively,[3] the short-term hired labourers (*sula* 蘇拉), eunuchs (*taijian* 太監) and palace maids (*gongnü* 宮女). Apart from these groups, less dependent, temporary employment of 'outsiders', for instance in the palace workshops, occurred as well.

Among the workers assigned to the service of the emperor, the bondservants probably made up the biggest group. In the early Qing dynasty they were combined into 29 or 30 companies (*zuoling* 佐領),[4] with about 300 men in each company.[5] This results in a figure of at least 8,700 to 9,000 men. Qi Meiqin cites a total number of 35 companies for the Kangxi period.[6]

Rawski asserts that the most numerous group of workers in the palace were the *sula*, casual or precarious labourers who were hired for short-term tasks and were not given official posts. According to a document in the Imperial Household Department palace memorial archives (*Neiwufu zou'an* 內務府奏案),[7] they numbered between several hundred and several thousand per month. Other references quote 150 *sula* posts under each company, or a total of 4,500 in 1708, but half of that in 1735. In 1774, the Qianlong emperor decreed

3 Rawski, *The Last Emperors*, 67. For a finer categorization of the terms *booi*, *aha*, and *booi aha*, see Qi Meiqin 祁美琴 and Cui Can 崔灿, 'Baoyi shenfen zaibian 包衣身份再辨', in *Qingshi yanjiu* 清史研究 = The Qing History Journal, no. 1 (2013): 117-128, 118.

4 *Guanling* 管領 and *zuoling* 佐領; Torbert, *The Ch'ing Imperial Household Department*, 61.

5 Torbert, *The Ch'ing Imperial Household Department*, 19, 61 (based on *Baqi tongzhi chuji* 八旗通志初集, which explains that additions were made to the companies during the 17th century).

6 Qi, *Qingdai neiwufu*, 73, calculated also on the basis of *Baqi tongzhi chuji*.

7 *Neiwufu zou'an*, no. 446-5-55. See Rawski, *The Last Emperors*, 168f, with reference to figures between 1708 and 1781.

that a figure of 50,000 *sula* working days per year should not be exceeded.[8] Rawski's tabulation of *sula* hired between 1760 and 1848 shows a decrease in the daily average, from 97.6 in the highest period (1760-1761) down to 50.4 in 1848.

Torbert explains that *sinjeku* or bondservants' slaves (Chinese: x*inzheku* 辛者庫), held lowest, slave-like position in the palace hierarchy.[9] Above them was another group of indentured house-servants, who were also subordinated to the bondservants.[10] A total number of *sinjeku* are given by Rawski, for instance, who cites a document of the late 17th century listing 170 individuals who fell into this lowly category, and who – together with their 'wives, sons, sons' wives, and other family members – a total of 762 dependents were converted into *sinjeku*'. *Sinjeku* women were mainly employed in menial tasks in the palace, numbering more than 5,000 in the 18th century. For 1681, a figure of over 4,000 *sinjeku* women were reported as working at the imperial mausolea and the Shenyang palace (*Shengjing xinggong* 盛京行宮) in Shenyang/Mukden.[11]

Palace maids were girls who mainly came from the banners. They were usually daughters of *sula* or lower-ranking soldier families, who began serving in the palace at the age of thirteen to fifteen and left their duties at thirty; later the age limit was reduced to twenty-five.[12] Their numbers reached a peak in 1734 at 500, reducing to slightly over 100 in around 1790, and rose again to between 150 and 200 under Empress Dowager Cixi, i.e. 1861 to 1908.[13]

For the number of eunuchs in the palace, Rawski cites archival documents which state that their numbers decreased from 3,107 in 1750 to 2,740 in 1800.[14] Wang Shuqing stresses that, at all times inspected after 1793 (i.e. Qianlong 58), the actual numbers of eunuchs were lower than the norm – in 1842 the

8 All figures from Rawski, *The Last Emperors*, 168/9, referring to reports of the ministers from the Imperial Household Department found in archival material at the First Historical Archives of China, Beijing.

9 Torbert, *The Ch'ing Imperial Household Department*, 65.

10 Ibid., 66.

11 Rawski, *The Last Emperors*, 172. More recently, Lai Huimin 賴惠敏 has challenged the conception of *sinjeku* as lowly slaves, pointing out that they still stood above the Han Chinese population, that a few *sinjeku* rose to high official positions, and that overall they had just as good an opportunity to earn a reasonable income as if they had obtained hereditary assignments as craftspeople such as smiths, or as storehouse keepers. See her 'Tiegan zhuangjia: Qingmo Neiwufu xinzheku ren de jiahu yu shengji 鐵杆莊稼—清末內務府辛者庫人的家戶與生計', *Zhongyang yanjiuyuan Jindaishi yanjiusuo jikan* 中央研究院近代史研究所集刊, no. 38 (Dec. 2002): 71-128.

12 Rawski, *The Last Emperors*, 170-1.

13 Ibid., 170.

14 Ibid., 165, quoting from archival documents of the Imperial Household Department (*Neiwufu zou'an* 內務府奏案).

norm quota for eunuchs was 3,600 but the actual number 2,216; in 1887 the norm quota was lowered to 2,500, but again, the actual number was only about 67% of that, namely 1,693. In the last years of the Guangxu reign (1875-1911) the actual number of eunuchs employed was 1,200.[15]

The artisans who worked in the palace workshops came from different backgrounds and status groups. Most important among these were the 'banner artisans' (*qijiang* 旗匠) and the mostly Han Chinese civilian artisans (*minjiang* 民匠) who were hired from outside the palace for permanent or temporary assignments. Moreover, the Inner Palace 'Regulations and Precedents' prove that, in addition, eunuchs were assigned to the imperial workshops and storehouses, such as the Workshops at the Yangxin Palace (*Yangxindian zaobanchu* 養心殿造辦處) of the Imperial Household Department, as well as to the Board of Work's (*Gongbu* 工部) Office of Construction (*Yingshansi* 營繕司).[16] This indicates that the lines between the Board of Work and the Imperial Household Department with respect to the deployment of workers and craftspeople were sometimes blurred.[17] However, it is clear that craft production for and in the palaces in the Forbidden City, the Yuanmingyuan 圓明園 and the Shenyang Palace was first and foremost organized by the Imperial Household Department. The other government institution that also produced goods for the palace, like the Board of Work, had only about half of the workforce of the Imperial Household Department.[18]

To give a very rough estimate, the overall number of people who jointly made the palace work in administrative, service, and production positions might have been about 1,500 officials, 9,000 bondservants, 3,000 *sula*, 4,000 *sinjeku*, 500 palace maids, and 3,000 eunuchs, plus a range of 1,000 external artisans, which adds up to a tentative number of 22,000 individuals all working for the palace at the same time in the mid-18th century. Although this figure reduced over time, it shows the volume of personnel needed to keep the palace tidy, maintain the buildings, guard, feed the inhabitants, and provide specialized, often luxury goods. This only includes the immediate

15 Wang Shuqing, 7-8. For the number of 2,216, see also *Qinding Gongzhong xianxing zeli di er zhong* 欽定宮中現行則例第二種 (hereafter *Gongzhong xianxing zeli* (2)), *juan* 4, 4a (c. 1883), reprinted in *Gugong zhenben congkan* (Haikou: Hainan chubanshe 2000), vol. 280, 337, with reference to the memorial of Daoguang 22/1/17 which specifies this number of eunuchs in the palaces in the Forbidden City, Yuanmingyuan and Southern Gardens (*Shengping shu* 昇平署).

16 *Gongzhong xianxing zeli* (2), *juan* 4, 7b-12b (*Gugong zhenben congkan*, vol. 280, 345-348).

17 For further statistics, see Christine Moll-Murata, *State and Crafts in the Qing Dynasty* (Amsterdam: Amsterdam University Press, 2018), 72-84.

18 Ibid., 106-107. In around 1800, the Imperial Household Department had approximately 3,000 workers in its service, and the Board of Work had about 400 (not counting the workers in the Imperial Mint).

service that is mentioned in the documents directly related to doing the palace's manual and administrative work. In a broader sense, the taxpaying agrarian population in the empire also contributed to maintaining the functions of the palace and its inhabitants – so every individual in the empire could be considered part of the palace's workforce. This is one example that illustrates how investigating the mechanism of the palace machine provides a fresh insight into how the empire worked.

Labour Relations at the Palaces: Ranges and Limits of Control

Palaces worldwide may not seem the obvious places to study labour relations. In the social sciences, labour relations are mostly conceptualized by observing entire populations or large parts of them, rather than small elite groups and the system that directly supports them.[19] Nevertheless, aristocratic elites have exemplary functions for societies at large. For instance, as Norbert Elias remarked in his study on the French court of the Ancien Régime, the bourgeois societies that came to prominence during 19th-century Europe took an antagonistic stance towards the luxury consumption of the courts before the French Revolution, but nevertheless also admired the court culture of bygone days and tried to emulate it.[20]

As shown above, a consideration of the actual number of people involved in management or production of the sources of political power and its representation has revealed that many more people may be involved than only those in direct contact with the rulers. In order to conceptualize their specific labour relations, this chapter applies the definitions developed by the Global Collaboratory on the History of Labour Relations, a group of researchers that intends to provide statistical insights into the global distribution of all types of labour relations in historical cross-sections. For this purpose, the Global Collaboratory has developed a classification of labour relations divided into four large groups: non-working, reciprocal, tributary, and commodified labour.[21]

19 See, for instance, the activities of the Global Collaboratory on the History of Labour Relations, http://socialhistory.org/en/projects/history-labour-relations-1500-2000, accessed September 2020.

20 Norbert Elias, *Die höfische Gesellschaft – Untersuchungen zur Soziologie des Königtums und der höfischen Aristokratie* (Frankfurt: Suhrkamp Taschenbuch Wissenschaft, 1983), 67, 74.

21 For the following definitions, refer to Christine Moll-Murata et al., 'Conquerors, Employers, and Arbiters: States and Shifts in Labour Relations 1500-2000: Introduction', in *International Review of Social History*, no. 61 (Special Issue: *Conquerors, Employers, and Arbiters: States and Shifts in Labour Relations 1500-2000*): 23-26.

'Non-working' includes people who cannot work or cannot be expected to work for reasons of age, physical constraints, or because they are studying; but also those who are affluent and those who are unemployed.

'Reciprocal' implies work done for other members of the same family, household, or community. It includes kin as well as subordinate non-kin (men, women and children) who work as reciprocal household servants and slaves and contribute to the maintenance of self-sufficient households. Servants of this type neither earn a salary nor are they free to leave their employer/relative of their own volition. They work as servants in autarchic households, monasteries, and palaces, and under all shades of conditions – from enforcement (including pawnship) to a desire to receive patronage.

'Tributary' labour relations, according to the Global Collaboratory's concept, refers to work that is provided as a tribute payment and is based on obligations vis-à-vis a polity that actually owns the entitlement to have work service rendered. In the case of the Qing palace and banner system, 'polity' and 'family' might be considered to overlap, thus blurring the distinction between 'reciprocal' and 'tributary' labour. The differences between the conceptions of 'reciprocal' and 'tributary' lie in two points: for reciprocity, self-sufficiency is key. This type of work is most often unpaid, and a kind of intergenerational trade-off ensures that work is done and service is rendered among members of nuclear and extended families. For tributary labour, the relationship between a worker and their employer is not necessarily kin-based. As a rule, the 'polity' is more abstract and transcends family configurations such as nuclear or extended families. Size is also a distinctive criterion: the 'family' formed by the banners consisted of hundreds of thousands, if not millions of people. More abstract obligations for this larger community applied, including monetary remuneration. Obligatory labour such as corvée, conscripted military service, and convict labour belong to this category. Yet obligatory work can also be an entitlement that confers a middle or high social standing, such as those of the banner people in Qing China. This type of work can also be indentured, either in the form of permanent tied labour for the polity or as service for a specific period of time (such as the palace girls mentioned above). Tributary serfs work for the polity because they are bound to its soil and obliged to provide specified tasks for a specified number of days. In the case of the Qing palace, this would include the people working on the Imperial Domains (*Huangzhuang* 皇莊). In a gradation of even greater restraint, tributary slaves are owned by and work for the polity indefinitely. They are deprived of the right to leave or refuse to work, and do not receive monetary compensation for their labour beyond food, clothing, and accommodation.

'Commodified' labour is where workers sell the products of their work or their labour in market exchange settings. This type of work either produces something for direct sale in the market, or for public institutions which may service the market, although not for the gain of private individuals.

As will be explored in the following subsections of this chapter, most of these definitions seem to be applicable to different types of labour that were administered in the Qing palaces at particular points in time.

There had been continuous labour at the Qing palace since the early 17th century, with some of its institutions going back even further than this. The largest share of labour the government resorted to was tributary labour. Yet, especially during the last phase of the Qing, although much work remained tributary or was of the reciprocal type – as could be argued in the case of the Manchu nobility – commodification was also possible and present.

Since I have explored the various types of labour relations of the late Ming and early Qing elsewhere,[22] I will focus here on the wider trends and biggest issues. I will investigate how they applied to the workspace of the Qing palace, in particular with respect to the highest echelon of the palace, i.e. the emperor, and those performing the physical work, i.e. craftsmen, servants, maids, eunuchs, and day labourers. This chapter does not cover the hoards of administrative staff doing investigations, filing reports etc.

Did Emperors and Empress Dowagers Actually Work?

Monarchic rulers and sovereigns mostly legitimize their rule by way of metaphysical appointment, at least of their first ancestor – or priests, officials, and historians legitimize it for them. Thereafter, the (usually patrilineal) heredity of the title and regency ensures that the rule of one predestined line will continue, according to the ideology of the ruling dynasty, for all eternity. Can we apply the term 'work' to these people at the top of the hierarchy of the palace, or would the purportedly affluent and idle Qing emperors qualify for the category of 'non-working' according to the taxonomy of the Global Collaboratory?

The basic definition of 'work' applied by the Global Collaboratory's taxonomy for the period 1500-2000 was taken from Charles and Chris Tilly. It

22 Christine Moll-Murata, 'Tributary Labour Relations in China During the Ming-Qing Transition (Seventeenth to Eighteenth Centuries)', *International Review of Social History*, no. 61 (Special Issue *Conquerors, Employers, and Arbiters: States and Shifts in Labour Relations 1500-2000*): 27-48, 36-38.

includes '... any human effort adding use value to goods and services'.[23] If the implementation of government rule, securing property rights, and ensuring the population's livelihood is seen as a service rendered by the monarch, then an emperor's activities can be considered as 'work'. Technically, the king was also the employer of a multitude of people, another category in the Global Collaboratory's taxonomy. Therefore, the category 'non-working' seems to be inadequate.

While it may seem arbitrary to judge monarchs who lived several hundred years in the past by 21st-century yardsticks, their contemporaries and immediate successors were well aware of whether the emperors, empress dowagers, and regents fulfilled their duties or not. Among the Qing monarchs, the Kangxi and Yongzheng emperors and Empress Dowager Cixi are famous for their continued devotion to administrative matters, and especially to the attendance of audiences with their officials. From this perspective, one could argue that some monarchs worked hard while others simply lived in the luxurious settings of the palaces and gardens of Qing Beijing. Yet, viewed from another angle, even those who did not attend audiences, comment and sign throne memorials, lead armies into wars, or administer the law, but mainly enjoyed a life of leisure, by definition also fulfilled the function of representing imperial rule – which constituted their 'work'.

Tributary and Commodified Labour in the Palace

Below the emperor in the palace there was an elaborate hierarchical structure of higher and lower officials in the central office and the various subsections of the Imperial Household Department that administered, supervised, and audited those who rendered work in tributary and commodified labour relations. The main pillar of service for the court was the 'tributary' labour provided by banner people, that is, the serfs (*zhuangding* 莊丁) who worked in the fields of the imperial agricultural estates in the vicinity of the capital, and the palace maids. They had to work for the polity because it had the right to make use of their labour.[24] Moreover, the number

23 Chris Tilly and Charles Tilly, *Work under Capitalism* (Boulder, CO: Westview Press, 1998), 22.

24 See Lai Huimin 賴惠敏, *Qianlong huangdi de hebao* 乾隆皇帝的荷包 (Taipei: Zhongyang yanjiuyuan jindaishi yanjiusuo, 2014), 51, for a table of the nominal labour force (*guanzhuang huding* 官莊戶丁) of 12,971 who were believed to do agricultural work on the Imperial Domains during the Qianlong era.

of palace maids assigned to a woman in the imperial harem indicated her status within the imperial hierarchy: twelve maids were required to serve the empress dowager(s), ten for empresses, eight for each imperial noble consort (*huang guifei* 皇貴妃) and the two noble consorts (*guifei* 貴妃), and six for each of the consorts and concubines (*fei* 妃 and *pin* 嬪).[25] The work of the palace maids and, according to Hsieh Bao Hua's study, also that of wet nurses, was bonded, in the sense that the women were drafted into their roles and could not (but usually also would not) refuse the service or leave their workplace at will once they were engaged.[26] The same status certainly applied for other assignments listed in the *Imperially Commissioned Current Regulations and Precedents for the Central Administration of the Imperial Household Department* (Qinding Zongguan Neiwufu xianxing zeli 欽定總管內務府現行則例), such as the many women employed in the Imperial Wardrobe (*Yiku* 衣庫), who were responsible for making boots and socks (450 workers), saddles (51), embroidered purses (84), knots for prayer bead rosaries (51) and armour linings (13), and the women in charge of rice (88), tea (60), and fruit (300). This constituted a total workforce of 1,126 craftswomen (*zuo huoji furen* 做活計婦人). They were also attached to the harem ladies in numbers that corresponded to the palace hierarchy, namely 88 people for the highest ranking imperial noble consort, 77 for the noble consorts, 66 for consorts, 55 for concubines, 30 for noble ladies (*guiren* 貴人), 25 for permanent female attendants (*changzai* 常在), and 44 for the imperial princes (*age* 阿哥) and their wives (*fujin* 福晉).[27]

Although tributary employment implied force and subjugation, it also provided a certain degree of entitlement and prerogatives. As He Rong'er 何榮兒, a palace maid who served the Empress Dowager Cixi recalled, 'we people under the banners' (*women qixiaren* 我們旗下人) received grain rations (*kouliang* 口糧) from the court as a 'favour from the emperor' (*huangshang gei de endian* 皇上給的恩典). In return, with reference to the lists of banner people, palace maids were hired at the age of thirteen or fourteen. This was considered an assignment based on the due respect and filial piety (*xiaojing* 孝敬) of the servitors (*nucai* 奴才).[28] The self-depreciatory term *nucai* for

25 *Gongzhong xianxing zeli* (2), *juan* 3, 5b (*Gugong zhenben congkan*, vol. 280, 286).

26 Hsieh Bao Hua, *Concubinage and Servitude in Late Imperial China* (Lanham: Lexington Books, 2014), 144.

27 *Qinding Zongguan Neiwufu xianxing zeli* (*yi*) 欽定總管內務府現行則例 (一) (Xianggang: Fuchi shuyuan chuban gongsi, 2014), 505.

28 Jin Yi 金易 and Shen Yiling 沈義羚, *Gongnü tangwang lu – Chuxiugong li suishi Cixi banian* 宮女談往錄宮女談往錄——儲秀宮裡隨侍慈禧八年 (Beijing: Zijincheng chubanshe, 1991), 5; an abridged translation is contained in *Mémoires d'une dame de cour dans la Cité interdite*

servitor or slave (the Chinese equivalent to the Manchu term *booi aha*) refers to a tied, master-slave-like relationship. If banner families were of a higher status or had connections to somebody in the Imperial Household Department, they could avoid this assignment. However, for poorer families, it brought an income of several taels of silver per month and monetary rewards (*shangqian* 賞錢) that were distributed every season. Moreover, if the girls learned some rules of etiquette in the palace, they would gain social prestige, and increase their chances of upward mobility in the form of an advantageous marriage to a person of higher status.

These are examples of tributary workers who were drafted into position and could not refuse to work. Their work had always been remunerated, since the beginning of the dynasty. As the palace machine was an open structure, labour in the palace was also recruited from the labour market in commodified-type arrangements. Some of those work arrangements stood in a transitional situation between tributary and commodified labour relations: for instance, the *sula* were paid for the days of service rendered. Since they belonged to the Eight Banner system, their banner units should have formally provided for them; the *sula* should have had the duty to serve and the right to be fed.[29] Yet it is well known that the banner people did not always receive the allowances they were entitled to, or were impoverished for other reasons, so had to do menial physical labour to make ends meet.[30] Eunuchs' employment situation was a clearer case of commodification, since they normally came from outside the Eight Banner system[31] and entered

(He Rong'er 何榮兒), collected by Jin Yi, transl. by Dong Qiang (Arles: Philippe Picquier, 1996). Compare also the account of Empress Dowager Cixi's Lady-in-Waiting Der Ling, *Two Years in the Forbidden City* (London: T. Fisher Unwin, 1912), 216, which also mentions that the sons of Manchu officials were called to court for service, including her two brothers Xunling and Xinling, who were responsible for the palace's electrical installations and for the operation of the empress dowager's steam launch. Xunling took the photographs that offer a close view of court life at the end of the dynasty. See David Hogge, 'The Empress Dowager and the Camera. Photographing Cixi, 1903-1904' http://vcarchive.com/mitvc/empress_dowager/index.html, accessed September 2020. On Der Ling's family, see Grant Hayter-Menzies, *Imperial Masquerade: The Legend of Princess Der Ling* (Hong Kong: Hong Kong University Press, 2008), 366. For the complex and dynamic system of remuneration of the banner people, see Mark C. Elliott, *The Manchu Way: The Eight Banners and Ethnic Identity in Late Imperial China* (Stanford, CA.: Stanford University Press, 2001), 191-193, who coined the term 'The Iron Rice Bowl of Privilege' for the preferential treatment banner people received from their status organization.

29 Moll-Murata, 'Tributary Labour Relations', 37.

30 Elliott, *The Manchu Way*, 201, 316.

31 Although *Da Qing huidian shili* (Guangxu ban), *juan* 1216, 2a, cites evidence from 1724 (Yongzheng 2) implying that the Qing tried to prevent banner people from becoming eunuchs, the castration of Manchu and banner people could not be completely forbidden (see also Torbert,

service in the palace from the external labour market. It would be cynical to qualify their joining the palace administration as based on their own or their parents' 'free volition'[32] but, in a legal sense, they were not forced to take the irreversible decision this entailed. As the literature on eunuchs describes, boys who hoped to join the palace service underwent castration at their own risk, aware that the operation alone did not guarantee employment.[33] After their capacities had been evaluated, eunuchs were either assigned to particular palaces or specific employers of the Qing nobility. Their salaries were on the payrolls of the Imperial Household Department.

Incentives and Sanctions

Both tributary and commodified labour were remunerated and 'rewarded'. The palace regulations from the Guangxu era were very explicit about who received how much and in which rhythms of payment, suggesting that money was of paramount importance and the major incentive for most of the people working in the palaces. Was money the oil that lubricated the palace machine?

The palace regulations, namely the *Imperially Commissioned Current Regulations and Precedents of the Court*, had been compiled since the Qianlong emperor ordered it in 1742. Their purpose was to record in detail all matters pertaining to the administration of the living quarters of the palace for later reference. As with regulations and precedents in general, they were supposed to be updated periodically.[34] After the first edition,

Ch'ing Imperial Household Department, 40). The *Gongzhong xianxing zeli* (2), *juan* 4, 4a (*Gugong zhenben congkan*, vol. 280, 337) states in a commentary that, when they first entered the palace, boys and young men between 10 and 20 years of age received a bonus of six taels in the case of (Han) civilian household registration and four taels for banner household registration. This suggests that banner people also could become eunuchs, even though it was generally believed otherwise and clearly was not encouraged by the authorities.

32 Compare the fictitious dialogue in which a father asks his son whether he really has decided to become a eunuch, and the son answers that he has, because of the family's financial constraints at the time. See the biography of Sun Yaoting 孙耀庭 (1902-1996) compiled by Ling Haicheng 凌海成, Yu Binhua 余斌華 et al., *Zui hou yige taijian* 最後一個太監 (Changchun: Jinlin wenshi chubanshe, 1991), 13-14; for a translated, abridged version, see Sun Yaoting and Ling Haicheng; Uwe Frankenhauser (transl.), *Der letzte Eunuch: Das Leben Sun Yaotings, letzter Eunuch des Kaisers Puyi, erzählt von ihm selbst* (Leipzig: Gustav Kiepenheuer, 1993), 39.

33 Ibid.; George Carter Stent, 'Chinese Eunuchs', *Journal of the North China Branch of the Royal Asiatic Society*, New Series, no. 11 (1877): 143-184.

34 *Qingdai Neifu keshu mulu jieti* 清代内府刻書目錄解題目, eds. Gugong bowuyuan tushuguan 故宮博物院圖書館 and Liaoning sheng tushuguan 遼寧省圖書館 (Beijing: Zijincheng chubanshe, 1995), 234.

which is no longer extant, the annotated bibliography of the works issued by the Imperial Household Department lists editions of Jiaqing 15, Daoguang 21 and 28, Xianfeng 6, Tongzhi 9, Guangxu 6 and Guangxu 10 (i.e. 1810, 1841, 1848, 1856, 1870, 1880 and 1884). They typically do not state the names of the editors and compilers, and the bibliography assumes that they were compiled by officials of the Bureau of Eunuchs (*Jingshifang* 敬事房), the office responsible for the private life of the emperors and the harem.

The court regulations were similar in content and structure to other regulations and precedents in the Imperial Household Department. Yet, while the Imperial Household Department's regulations focused more on the wide array of details concerning the administration of imperial estates and the financial and disciplinary supervision of the department and its large echelon of officials, the palace regulations laid out the actual services required in the various palaces. Taking the last edition of 1884 as an example,[35] the first of its four chapters (*juan* 卷) is concerned with previous imperial instructions and edicts concerning the management of the palace (*xunyu* 訓諭). The second chapter presents designations and titles of the harem women (*minghao* 名號), rules for entry into the imperial genealogy (*yudie* 玉牒) and regulations for participants, places and activities during festivals and birthdays, and rites during ceremonial banquets (*liyi* 禮儀, *yanyi* 宴儀), conferment of titles (*cebao* 冊寶), and sacrificial directions (*diangu* 典故), including dress and accessories of the harem ladies (*fuse* 服色). The third chapter is dedicated to the general rules and regulations of the palaces (*gonggui* 宮規), starting with people, namely the annual allowances for every harem lady, from the empress dowager down to the wives of imperial great-great-grandsons, rewards for servants (*gongfen* 宮分) and rewards after childbirth (*yuxi* 遇喜); then specifying the material equipment of the palace buildings (*pugong* 舖宮), such as carts and carriages (*cheyu* 車輿), furniture, vessels, metal lanterns, ceramics and lacquer, as well as interior decorations such as lanterns and paravents (*anshe* 安設). The spring rites (*jinchun* 進春) and the rites to express thankfulness for imperial gratuities (*xie'en* 謝恩) form the end of the chapter. Chapter four is the most relevant to the question of labour. It lists the monetary and food allowances (*qianliang* 錢糧) for the palaces, for princes and princesses, palace maids, nurses, and eunuchs, along with their food allowances according to rank, and in cases of sickness or disability. The chapter also specifies the end-of-year bonuses for eunuchs, the amount of which was defined according to which palace building they were engaged

35 See *Gongzhong xianxing zeli* (2).

at, as well as the eunuchs' clothing (*taijian fuse* 太監服色). In parallel to the financial aspect of maintaining all these people, the chapter also has a subsection on the money needed for the annual maintenance of the buildings (*suixiu* 歲修). The final subsections cover guarding the gates to limit access (*menjin* 門禁), fire-fighting provisions, and punishments (*chufen* 處分) for palace maids and eunuchs in cases of insubordination or mistakes of any kind. As other chapters in this volume demonstrate, the 'Regulations and Precedents' focused on a combination of the monetary and material sides of the workings of the palace machine, since the Imperial Household Department was responsible for disbursing and accounting of all funds. The court regulations in addition considered social hierarchies, work duties, and rules of conduct of the people in the palace machine, thereby adding remunerations and rewards to the construct.

Money was needed to acquire materials, maintain existing structures and create new symbols of power, and to obtain the manifold types of labour services. But money also motivated the personnel, from the lowest strata of the palace workforce up to the concubines and members of the imperial family. An interesting aspect regarding the monetization of the palace economy were the allowances paid to imperial wives and concubines for the birth of imperial offspring. The highest value, 1,000 taels of silver, was assigned to a childbearing empress while, at the other end of the hierarchy within the harem, 50 taels were granted to the wife of an imperial grandson. The money was disbursed one month after childbirth, together with a certain quantity of cloth, ranging from 300 to 10 lengths.[36] The annual endowments for the same harem women outnumbered even these gratuities. In addition to money, they included long lists of gold, silver, jewellery and clothing for the ladies,[37] as well as items to furnish their palaces (*pugong* 鋪宮) in decreasing quantity from 67 items for the empress dowager down to 15 for the wife of an imperial great-grandson.[38] The empress dowager's yearly apanage, for instance, started out at 20 taels of gold and 2,000 of silver, and included about 70 more items.[39]

How can this type of labour arrangement between harem women and the emperor – or rather the court as the employing institution – be conceptualized, in light of these monetary and material rewards? As a family bond

36 *Gongzhong xianxing zeli* (2), *juan* 3 (yuxi 遇喜), 1a-2b (*Gugong zhenben congkan*, vol. 280, 329).

37 Ibid., *juan* 3 (gongfen 宮分), 1a-52a (*Gugong zhenben congkan*, vol. 280, 288-314).

38 Ibid., *juan* 3 (pugong 鋪宮), 1a-26b (*Gugong zhenben congkan*, vol. 280, 316-328).

39 Ibid., *juan* 3 (gongfen 宮分), 1a-5b (*Gugong zhenben congkan*, vol. 280, 288-290).

was created for the harem women through their reproductive service for the dynasty, it could be argued that these were reciprocal labour relations. From the perspectives of organization and remuneration, the mix of commercialization and dependence on the court can lead to the conclusion that these relationships belonged to a transitory state between tributary and commodified labour. This tallies with, for instance, the observation made by Norbert Elias on the general situation of the French lower and middle court nobility of the early 17th and late 18th centuries, and even for the higher-ranking nobles, if their income relied on the royal treasury. He stated that their lifestyle and role in the large-scale organization of the court, all obvious differences notwithstanding, resembled that of workers and employees in a large industrial enterprise.[40] Elias considered this relationship from the perspective of the complexity of the organization and the actors' dependence upon their employer(s), namely the monarch(s).

In the Qing court, eunuchs' income consisted of a mixture of allowances for food, for expenses when on official errands, rewards at the year's end and for specific events such as funerals. The palace regulations contained finely-detailed lists itemizing these various revenues, along with the rules for continuing paying for them in cases of illness, and compensating the individuals' families in the event of death, again according to the rank of the personnel (eunuch ranks were more diverse than those of palace maids). Again, this raises the question about the relative importance of commodification.[41] One distinct aspect of the monetization of eunuch labour relations was that eunuchs had to pay rent for their housing. Moreover, they could borrow money from their employer,[42] but this often led to a dependency beyond that of merely employer and worker. One potential reason why eunuchs needed to borrow money from their employer was to pay for fines imposed as punishments. The palace regulations prescribed monetary fines worth between two months' and a year's wages, and chief stewards and other high-level eunuchs were often faced with these because,

40 Elias, *Die höfische Gesellschaft*, 402.

41 An argument against monetization was made in a statement by Der Ling, *Two Years in the Forbidden City*, 278, regarding the remuneration of the painter Rebecca Carl, who portrayed the empress dowager. Der Ling claimed that the empress dowager would prefer not to pay the artist, but decorate her in reward for her efforts, claiming that in China monetary remuneration for such work would be regarded as an insult. Cixi would only consider making a monetary payment after repeated entreaties by Der Ling (ibid., 347-8).

42 *Gongzhong xianxing zeli* (2), *juan* 4 (taijian 太監), 4a, 6b (*Gugong zhenben congkan*, vol. 280, 337). The allocation of fines for offences goes into great detail and even specifies the exchange rates between several units of copper cash (*zhiqian* 制錢), namely that 350 strings of 'big cash' (*daqian* 大錢) converted into 525 strings of 'small cash' (*xiaoqian* 小錢).

in their positions as superior personnel, they were held responsible whenever any of their subordinates committed an offence.[43]

The regulations stipulated harsh punishments for palace maids and eunuchs for even minor offences, or for wrongdoings which present-day readers might not consider to be crimes at all. Severe penalties were applied to palace maids and eunuchs who attempted or succeeded in committing suicide, with gradations based on whether they did this in the palace or in the gardens, with or without metal weapons – which they were severely forbidden from possessing. The sentence for a person who survived such an attempt was exile as a military slave in Yili, a garrison in the far western province of Xinjiang. If the act was committed in the palace and the person died, their corpse would be exposed to the elements in the wilderness and their family members would be exiled as military slaves to Urumqi, the newly-built administrative centre of Xinjiang, which had been incorporated into the Qing empire by military force in 1755. If the suicide had been attempted using metal weapons, and the person could be saved, he or she was sentenced to strangulation on reprieve.[44]

Whereas the palace regulations specified in great detail the monetary remuneration for eunuchs, implying that their labour relations corresponded to a commodified form of labour, the section on 'punishments' (*chufen* 處分) makes it clear that eunuchs had very limited personal freedom. The Imperial Household Department and the Bureau of Eunuchs were supposed to be informed about the whereabouts of the eunuchs under their command at all times. Leaving their assigned workplace or walking about the palace at random was penalized with monetary fines for the supervising eunuchs and corporal punishments for the perpetrators.[45] Eunuchs were only allowed to ask for a specified number of days off to attend family events,[46] or for sick leave if their illness could not be treated in the palace, their return date was controlled, and sanctions applied in cases of delay. Sick leave outside

43 See the telling story of a theft from a workshop given in Chapter Four of this volume, where the higher the position of the people involved, the more they were fined.

44 *Gongzhong xianxing zeli* (2), *juan* 4 (chufen 處分), 9b-10a (*Gugong zhenben congkan*, vol. 280, 391-2). The gradations of punishments for attempted suicide using metal weapons meant that palace maids could at least redeem themselves if their life was saved. A corpse would be exposed in cases where an individual tried or actually committed the act in the gardens, not in the buildings, which was considered an even graver offence.

45 *Gongzhong xianxing zeli* (2), *juan* 4 (chufen), 7b (*Gugong zhenben congkan*, vol. 280, 390). For leaving the workplace and walking about the palace without authorization the supervisory eunuch (*shouling taijian* 首領太監) was punished by the deduction of two months' salary and the perpetrator given 20 strokes of the lash.

46 Ibid., *juan* 4 (chufen), 10b (*Gugong zhenben congkan*, vol. 280, 392).

the palace was granted for a maximum of one year. After that, if the ailing person had passed the age of 65 or could not get up from their sickbed, he was allowed to return to civilian status (*weimin* 爲民); otherwise he had to remain in service.[47] In other words, if an individual was suspected of feigning illness, he was forced back to duty. Thus, although entry into service as a eunuch was more or less based on free volition, usually forced by economic necessity, the work situation they found themselves in thereafter was a long-term indenture.

People, their bodies and their work, were crucial aspects of the palace as a machine – with the palace regulations taking special care to direct, monitor and control the duties, behaviour, rewards, and punishment of these bodies. Other chapters in the regulations focused on the material parts of the machine, i.e. the value and meaning of goods and their display. Where these two spheres clashed two further aspects of the hierarchies and value system in the palace became clear. First, negligence by palace maids and eunuchs in their treatment of material goods was severely punished by monetary fines imposed on the supervising eunuch officials and flogging for the perpetrators.[48] Second, illegal use of facilities in the palaces and gardens was strictly forbidden but obviously did happen, such as fishing in the ponds and lakes on the imperial estates. The latter was punished by wearing a cangue for one month, then the perpetrator was relegated to cutting grass for three years, and after that period he was only eligible to work as servant (*dangchai* 當差), either in the Office of Insignia (*Zhangyisi* 掌儀司) or in the Inner Department of Works (*Yingzaosi* 營造司).[49] This reveals that work in the mainly production-related units was considered lower status than direct service for the court nobility.

In sum, the palace relied on the Chinese legalist tradition of rewards and punishments as a means to discipline the subaltern workforce. Work in the palace could yield monetary, material and immaterial rewards, such as status and prestige; but it could also entail punishing monetary fines. More demeaning

47 Ibid., *juan* 4 (chufen), 17b (*Gugong zhenben congkan*, vol. 280, 395).

48 Ibid., *juan* 4 (chufen), 4a, 5b, 6b. These rules were only formulated for eunuchs. Two conflicting statements from the early 20th century in the service of Cixi are those by He Rong'er, who claimed that, unlike eunuchs, palace maids were never hit in the face (Jin Yi, *Gongnü tanwang lu*, 8) and by Der Ling, who described seeing a maid slap a colleague's cheeks on Cixi's orders as a punishment for bringing her two socks that did not match (Der Ling, *Two Years in the Forbidden City*, 119). Der Ling also claimed that Cixi always carried a yellow bag containing bamboo sticks for flogging eunuchs, palace maids and older female palace servants wherever she went, to have it at hand when she thought that staff needed to be punished immediately (ibid., 117).

49 *Gongzhong xianxing zeli* (2), *juan* 4 (chufen), 24a/b (*Gugong zhenben congkan*, vol. 280, 399).

and hazardous to health were the corporal punishments, usually in the form of carrying a cangue or being lashed with a bamboo cane. The demotion from serving nobles inside palaces to workplaces that required physical labour can also be considered a form of corporal punishment. In addition, it offered reduced or no opportunities for personal contact with the ruling classes, so was intended as a further means to discipline the palace workforce.

Palace Personnel's Self-Perception

There are relatively few accounts of everyday life in the palace – the most famous being three from the end of the monarchical system. There is firstly one by the palace maid He Rong'er, who lived in the palace between ca. 1890 and 1901 and accompanied the court's flight to Xi'an in the wake of the Boxer Rebellion in 1900;[50] then there is the narrative by Der Ling (Yu Deling 裕德齡, 1885-1944), who served as a lady-in-waiting for the Empress Dowager Cixi 慈禧 (1835-1908) between 1903 and 1905, and finally the biography of Sun Yaoting 孫耀庭 (1902-1996), who became an eunuch after the demise of the Qing dynasty in 1911 to enter the service of the princely palace of the emperor's uncle Zaitao 載濤 (1887-1970). Sun later moved to the Forbidden City to work for the Imperial Consort Dowager Duankang 端康皇太妃 (1873-1924), for the abdicated Emperor Pu Yi 溥儀 (1906-1967) and for his Empress Wanrong 婉容 (1906-1946) until the imperial clan was expelled from the Forbidden City in 1924. All three narrators depicted the palace in different ways and each had particular motives in how they explained their own roles in its organization. He Rong'er and Sun Yaoting did not find sympathetic listeners and authors to record their accounts until after the foundation of the People's Republic in 1949. Moreover, the publication of both narratives had to wait until after the more rigid class-war period of the first thirty years of the People's Republic, and thus only appeared long after the events described. In the case of He Rong'er, at least, the information had already been gathered in the 1940s by the historians Jin Yi and Shen Jinling. He and Sun never left China, and their intended audiences were first and

50 No precise dates are given for He Rong'er's birth and death in the text. According to the preface she entered the palace aged 13, was married at 18, returned to service shortly thereafter, and accompanied the empress dowager into exile in 1900. After the court returned in 1901, she was forced to resign from service because palace maids were not supposed to serve after the age of 25 (Jin Yi, *Gongnü tanwanglu*, 2). The figures do not quite tally, but suggest a birth date of 1876 and entry into service at about 1889. The preface states that He Rong'er returned to her hometown and died there in 1950 (ibid., 3).

foremost Chinese readers. Der Ling's story was different. Her father was a high-ranking bannerman in the Qing diplomatic service in Japan and France, and she had acquired a Western education while in France. Der Ling published her book, and several subsequent memoirs, in English for a non-Chinese audience, with the intention of correcting the exaggeratedly negative image depicting the empress dowager.[51]

Comparing the view of the palace portrayed in these three narratives, the smallest common denominator is that of an orderly, highly regulated, hierarchical organization that enforced its rules strictly. It is striking that He Rong'er's recollection contains little criticism, but rather nostalgia for the days of yore. She presents a vivid picture of the harshness of life outside the palace walls for out-of-work palace girls and eunuchs, especially after the demise of the Qing. He Rong'er considered it a great honour and a special privilege to be allowed to return to serve the empress dowager again after having left the palace to get married.[52]

Nevertheless, He Rong'er's account also includes the aspect of extreme self-restraint that was demanded of all palace servants. Private feelings should never be shown, and palace girls and eunuchs were supposed to speak as little as possible and in a quiet voice. The metaphors she used were that 'everybody was as if covered in a layer of wax skin' (*meiren you yi ceng lapi baozhe side* 每人有一層蠟皮包着似的), and that the 'inside of the palace resembled an ice vault, where people would shrink back everywhere' (*gongli jiu xiang bingkao yiyang, rang renmen chuchu dou yao suoshou suojiao de* 宮裡就像冰窖一樣, 讓人們處處都要縮手縮腳的).[53] The 'hardest chore' (*zuikude chaishi* 最苦的差事), in her eyes, was depicted in the machine-like movements of the eight palace girls whose task it was to constantly sweep dust from the tiled floors that were warmed with an underfloor heating system. The empress dowager's entire residence needed to be kept meticulously dust-free, being washed with water three times every morning. The eight women wiped the floor on their knees, working in four two-person teams. They were under enormous time pressure, since they had to wait until Cixi had risen and gone to her morning audience. The cleaners had to finish their task, without leaving any footprints, by the time the empress dowager returned. He Rong'er explained that 'those eight palace girls rose at five in the morning and toiled hungrily until after nine. Crouching on the floor,

51 Li Yuhang and Harriet Zurndorfer, 'Rethinking Empress Dowager Cixi through the Production of Art', *Nan Nü*, 14/1 (2012): 1-20, here 6-8; Hayter-Menzies, *Imperial Masquerade*, 275-278.

52 Jin Yi, *Gongnü tangwang lu*, 2.

53 Ibid., 15.

they couldn't eat. In the midst of winter, those eight would still have their faces covered in sweat'.[54] But, according to He Rong'er, the lives of those at the top of the echelon was also miserable and bound by conventions. The empress dowager in particular did not 'enjoy herself', but 'suffered hardships' (*shouzui* 受罪) because there was nobody she could trust. The entire court life basically resembled an opera performance (*xiang changxi yiyang* 像唱戲一樣) which involved concealing one's actual emotions and opinions.[55] For the empress dowager, said He Rong'er, the only way to dispel her feeling of isolation was to study the throne memorials submitted by high-ranking officials and respond to these: in other words, to exert power.[56] From a subaltern perspective the framework of the palace might have felt like a mindless machine. Everybody, from the very highest to the lowest echelons of the hierarchy, formed part thereof and was subdued by its mechanisms. Nevertheless, those in higher positions had the power to change the rules by which it ran, while the lower ranks could try to change their position or movements to possibly evade those rules and conventions.

The 'punishments' section of the palace regulations cited a famous example of misconduct to prove how severely illegal behaviour was punished. In the case in question, An Dehai 安德海 (1837-1869), a high-ranking eunuch official previously favoured by Empress Dowager Cixi, was accused of acting inappropriately while travelling on an official mission. Mr. An was beheaded for this abrogation of imperial prestige.[57] Another example shows that the rules of conduct were not always maintained. The regulations forbade female palace servants and eunuchs from even speaking to each other.[58] Yet, the cohabitation of eunuchs and elderly seamstresses and embroiderers in a peripheral part of the Yuanmingyuan, called Suzhou Street[59] was an issue that, as He Rong'er expressed, 'the civilians did not talk about and the officials did not investigate'.[60]

54 Ibid., 52; the French translation is incomplete here (Jin Yi, *Mémoires d'une dame de cour*, 84).

55 Ibid., 58.

56 Ibid.

57 *Gongzhong xianxing zeli* (2), *juan* 4 (chufen), 31a/b (*Gugong zhenben congkan*, vol. 280, 401); also mentioned in section 'taijian', 64a/b (*Gugong zhenben congkan*, vol. 280, 374), with the date of the memorial of Tongzhi 8/8/11 (i.e. 16 Sept. 1869) on this incident by Shandong provincial governor Ding Baozhen 丁寶楨. See also Rawski, *The Last Emperors*, 189-90. Accordingly, An Dehai probably fell victim to a power struggle between Cixi and the regent, Prince Gong. For a popular version of the event, see Jin Yi, *Gongnü tanwang lu*, 40-41.

58 *Gongzhong xianxing zeli* (2), *juan* 3 (gonggui 宮規), 4a-5a (*Gugong zhenben congkan*, vol. 280, 287).

59 Jin Yi, *Gongnü tanwang lu*, 174.

60 Ibid.

After the abdication of the Qing in 1912, the rules by which the palace machine was run remained in force for a certain time.[61] The Qing court continued to be under the command of the Imperial Household Department with a still considerable number of eunuch servants on duty. Two well-known interruptions of its smooth workings were first, President Yuan Shikai's 袁世凱 (1859-1916) aspiration to monarchical rule in 1916 and second, General Zhang Xun's 張勳 (1854-1923) failed attempt to restore the Qing monarchs. Even under Republican China's command over the palace, neither the ex-emperor Pu Yi nor his consort Wanrong had freedom of movement or could liberate themselves from their function as part of the machine. This remained the case even after they were expelled from the palace in 1924 along with most of the eunuchs. For the eunuchs, the rules of conduct became more relaxed as, in the anti-monarchic public opinion of the period, they were seen as victims of an inhuman Manchu demand for mutilated servants who had deprived themselves of the chance to reproduce. They were forced to suffer extreme, lifelong health hazards, simply to ensure that all the children born in the palace were offspring of the Qing emperors.

The biographical narrative of one of the last eunuchs, Sun Yaoting, is an interesting source of information about the situation of the court and the scattered 'human' parts of the palace machine after the demise of the Qing. Sun's oral testimony (*koushu* 口述) has been published in several versions and translations, and the content has been adapted by various editors. Differing from He Rong'er's account, Sun Yaoting's experiences are embedded into a novel-like historical narrative and are not meant to be taken literally. Rather, these adaptations partly reflect the convictions of the late 20th-century editors.[62] According to Sun, the obvious incentive for eunuchs were the money, food, and material income they received for their service. Working in the imperial palace was the ultimate goal, because

61 The copy of the *Gongzhong xianxing zeli* in the holdings of the School of Oriental and African Studies, London, carries a dedication from Puyi to Sir Reginald Johnston, his tutor between 1919 and 1924. Since Johnston was formally employed by the Imperial Household Department and had access to the palace interior, the ex-emperor may have considered this to be a useful work of reference.

62 I follow Ling Haicheng's version of Sun Yaoting's narrative, see Ling Haicheng et al., *Zui hou yige taijian*, 13-14; an earlier version is the shorter arrangement of the narrative by Zhang Wenhui 張文惠, *Zui hou yige taijian* 最後一個太監 (Hangzhou: Zhejiang shaonian ertong chubanshe, 1984), which was obviously intended for a young audience; a later version is Jia Yinghua 賈英華, *Modai taijian Sun Tingyao* 末代太監孫耀庭 (Hong Kong: Zhonghua shuju, 2014), which expands the narrative up to the period after Sun Yaoting's death. See also footnote 32.

service in one of the princely residences was much less well paid.[63] In Sun's view the Qing court system had fed, clad, and housed eunuch servants for more than two centuries, thus the most important thing for them was to remain inside the system rather than to be liberated from it. In contrast, Sun asserted, for Pu Yi's Empress Wanrong, the palace resembled a prison. Wanrong had been fifteen years old when she got married in 1922. Until the imperial court was expelled from the palace two years later, she had been forced to cease all her previous leisure activities outside the palace, such as going to the cinema with friends and clothes shopping, and had to restrict herself to reading and playing the piano, or occasional games of carom billiard with eunuchs and palace ladies inside the palace.[64] Sun also described how the image and function of a eunuch had been challenged by societal changes, since Pu Yi was aware that other royal courts around the world, such as in Britain, Japan, and Sweden, did not use eunuchs. He was also afraid of being attacked or killed by disgruntled eunuchs and wanted to dismiss them all.[65] However, his uncle Zaitao had argued that the female servants would not be able to shoulder all the tasks in the palace that required physical strength, and that it would be even more dangerous to hire male servants instead of eunuchs.[66] Eventually, 800 to 900 eunuchs were dismissed on 16 July 1923, with only about 175 being allowed to stay on. Sun Yaoting was not among them[67] and, having already risen to a certain position within the eunuch hierarchy, had to bury his dreams of becoming chief eunuch. Thereafter he led a life with intermittent assignments serving the abdicated imperial court. For a short period in about 1940 he took up a post in Changchun as a personal servant to Pu Yi, then the Kangde 康德 emperor of Manchuria (*Manzhouguo* 滿洲國).[68]

Sun Yaoting had begun his service when the dynasty and its ability to enforce the court rules was already defunct. This must surely have influenced his account and may explain the greater freedom of movement and contacts that seemed possible for him than for other palace employees.

63 In a dialogue – which presumably did not take place precisely as it is recorded in Ling's book – an old eunuch explains to the novice the riches a eunuch could generate by working in the imperial palace. Ling Haicheng et al., *Zui hou yige taijian*, 25-26, 65.

64 These pastimes inside and outside the palace show the interest the (ex-)imperial couple had in Western culture. Ling Haicheng et al., *Zui hou yige taijian*, 174-5, 179.

65 Ling Haicheng et al., *Zui hou yige taijian*, 160.

66 Ibid.

67 Ibid., 159-160.

68 Ibid., 311. The text mentions that this occurred after Sun's 39th birthday, thus circa 1940 or 1941.

Der Ling's book recounting her experiences at court differed from He Rong'er's and Sun Yaoting's stories in several respects. She wrote it herself shortly after the event, instead of telling her story to a much younger author who had no first-hand knowledge of Qing palaces; and her intention was to defend the deceased Empress Dowager Cixi against accusations which she deemed unfair. It is safe to assume that both He Rong'er's and Sun Yaoting's biographers knew about or had read Der Ling's book, which was translated into Chinese from English in the 1920s[69] and serialized in the magazine *Dongfang zazhi* 東方雜誌 (*Eastern Miscellany*). A more important difference resulted from the fact that Der Ling stood in a socially more elevated position than He and Sun, as a lady-in-waiting rather than a servant, and thus had other, more highly respected tasks in the palace. For example, she looked after the empress dowager's jewellery and acted as an interpreter for foreign guests who paid visits to Cixi.[70] Just like He Rong'er and Sun Yaoting, she described the rules of the palace as being robustly enforced, with corporal punishment for mistakes made by staff, and as a place humming with tasks which kept workers busy from dawn till dusk. Due to her position in the palace's hierarchy and her particular assignment as a cross-cultural mediator (a role she also maintained during her later life when she moved to the United States with her American husband in 1928),[71] her portrayal of the palace is brighter than that of the other two books. Der Ling was very critical of the eunuchs, describing them as 'brainless people' in a continuous fight for survival: 'Whenever Her Majesty gave an order they always said "Jer" (Yes) and as soon as they got to our waiting room they would say to each other: "What was the order? I have forgotten all about it," asking the palace ladies to inform them, because they had not listened to the order'.[72] Der Ling thought that the female palace servants were 'a much better class of people than the eunuchs', since they were recruited from the daughters of Manchu soldiers.[73] In fact, they had to be daughters of banner people (such as Der Ling herself), who were not necessarily all of Manchu descent. Of course, ethnic affiliation was a central issue in Qing China, since it defined and legitimized the tributary labour relations. Later historiographers pointed out

69 Deling 德齡, *Qing gong ernian ji* 清宮二年記, translated by Dongfang zazhishe 東方雜誌社 (Shanghai: Shangwu yinshuguan, 1924); eadem, *Qing gongjin ernian ji* 清宮禁二年記 (Chengdu: Changfu gongsi, 1920).

70 Der Ling, *Two Years in the Forbidden City*, 61.

71 Hayter-Menzies, *Imperial Masquerade*, 259-262. In the United States, she worked as a freelance writer and a Chinese teacher at the University of California at Berkeley. Ibid., 330.

72 Der Ling, *Two Years in the Forbidden City*, 121-122.

73 Ibid., 120.

that Der Ling herself was not a 'princess' of Manchu nobility, but a banner woman, probably of Han Chinese descent.[74]

None of the above accounts literally described the palace as a 'machine', but as a conglomerate of people and things which had manifold restrictions on their movement and behaviour. The narrations are full of statements about hierarchy and regulation, and on the material wealth it represented, such as the jewels in the boxes Der Ling was entrusted with. Describing the empress dowager's and the emperor's enthusiasm for costly new gadgets such as electricity and a steamboat for the lakes, her account shows the empress dowager and the emperor as being not too dissimilar from contemporary monarchs elsewhere in the world.

Reconsideration of the Model of a Machine for the Qing Palace

How can the model of a machine be applied to the everyday life and production within a palace? Michael E. Meeker's portrayal of the Istanbul palace buildings in the epoch of the Ottoman Sultan Mehmet the Conqueror (reg. 1451-1481), for instance, conceives of the palace building as a machine which enabled and represented the ruler's gaze from inside to outside, while he was enabled to remain unobserved.[75] Meeker's machine as a thinking model thus places the elements of control and line of authority on centre stage.

Present-day specialists on organizations and management hold that the organizational metaphor of a machine tends to disregard human agency.[76] In their view, the term 'machine' implies specialization in the sense of a division of labour, standardization, replaceability, and predictability, with an emphasis on orderliness, rather than dynamics and adaptability.[77] Yet, for the labour conditions described above, the machine as a thinking model seems appropriate. Within the palace and its several thousand-strong staff working in a sensitive, iconic space, there was a high degree of specialization with a meticulously fine-grained division of labour; standardization was intense and dictated by written-down rules, and although palace maids and

74 Hayter-Menzies, *Imperial Masquerade,* preface, xvi-xvii.

75 Michael E. Meeker, *A Nation of Empire* (Berkeley: University of California Press, 2001), 121.

76 Kathy Milhauser, 'Organizational Metaphors', PPT presentation on YouTube, https://www.youtube.com/watch?v=Xt_1qCHH_Qk, uploaded 24 Feb. 2014, accessed 28 Aug. 2020, 0:49 ff.

77 Kate Dieffenbach Wall, 'The Machine Metaphor in Modern Organizations', video presentation on YouTube, https://www.youtube.com/watch?v=k6OXf_A99KU, uploaded 20 Apr. 2012, accessed 28 Aug. 2020, 1:02ff., referring to Katherine Miller, *Organizational Communication, Approaches and Processes*. Belmont, CA: Wadsworth, 2006.

the lower ranks of the eunuchs and other palace personnel often were not literate enough to read the palace regulations themselves, they were aware of, and alert to, what He Rong'er frequently referred to as 'the family law of the palatial ancestors' that set the standards for usages and practices in the palace.[78] Furthermore, as all the functions were clearly defined the palace women and other residents were easily and frequently replaced – whether that was due to age restrictions or because they had violated the rules. Predictability and continuity at the palace were ensured not only by its role as the iconic centre of Qing rule, but also by its image as a long-established institution as an 'employer' that offered rich rewards to the higher echelon officials and a decent living to the mass of the personnel. The palace held onto this function right until the end of the monarchy, and even for some time afterwards.

This did not mean that the palace could not cope with change. Despite the monarchic system that was intended to function by continuous, stable patrilineal succession, the practices of everyday life, production and maintenance in the palace afforded continuous dynamic adjustments. This is evident from the periodic revisions of the rules and regulations, from the use of new techniques and materials, and from updates in the costs of certain work assignments.

As to labour, labour relations, and labour conditions, from the very beginning of the Qing dynasty, the court had applied a remuneration and a tributary system to the Manchus and the banner people, as opposed to the indenture system imposed on its eunuchs and selected specialists in craft production. This situation can be observed until the very end of the Qing. External viewers were brought into the court by receiving foreign visitors, mediated by persons such as Der Ling, or by hiring a British tutor for the abdicated emperor.

The dismissal of almost the entire eunuch workforce in 1923 was a sharp rupture, a precursor to the court's expulsion from the Forbidden City in 1924. The Qing palaces were not perpetual motion machines, even though attempts to save the monarchy were made (see Yuan Shikai 1916, Zhang Xun 1917) and realized in the ephemeral Manzhouguo court, which was, in many ways, modelled on the Japanese court.

Its many thousands of servants had diverse incentives for making the Qing palace machine work. For the banner people, their ethnic identification with the ruling house also motivated those who propelled the machine from below, such as the palace maids, casual labourers, or palace guardsmen.

78 Jin Yi, *Gongnü tangwang lu*, 62.

This implied obligations and entitlements of a tributary nature. The Han Chinese eunuch servants were motivated by the prestige and economic advantages that an appointment in the palace offered. In their case, labour was commodified and indentured. The extent to which people would go in order to realize the advantages the palace institution might provide, were quite tragic, including self-maiming and acceptance of the rigid and unjust behavioural regulations. Even after the demise of the Qing, the lack of other options made Sun Yaoting and colleagues return to the already largely dysfunctional court. Yet He Rong'er's life outside the palace, when she worked as a housekeeper in Beiping 北平 (as Beijing was called at that time), was no easier than within.

As long as they were serving and functioning, the palace machine provided well for the people that made it run. Yet it did not always work as the mechanisms of palace regulations and ancestral laws stipulated – and this can be seen as one of its core characteristics. The machine was designed to realize abstract goals, whereas the people operating it did not only function according to the prescribed mechanisms, but also harboured their own agendas and agency.

About the Author

Christine Moll-Murata is Chair professor at the Department for History of China, Ruhr University Bochum, Germany, and honorary fellow of the International Institute of Social History, Amsterdam. Her research focuses on labour history of China, Taiwan, and Japan, the history of crafts and early industrialization in China and East Asia.

2 Manager or Craftsman

Skillful Bannermen of the Qing Dynasty (1644-1912)

Kai Jun Chen

Abstract

The chapter investigates how the 'palace machine' of the Qing dynasty reproduced (or systematically trained) particularly skilled bannermen as ethnically-marked official experts. By mapping out these bannermen's education, training process, and official appointments, I explain how the court system perpetuated the administrative privilege of bannermen families and how specific skills of different generations matched the particular demands of empire building projects of the Qing dynasty in different stages. I focus on a representative family, the Wanggiyan/Wanyan clan, generations of which served the court within the institutional framework of the Imperial Household Department. Placing this extended family in the context of peer bannermen equipped with specialized skills allows me to shed light on the larger issue of the relationship between hereditary status and specialized skills in the Qing palace machine.

Keywords: expert, specialized skill, family, Wanyan clan

This chapter investigates the mutual formation of the 'palace machine' of the Qing dynasty and the ethnically-marked official experts, that is, the bannerman specialists involved in military campaigns, finance, and large-scale construction, all aspects of statecraft that required specialist knowledge outside the purview of Confucian classics. In addition to capturing the regulated mechanism of several innovative court bureaus established by the Manchu rulers, as outlined in the Introduction, the concept of a machine, *ji* 機, illustrates the pivotal role of skilled labourers

Siebert, Martina, Kai Jun Chen, and Dorothy Ko (eds), *Making the Palace Machine Work: Mobilizing People, Objects, and Nature in the Qing Empire*. Taylor & Francis Group, 2021
DOI: 10.5117/9789463720359_CH02

in practical statecraft.[1] Together with the lower-ranked workers examined in Chapter One, the princes and bannerman experts working in the Imperial Household Department comprised the human operators who drove the 'palace machine'. In the very process of operation, the mechanism reproduced qualified workers. The official *Regulations and Precedents* (*zeli* 則例) and the constantly-updated *Precedent Cases of Statutes of the Great Qing* (Da Qing huidian shili 大清會典事例) documented the bureaucratic structure and its transformation in principle. However, that schematic framework, by its very nature of explaining how things *should* be done, left out the detailed process of actual production and the workers' specialist skills, which is the focus of this chapter.

By scrutinizing a few representative bannermen specialists' education and training processes, career patterns, and cultural output, this chapter examines how the palace machine perpetuated a group of officials' administrative privilege, and how the skills of different generations of a bannerman lineage met the particular challenges of the empire-building project in different phases. Scholars have clarified the evolving Manchu bureaucratic system in which bannerman bondservants elevated their status from house slaves over time.[2] After the establishment of the Eight Banner system, the term 'bondservants' came to refer to servants registered in banner companies (Chinese: *zuoling* 佐領; Manchu: *niru*). The bannerman bondservants were trusted accountants, managers, or master technicians who occupied the top tier of the pyramid of workers.[3] Through a carefully-channelled memorial system, they responded

1 The word *ji* even appears in the name of the Grand Council (*Junjichu* 軍機處), one of the three innovative bureaus created by the Qing ruling house. See Beatrice S. Bartlett, *Monarchs and Ministers: The Grand Council in Mid-Ch'ing China, 1723-1820* (Berkeley, CA: University of California Press, 1991). The other two were the Imperial Household Department and the Ministry for Ruling Outer Provinces (*Lifanyuan* 理藩院). See Dittmar Schorkowitz and Chia Ning, eds., *Managing Frontiers in Qing China: The Lifanyuan and Libu Revisited* (Leiden: Brill, 2017). For an explanation of the concept of *ji* 機 in traditional military thought in China, see Earnest Caldwell, 'Opportune moments in early Chinese military thought: the concept of *ji* 機 in the Warring States period manuscript Cao Mie's Battle Array', in *Chinese and Indian Warfare – From the Classical Age to 1870*, eds. Kaushik Roy and Peter Lorge (London: Routledge, 2014), 17-31.

2 Evelyn S. Rawski, *The Last Emperors: A Social History of Qing Imperial Institutions* (Berkeley, CA: University of California Press, 2001), 166-71. Mark Elliott, *The Manchu Way: The Eight Banners and Ethnic Identity in Late Imperial China* (Stanford, CA: Stanford University Press, 2001), 81-84. Liu Xiaomeng 劉小萌, *Manzu cong buluo dao guojia* 滿族從部落到國家 (Beijing: Zhongguo shehui kexue chubanshe, 2007). See also Qi Meiqin 祁美琴 and Cui Can 崔燦, 'Baoyi shenfen zaibian 包衣身份再辨', *Qingshi yanjiu* 清史研究, no.1 (2013): 117-28.

3 For a brief comment on powerful bondservants, see Rawski, *The Last Emperors*, 171; for a detailed case study of an illustrious bondservants' clan, see Jonathan Spence, *Ts'ao Yin and the K'ang-hsi Emperor: Bondservant and Master* (New Haven: Yale University Press, 1966).

directly to – or even negotiated in person with – emperors about details of empire-building projects, bypassing the regular bureaucracy.

The character of the bannerman bondservants' remits reveals a conflation of state institutions and household organizations. As the pre-dynastic royal clans built the Qing dynasty and the Aisin lineage managed to reduce peer aristocrats' power, bondservants witnessed an expansion of hierarchy among themselves. Only bondservants from the upper three banners which directly responded to the ruling house would function as the rulers' confidantes in the Imperial Household Department. This chapter focuses on bannerman bondservants from the companies in the upper three banners, who formed the core personnel of the Imperial Household Department. Meanwhile, some bannerman specialists from prestigious clans did not register in bondservants' companies and worked outside the Imperial Household Department.[4] Others, such as the pre-dynastic keepers, hunters, herders, gatherers, soldiers, farmers, and petty traders accompanying the Aisin warriors to the court in Beijing, became assistants in the project of ruling the empire by translating documents and the classics, collecting taxes, running manufactories, and intervening in various ministries in the civil government. One vivid illustration of such transforming responsibility was found in a memorial from the 16th year of Kangxi reign (1677),[5] concerning changing the names of subdivisions in the Imperial Household Department from tribal groupings to more bureaucratic offices. To date, scholars have mapped out the intricate organization of bondservants in various offices which were in charge of practical statecraft, but have left one question unanswered: how did bondservants acquire their specialist skills at court and keep the palace machine working? The mechanism of assessment and promotion sheds some light on bondservants' career patterns, as I will discuss below.[6] Official genealogies provide information on the extended lineage of bannermen, including bondservant companies, and list their

4 The Imperial Household Department is best described as the emperor's private treasure vault, business company, and manufacturing facilities rolled into one. See Qi Meiqin, *Qingdai Neiwufu* 清代內務府 (Beijing: Zhongguo renmin daxue chubanshe, 1998); Preston M. Torbert, *The Ch'ing Imperial Household Department: A Study of Its Organization and Principal Functions, 1662-1796* (Cambridge, MA: Harvard University Press, 1977); Chang Te-Chang, 'The Economic Role of the Imperial Household in the Ch'ing Dynasty', *The Journal of Asian Studies*, vol. 31, no. 2 (1972): 243-273.

5 'Ga-lu deng wei zunzhi huiyi Zongguan Neiwufu xia yashu gengming jiyin de tiben 噶魯等為遵旨會議總管內務府下衙署更名給印的題本', in *Qingdai neige daku sanyi manwen dang'an xuanbian* 清代內閣大庫散佚滿文檔案選編, edited by Liaoning shehui kexueyuan lishi yanjiusuo 辽宁社会科学院历史研究所 et al. (Tianjin: Tianjin guji chubanshe, 1991), 1-3.

6 For an extensive discussion of the career patterns of Manchu and Han Chinese at the Qing court, see Raymond Chu and William Saywell, *Career Patterns in the Ch'ing Dynasty: The Office of Governor-General* (Ann Arbor: Center for Chinese Studies, University of Michigan, 1984).

official positions. Even more detailed information on bondservants' skills and their agency in the palace machine is available in individual bondservants' collected writings, which are seldom examined by historians.

Family Resemblance

One of the key characteristics of bondservants as the personnel of the palace machine was the family resemblance of their career pattern. A father and a son might not start from the same position at court and might not progress through the same series of posts. Yet the members of six or more generations of a lineage tended to recurrently hold court and regional posts reserved for bannermen, or even civil posts frequented by bannermen. Moreover, the phenomenon of recurrent occupation across generations in a lineage also took place in peer lineages of bondservants, probably because bannerman bondservants were constantly trained and appointed to a set of bureaus which were central in the palace machine. Wang Zhonghan and Liu Xiaomeng have identified more than twenty lineages of bondservants who regularly occupied posts in custom houses, manufactories, salt administration, and river conservation (*hefang* 河防)[7]. This chapter examines the work of one especially extended lineage, the Wanggiyans (Chinese Wanyans 完顏), to illustrate a general career pattern which was common among other lineages like them, thereby exploring the co-productive relationship between bannerman specialists' skills and the development of court institutions in different phases of the Qing dynasty (see Figure 2.1).

Military Allies and Translators

The roots of banner organization lay in military service.[8] A company, literally meaning arrow in Manchu, was the unit by which bannermen, including bondservants, were integrated into the privileged banners. Many bondservants' families traced their pre-dynastic ancestors back to loyal warriors. Wanggiyan Huqiha 完顏乎齊哈 (active 1642) and Wanggiyan

7 Wang Zhonghan, 'Neiwufu shijia kao 內務府世家考', in idem, *Qingshi xukao* 清史續考 (Taipei: Huashi chubanshe, 1993), 524-541. Liu Xiaomeng 劉小萌, 'Guanyu qingdai neiwufu shijia 關於清代內務府世家', in *Mingqing shi luncong* 明清史論叢 (Shengyang: Liaoning daxue chubanshe 2004), 138-158.

8 The kinsmen in the Aisin clan and their allied friends (*gucu*) formed the cornerstone of the early military organization, and their offspring carried on serving in army.

Figure 2.1 Wanggiyan clan's family tree and genealogy

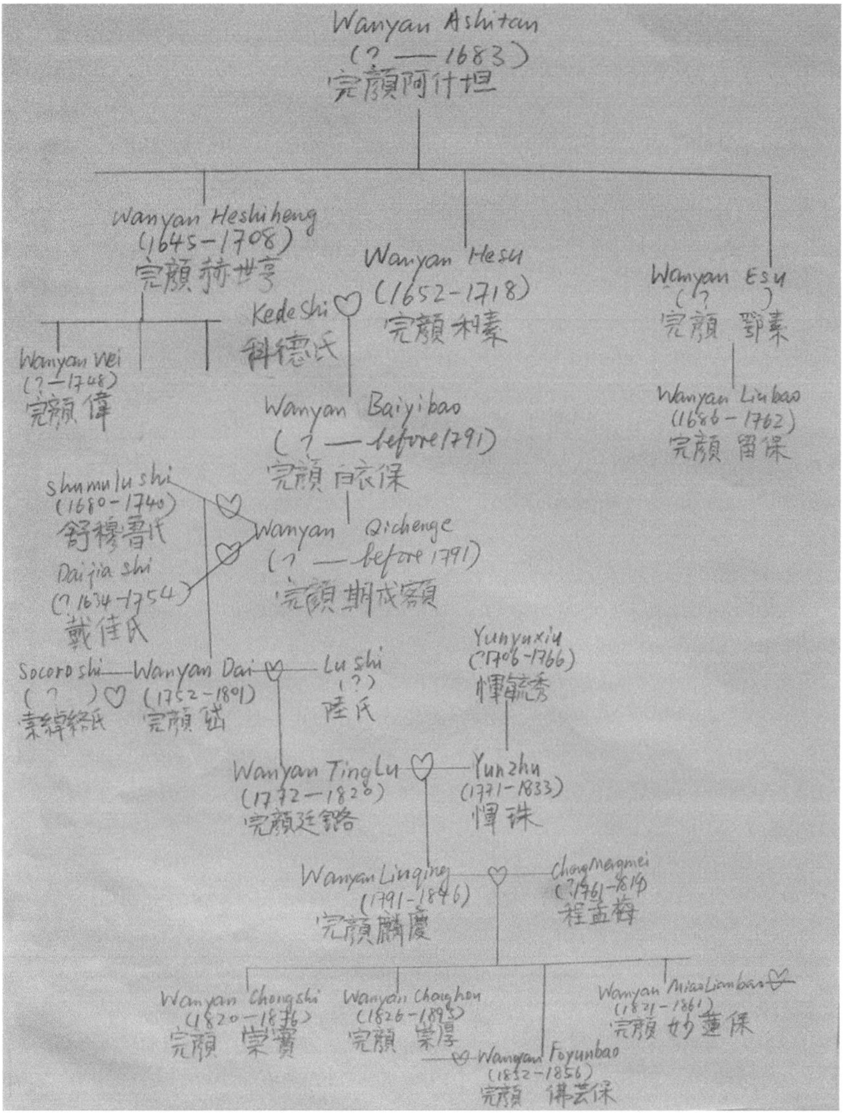

Reconstruction and hand drawing by Kai Jun Chen

Daqiha 完顏達齊哈 (active 1644) were among Aisin Gioro Nurgaqi's 努爾哈赤 (1559-1626) earliest military allies.[9] They led one of the nine Manchu

9 Ding Yizhuang 定宜莊, 'Neiwufu Wanyan shijia kao 內務府完顏世家考', in *Qingshi luncong* 清史論叢 (Shenyang: Liaoning guji chubanshe, 1995), 133-34; E'ertai 鄂爾泰 (Manchu:

companies in the Imperial Household Department in 1644. In the conquest wars, they were able to command battalions equipped with cannons, which primarily consisted of surrendered Chinese soldiers. This suggests that the Wanggiyan ancestors commanded at least rudimentary skills in communicating with Chinese soldiers. The early Wanggiyans' bilingual aspirations is further evidenced by a copy of the *Rules of Composition* (Wenzhang guifan 文章軌範), a Chinese anthology of model essay compositions dated from the Song dynasty, held in their descendants' library.[10] Daqiha 達齊哈 is said to have grabbed the book during his raid on Shandong. Thus, it is not surprising that at least three members of the following two generations of the Wanggiyan (now sinicized to Wanyan) lineage made their name in translation and translingual communications.

Communicating information across language barriers must have been an extremely valuable skill for the ruling house to have, to appropriate knowledge and extract resources and labour in the expanding territory. Confucian classics and primers of various sorts took priority in the early translation projects, both as a source of ideology and a reservoir of practical knowledge. Daqiha's son, Wanyan Ashitan 完顏阿什坦 (?-1683) was the premier translator at the court, and rendered Confucian classics into Manchu. He also served as a tutor and high-ranking secretary, both at court and in the civil government.[11] It was not uncommon for bannermen to change jobs between the court bureaus, namely the Ministry for Ruling Outer Provinces (*Lifanyuan* 理藩院), the Imperial Household Department, and later, the Grand Council (*Junjichu* 軍機處). They could even serve in civil government bodies such as the six boards or provincial governments, as long as they passed the civil service examination. Although Ashitan's lineage is particularly remembered as degree-holders in the civil service examination, members of successive generations followed two career routes reserved for bannermen: scribe (*bitieshi* 筆帖式, sinicized from the Manchu term *bithesi*) or imperial guard (*shiwei* 侍衛).

A *bithesi* or scribe was the ubiquitous clerk in almost all bureaus across the empire, often in charge of copying, translating, and archiving official

Ortai) et al., *Baqi manzhou shizu tongpu* 八旗滿州氏族通譜 (Beijing: Wuying dian, 1744), *juan* 28, 9-10.

10 See chapter 'Langhuan changshu 嫏嬛藏書 (Langhuan Library)', in Wanggiyan Linkin (1791-1846; Chinese: Wanyan Linqing 完顏麟慶), *Hongxue yinyuan tuji* 鴻雪姻緣圖記 (Yangzhou: Private Print, 1847), 911.

11 Huang Li-chun 黃麗君, 'Qingchu manhanwen fanyijia Wanyan Ashitan shihuan shengping kao 清初滿漢文繙譯家完顏阿什坦仕宦生平考', in *Gongting yu haiyang de jiaohui* 宮廷與海洋的交匯, ed. Li Qilin 李其霖 (Taipei: Tamkang University Press, 2017), 91-117.

documents.[12] The job was reserved for bannermen. Some might inherit a low-ranking scribe role, while others would purchase a position, especially later on in the Qing dynasty. Yet the civil service and special Manchu examinations were clearly devised to ascertain a scribe's literacy skill – and the bilingual capabilities of candidates attending translation examinations.[13] Ashitan's second son, Wanyan Hesu 完顏和素 (1652-1718)[14] began his career as a scribe and went on to chair the Inner Bureau of Translation (Manchu: *Dorgi bithe ubariyambure boo*; Chinese: *Nei fanshu fang* 內翻書房) in 1688.[15] Although Hesu was believed to have translated vernacular novels such as *The Romance of the West Chamber* (Xixiang ji 西廂記) and *The Plum in the Golden Vase* (Jinpingmei 金瓶梅), his signed translation falls squarely within Confucian doctrines. His Manchu translation of Kangxi's *Imperial Discussion of Triangles and Computation* (Yuzhi sanjiaoxing tuisuan falun 御製三角形推算法論) demonstrates a basic mathematical knowledge as well as the craft of bending scientific treatises to fit ideological discourse.[16]

Later on, as an supervisor in the Imperial (Printing) Workshops (*Zaobanchu* 造辦處) of the Wuying Palace 武英殿, Hesu's mediating position between a Jesuit named Bouvet and Emperor Kangxi was continued by his elder brother Wanyan Heshihen's 完顏赫世亨 (1645-1708) management of Jesuit missionaries. Heshihen had probably commenced his career in a

12 Chen Wen-shih 陳文石, 'Qingdai de shiwei 清代的侍衛', in idem, *Ming Qing zhengzhi shehui shilun* 明清政治社會史論 (Historical Investigation of the Politics and Society during Ming and Qing) (Taipei: Taiwan xuesheng shuju, 1991), 623-650; Elliott, *The Manchu Way*, 150-51, 204-5.

13 Yeh, Kao-shu 葉高樹, 'Qingchao de Qixue yu Qiren de Fanyi Jiaoyu 清朝的旗學與旗人的繙譯教育', *Taiwan shida lishi xuebao* 臺灣師大歷史學報, no. 48 (2012): 129-49; Pamela Crossley, 'Manchu Education', in *Education and Society in Late Imperial China, 1600-1900*, eds. Benjamin Elman and Alexander Woodside (Berkeley: University of California Press, 1994), 340-78, especially 350-52.

14 Zhang Zhaoping 張兆平, 'Kangxi chao zhuming manwen fanyijia Hesu 康熙朝著名滿文翻譯家和素', *Gugong bowuyuan yuankan* 故宮博物院院刊, vol. 168, no. 4 (2013): 143-150. Zhao Binwen 趙炳文, 'Shi lunshu Qingdai fanyi shijia Hesu 試論述清代繙譯世家和素', in *Qingdai dangwen yu Qing gong wenhua – diyiqi Qinggongshi yantaohui lunwen ji* 清代檔案與清宮文化——第九屆清宮史研討會論文集 (Beijing: Zhongguo dang'an chubanshe, 2010), 489-99.

15 'In 1688 [the 27th year of Kangxi reign], [He] received imperial favour, which appointed him Grand Master for Governance [an honorary title for mid-ranking civil servants], Director of the Imperial Stud, and Chief Supervisor of the Bureau of Translation (康熙二十七年遇覃恩, 贈奉政大夫、上駟院郎中、翻書處總管)'. See E'ertai 鄂爾泰 (Ortai) et al., *Baqi tongzhi chuji* 八旗通志初集, *juan* 237, chapter 'Rulin zhuan xia 儒林傳下 (Biography of Scholars, part 2)' (Changchun: Dongbei shifan daxue chubanshe 東北師範大學出版社, 1985), 5341. See also Zhang Yajing 張雅晶, 'Qingdai Wanyan shi Hesu shiji kao 清代完顏氏和素事蹟考', in *Manxue luncong diyi ji* 滿學論叢第一輯 (Shenyang: Liaoning minzu chubanshe, 2011), 254-65.

16 Huang Li-chun 黃麗君, 'Hesu jiqi yizuo Qibentou yanjiu 和素及其譯作七本頭研究', *Taiwan Shida lishi xuebao* 臺灣師大歷史學報, no. 58 (2017): 91-150.

menial job in the Imperial Household Department, as he was not recorded as being either a scribe or an imperial guard.[17] Yet, with his skill in Manchu and Chinese, as well as the emperor's fondness for the family, Heshihen was trusted with several major publishing projects, such as the *Imperial Anthology of Manchu Essays* (Yuzhi Qingwen jian 御製清文鑒) and *Imperially Commissioned Survey of Territory and Routes* (Qinding fangyu lucheng kaolüe 欽定方輿路程考略), and ended up as a chief supervisor in the Imperial Workshops. Differing from the handicraft ateliers in the Yangxin Palace (*Yangxindian* 養心殿), the Imperial Workshops of the Wuying Palace mainly housed publication projects, putting these early polyglots to work, primarily to appropriate textual knowledge from the Chinese and Europeans. Heshihen built up such a strong bond with the Jesuits in China at that time that he converted to Christianity in 1707, which may have ultimately made Kangxi doubt his loyalty to the empire. Heshihen died in disgrace, like many other bannerman bondservants who had risen from humble *bithesi* to become chief supervisors at the emperor's whim.

The position of scribe offered bondservants good opportunities for career advancement for several reasons. First, the system had been designed to provide more openings for bannermen and allow promotion across martial and clerical tracks; second, that work experience honed scribes' bilingual or multilingual skills; and third, scribes were needed to manage the flows of information in the complex forms of correspondence among myriad bureaus. In contrast to the more abstract philological and verbal adroitness which was assessed by the civil service examination, a scribe on the ground was required to demonstrate a practical ability to use languages, comply with formats and meet deadlines, which could be regarded as an alternative system of clerical training.

Yet between the roles of humble scribe and exalted supervisor, Hesu also held a couple of curious menial positions where he ran errands between the Vaults of Money and Textile of the Imperial Household Department (*Biboku* 幣帛庫) and the Imperial Stud (*Shangsiyuan* 上駟院). Evidently, his satisfactory service at the horse stable rendered him suitable to tutor young princes straight afterwards. The constant factor required for this diverse string of positions was good character, on top of literacy. Being trusted not to steal seemed to be the basic – in fact crucial – criterion, because errand runners looked after a large number of valuable artefacts in these jobs. Myriad goods including silver and copper, manufactured

17 Chen Kuo-tung 陳國棟, 'Wuying dian zongjianzao He-shi-hen 武英殿總監造赫世亨', *Gugong xueshu jikan* 故宮學術季刊, 30-1 (2012): 87-134.

textiles, porcelain, and weapons had to be moved within the palace and across remote provinces. Bannerman bondservants were trusted to handle this empire of materials.

Imperial Guard-Supervisor

No personnel could compete with the imperial guards in terms of mobility and imperial trust. The institution of the imperial guard often recruited young descendants from old military allies of the ruling house[18] with mainly those from the upper three banners, the direct attachés of the Aisin clan before conquest, being considered eligible. The diachronic development of the intricate hierarchy in the institution of the imperial guard has been meticulously detailed in previous scholarship.[19] The most pertinent point here is that some imperial guards were simultaneously appointed to many seemingly menial divisions, such as the Imperial Stud in the Imperial Household Department, which implies overlapping functions and possible skill acquisition. Once favourably assessed in such low-level roles, those imperial guards might be rapidly promoted along diverse tracks, either at court or to major custom houses, manufactories, and other vital posts in the provinces.

Several famous specialist bondservants had once guarded the gates of the imperial residents, or closely accompanied emperors on their journeys to borderlands. This proximity often developed into a familial intimacy, and the trust earned would benefit their career and entire family for a long time. Cao Yin 曹寅 (1658-1712), eventually known as a famous textile commissioner based in Nanjing, had once served as a low-ranking guard and was promoted to the Imperial Procession Guard (*Luanyi wei* 鑾儀衛) before his illustrious and consuming career in Jiangnan.[20] Cao Yin's mother

18 Lei Bingyan 雷炳炎, *Qingdai shehui baqi guizu zhijia shili yanjiu* 清代社會八旗貴族世家勢力研究 (Beijing: Zhongguo shehui kexue chubanshe, 2016).

19 For a succinct summary of the structure, see Rawski, *The Last Emperors*, 82-87; Elliot, *The Manchu Way*, 81.

20 Note the phrase 'I recall once guarding the Irradiant Palace (*yi xi suwei Guangming gong* 憶昔宿衛光明宮)', in the poem 'Inscription on Night Talk in Beadtree Pavilion (Ti Lianting yehua tu 題楝亭夜話圖)', from Cao Yin 曹寅, *Lianting ji jianzhu* 楝亭集箋注 (Beijing: Beijing tushuguan chubanshe, 2007), 96. Spence provides a classic study of the iconic Cao clan's multiple roles in publication, textile manufacture, salt administration and the intelligence service during the Kangxi reign, especially 49-50. Cao Yin's father Cao Xi 曹璽 and one of his sons, Cao Qi 曹頎 definitely served as ranked imperial guards and the latter continued to enjoy the succeeding emperor, Yongzheng's favour, while the other lines of the family faced catastrophic confiscation.

had been Emperor Kangxi's wet nurse. The imperial guard system again confirms the conflated nature of bureaucracy and family organization at the early Qing court. Another telling case concerns Gao Bin 高斌 (1683-1755), who is remembered as a competent hydraulic specialist. His daughter was married to the heir prince Hongli and the entire Gao clan was elevated to the prestigious Manchu Bordered Yellow Banner when Hongli ascended to the throne as Emperor Qianlong.[21] Yet Gao Bin had served as an imperial guard for almost a decade before taking jobs in the Grand Storage Office (*Guangchusi* 廣儲司) and then the Imperial Textile Manufactory in Suzhou (*Suzhou zhizao* 蘇州織造),[22] both of which were annexed to the Imperial Household Department.

We are left wondering how much technical knowledge Cao Yin and Gao Bin had before they could oversee textile production, and how far their jobs as imperial guards had prepared them for this. Apart from an occasional erudite reference to elegant fabrics, there is no technical discussion about textiles or loom structures in either Cao Yin's literary anthology or official memorials. Gao Bin never wrote about textiles. This raises the question: Were these so-called specialist bondservants in fact no different from generalist civil officials who were rarely equipped with technical knowledge and mostly hired a team of private secretaries to accomplish the tasks? Gao Bin's detailed description of hydraulic tools, dams, and river estuaries in his poetry disproves this generalist hypothesis,[23] but shows that his expertise was more relevant for the post of a river conservator. If we look back at the extended Wanggiyan/Wanyan lineage, we can find Heshihen's son, Wanyan Wei 完顏偉 (?-1748), a hydraulic novice when appointed to the post of director-general of the Waterways in Jiangnan (*Jiangnan hedao zongdu* 江南河道總督) in the seventh month of 1741. When the water level of the Huai and the Yellow Rivers reached a critical height, he ordered the dams open to alleviate pressure, resulting in disastrous floods.[24] It is tempting to hypothesize that bondservants did not acquire specialist skills as imperial guards or errand runners at court. So what was the rationale behind appointing bondservants to senior-level practical regional positions?

21 Ding Yizhuang 定宜莊 and Hu Hongbao 胡鴻保, 'Qingdai Neiwufu Gaojia shijia de hunyin quan 清代內務府高佳世家的婚姻圈', *Qingshi yanjiu* 清史研究, no. 3 (2005): 25-32.

22 Gao Bin 高斌, *Guzai caotang shiji* 固哉草堂詩集, in *Qingdai shiwen ji huibian* 清代詩文集彙編, *juan* 1, 13 (rep. Shanghai: Shanghai guji chubanshe, 2010), vol. 273: 544.

23 Gao Bin, *Guzai caotang shiji*, *juan* 4, 4; see also poems concerning hydraulic techniques such as 'River survey (*Kanhe* 勘河)', ibid., 588; 'River brake (*Zha* 閘)', ibid., 592; 'Cloud ladders (*Yunti* 雲梯)', ibid, 596.

24 Ding Yizhuang, 'Neiwufu Wanyan shijia kao', 140.

A few cases from some other specialized fields may explain the apparent gap between bondservants' training at court and their service around the empire.

Errand Runner to Designer

Compared with the role of scribe or imperial guard, being appointed as an errand runner (*chai* 差) was a more humble start for bannermen.[25] Bondservants who were not from a high-ranking lineage were more likely to run menial errands (*dangchai* 當差) for a long time in various departments, in roles such as keepers in the vaults of tea and clothes, foremen in the Imperial Workshops, or pagers in the Account Bureau. As documents in the Imperial Archive show, one very common task for errand runners was carrying things around without damage or delay – either raw materials, livestock or finished products. When they attained a higher rank, such as assistant director of a bureau (*yuanwailang* 員外郎), they communicated official orders written on paper and oversaw imperial commissions. Imperial guards appeared to be suitable for such jobs because of emperor's trust and their familiarity with the security system at court, so were often given these roles as a joint appointment in addition to their guard duties. Whether such trusted functionaries were appointed to supervise manufactories of textiles and porcelain in Jiangnan, or to purchase jade from Xinjiang and European goods in Canton, their responsibility was essentially the same[26] – they kept track of the flows of fabric, porcelain, jade, silver, and imperial orders. Even if most of them were not involved in hands-on production, they still needed a basic knowledge of accounting and the materials involved in order to assess their quality and quantity. Moreover, transportation across long distance required having strategic information about regional territories and experience of coordination and logistics, essential factors for both military operation and trade. This probably explains why Tang Ying 唐英 (1682-1758), once Gao Bin's fellow guard and later an eminent porcelain supervisor, carefully noted down journey routes, dangerous reefs and turbulence, distances between neighbouring villages, and precipitation

25 The best Manchu term for this would be *alban*, which could mean public service, forced labour or duty. See Jerry Norman, *A Concise Manchu-English Lexicon* (Seattle: University of Washington Press, 1978), 12.

26 Chen Kuo-Tung 陳國棟, 'Neiwufu guanyuan de waipai, wairen yu Qianlong gongting wenwu gongji zhijian de guanxi 內務府官員的外派外任—與乾隆宮廷文物供給之間的關係', *Meishushi yanjiu jikan* 美術史研究集刊, no. 33 (Sep. 2012): 225-69, 346.

at particular locations and times of the year.[27] These managers probably developed routines and methods to keep organized records of their menial jobs in the palace. Equipped with fundamental skills in languages, numbers, and images, bondservants could efficiently acquire new knowledge in specialist fields under the urgent official obligation imposed by the court.

Some bondservants did go through a more specialized apprenticeship at court, especially in the Imperial Workshops. But I argue that, instead of obtaining an in-depth knowledge of tea, horse, or muskets, what was more critical to their career was the experience of training their bodies to work with certain media and organizational protocols. The practical skill of communicating qualitative and quantitative information by means of numbers and visualization was widely applicable, whether collecting taxes in custom houses, supervising porcelain production, or building dikes.

The 'study of numbers' (*shuxue* 數學, i.e. the modern term for mathematics) encompassed both arithmetic and some mantic arts. This is at least how a bannerman mathematician conceptualized it. Nian Xiyao 年希堯 (1671-1738) was the elder brother of the powerful general Nian Gengyao 年羹堯 (1679-1726).[28] In his preface to *Handbook for Calculating Area and Volume* (Mianti bili bianlan 面體比例便覽), a simplified selection of Kangxi's mathematical work, he construed a number as a thing (*wu* 物) which could communicate among heaven, earth and the material realm of ten thousand things. What he called 'numbers with form' (*youxing zhi shu* 有形之數) were used to measure across different physical dimensions, and yin-yang and the five phases were 'formless numbers' (*wuxing zhi shu* 無形之數).[29] Yet the handbook does not deal with anything metaphysical. It introduces practical methods for calculating the area and volume of various polyhedrons, and lists how much a cube would weigh if made out of 32 different kinds of material (see Figure 2.2).

This knowledge of polyhedrons must have been helpful for porcelain supervisors like Nian Xiyao and his successor, Tang Ying, when they estimated the quantity of clay needed to cast porcelain bodies.[30] Between 1728 And 1735,

27 It is also possible that he copied the information from merchants' handbooks. See Tang Ying's working journal. Tang Ying, *Tang Ying quan ji* 唐英全集, ed. Zhang Faying (Beijing: Xueyuan chubanshe, 2008), vol. 2, 536.

28 Elisabetta Corsi, 'Nian Xiyao de shengping jiqi dui yishu he kexue de gongxian 年希堯的生平及其對藝術和科學的貢獻', *Zhongguoshi yanjiu* = Journal of Chinese Historical Studies, no. 3 (2000): 155-65.

29 Nian Xiyao 年希堯, *Mianti bili bianlan* 面體比例便覽, in *Gujin suanxue congshu* 古今算學叢書, third collection, ed. Liu Duo 劉鐸 (Shanghai: Suanxue shuju, 1898), vol. 429, 1.

30 The word casting (*rongfan* 鎔範) is even used when the author explains the required precision. See Nian Xiyao, *Mianti bili bianlan*, 2b.

Figure 2.2 List of how much a cube would weigh in thirty-two materials

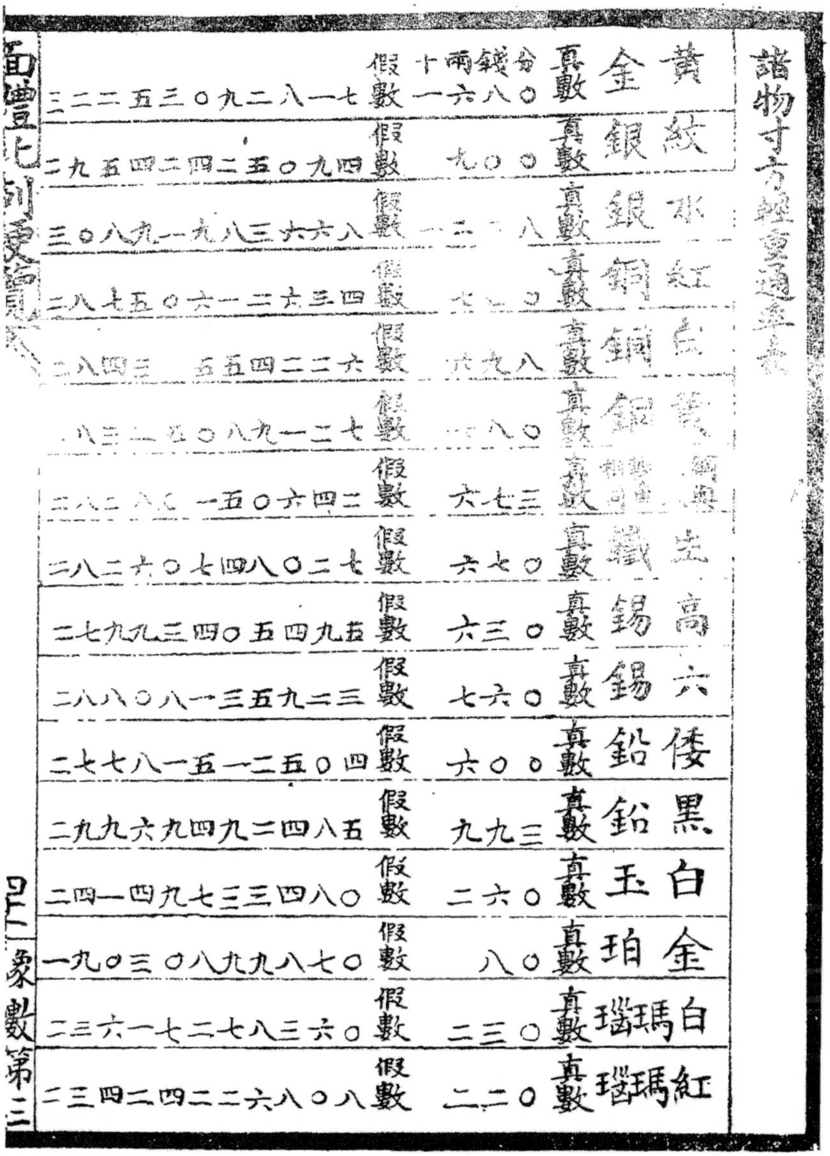

諸物寸方輕重通率表

	真數	假數
黃金	一六八〇 (十 兩 錢 分)	三二二五三〇九二八一七
紋銀	九〇〇	二九五四二四二五〇九四
水銀	一二[illegible]八	三〇八九一九八三六六八
紅銅	[illegible]	二八七五〇六一二六三四
白銅	六九八	二八四三[illegible]五五四二二六
黃銅	[illegible]八〇	[illegible]八三二[illegible]〇八九一二七
[illegible]	六七三	二八二[illegible]一五〇六四二
[illegible]鐵	六七〇	二八二六〇七四八〇二七
高錫	六三〇	二七九九三四〇五四九五
六錫	七六〇	二八八〇八一三五九二三
倭鉛	六〇〇	二七七八一五一二五〇四
黑鉛	九九三	二九九六九四九二四八五
白玉	二六〇	二四一四九七三三四八〇
金珀	八〇	一九〇三〇八九九八七〇
白瑪瑙	二三〇	二三六一七二七八三六〇
紅瑪瑙	二二〇	二三四二四二二六八〇八

面體比例便覽 [illegible]

[illegible] 象數第三

Source: Nian Xiyao, *Mianti bili bianlan*, 41

Nian Xiyao inspected Tang Ying in the Porcelain Manufactory, just as Tang Ying had inspected his successor, Laoge 老格 (active in Jingdezhen 1741-1768). The transmission of technical skills through collaborative mentorship seemed to extend to the workshop setting at court where highly-skilled southern artisans (*nanjiang* 南匠) and European 'technicians' created an extraordinarily dynamic but tightly-constrained learning environment. Imperial control was embedded in the massive standardization of expenditure, which materials to use, deadlines to meet, and production procedures. Both textual and visual blueprints abounded in recording plans and guides for realizing the plans.

The genres used for standardization were 'Regulations and Precedents' in written texts (*zeli* 則例) and painted or modelled design drafts (*yang* 樣) in the form of artefacts.[31] Metaphorically speaking, the 'Regulations and Precedents' were detailed maintenance reports of the palace machine, since the princes and bannerman supervisors compiled and regularly updated them for almost every official department. This matrix of assembled data connected the quantity of expenditure, raw materials, labour, products and consumers, to generate an ideal plan of how the palace machine should work, while the standardization regulated both the status quo and project management. On the one hand, the rationale underlying the sumptuary regulation of attire, daily supplies for royal households or imperial pomp, was mainly a social one. Pervasive examples of this can be found in the *Imperially Commissioned Current Regulations and Precedents of the Court* (Qinding gongzhong xianxing zeli 欽定宮中現行則例). On the other hand, planning for commissioned projects entailed a mixture of responses to both the forces of nature and social conventions. Thus, the skilful input of hands-on and locally informed bannerman managers is more clearly demonstrated in the 'Regulations and Precedents' concerning procedural projects.

When it came to constructing things – be it palatial gardens, large dikes or moveable type blocks – there were a few subgenres of 'Regulations and Precedents', such as 'Methods of Making' (*zuofa* 作法), 'Precedent Rules' (*chenggui* 成規), and 'Procedures and Models' (*chengshi* 程式). The 'Methods of Making' does not explain how to build things as if craftsmen needed manuals. As is shown in the manual *Methods of Making Architectonic Projects* (Gongcheng zuofa 工程作法), their primary concern was to systematically portray all the necessary materials in standardized measurements, along with – most important of all – the price tag of everything, including wages and rations required for labourers.[32] The information was explicitly organ-

31 Song Jianze and Christine Moll-Murata, *Regulations and Precedents* – 'Notes on Qing Dynasty "Handicraft Regulations and Precedents" (*Jiangzuo zeli*), with a Special Focus on Regulations on Materials, Working Times, Prices and Wages', *Late Imperial China*, vol. 32, no. 2 (Dec. 2002): 87-126.

32 Wang Puzi 王璞子, *Gongcheng zuofa zhushi* 工程作法註釋 (Beijing: Zhongguo jianzhu gongye chubanshe, 1995).

ized for auditing and to facilitate reimbursement.[33] An observant bannerman supervisor would take these material constraints into consideration when revising the precedent rules. In his preface to *Precedent Rules Concerning Five Metropolitan Circuits in Zhili Province* (Zhili wudao chenggui 直隸五道成規), for example, the capable hydraulic commissioner Gao Bin gave advice on such minute but crucial decisions as whether to increase or decrease the dredging workload based on the differing viscosity of soil under three river banks near Tianjin.[34]

The bannerman supervisors did not only reveal such perceptiveness in words, but in visualizations as well, and their attention to material specificity permeated both their work for the court and personal expression. Wanyan Linqing 完顏麟慶 (1781-1846), Hesu's grandson's grandson, who became another eminent hydraulic commissioner, compiled an *Illustrated Treatise on Estuaries* (Huangyun hekou gujin tushuo 黃運河口古今圖說, 1841) and an *Illustrated Treatise on Hydraulic Tools* (Hegong qiju tushuo 河工器具圖說, 1836) to disseminate knowledge of rivers and estuaries, as well as the instruments required for hydraulic constructions.

Many bannerman bondservants exhibited professional skills in image-making. Although a few bannermen were expert or demonstrated an ethnically-marked accomplishment in painting,[35] images fashioned by bannermen visual specialists predominantly served practical purposes. In other words, many pictures circulated around the palace machine were made to document technical processes, depict ethnic groups in borderlands or communicate designs. For instance, porcelain supervisor Tang Ying collaborated with a group of court painters to make an *Illustrated Manual of Porcelain Production* (Taoye tushuo 陶冶圖說).[36] This album depicts the production procedures in twenty scenes and was probably not created to instruct craftsmen but the object's commissioner, Emperor Qianlong, who

33 For the institution and process of financial operation, see Chapter Three in this volume.

34 Gao Bin 高斌, *Zhili wudao chenggui* 直隸五道成規 (pre-modern print, 1743. Starr Rarebook Collection at Columbia University), Preface, 2.

35 See brief introductions to Tang Dai 唐岱 (1675-1752), an eminent court painter, in Chou Ju-hsi, 'Tangdai: A Biographical Sketch', in *Elegant Brush: Chinese Painting under the Qianlong Emperor*, (Phoenix, Ariz.: Phoenix Art Museum, 1985), 133-140; Wang Yao-t'ing 王耀庭, 'Yuci huazhuangyuan tangdai 御賜畫狀元唐岱', *Gugong wenwu yuekan* 故宮文物月刊, no. 392 (Nov. 2015): 18-29. There was also a bannerman painter who was famous for his finger painting – see Klaas Ruitenbeek, *Discarding the Brush: Gao Qipei 高其佩 (1660-1734) and the Art of Chinese Finger Painting* (Amsterdam: Rijksmuseum, 1992).

36 Stephen Bushell, *Description of Chinese Pottery and Porcelain; Being a Translation of the T'ao Shuo* (Oxford: The Clarendon Press, 1910). The album is currently housed in a private collection in Taiwan.

would have felt empowered by gaining this knowledge. Combined with the explanatory breakdown of cost in the 'Regulations and Precedents' about porcelain production, the album might also have facilitated inspection and quality control. Thus, visual specialists employed pictorial forms in a rather limited way. Decisions about the composition, colour, brushwork, figures and details were all intended to foreground the information relevant for improving imperial projects. This agenda demanded a different visualizing skill from that of artists who painted for connoisseurs and explored the potential of pictorial media primarily to converse with the immense literati tradition.[37] The documentary representations served pedagogical and archival purposes, to ensure that the transmitted knowledge would remain stable for a while. Jin Jian 金簡 (?-1794), a bannerman specialist of printing, most clearly defined the genre of 'illustrated treatise' (*tushuo* 圖說, literally explanation with pictures) in his postscript to the *Manual of Moveable Type Prints in the Wuying Palace* (Wuying dian juzhen ban chengshi 武英殿聚珍版程式), where he instructed: 'Make pictures based on the project. Attach the explanation to the pictures. The project should be [explained] in detail. The explanation should be concrete. Even crafty minutiae should be thoroughly understood. In this way, the responsible person has [methods] to hold on to and those to come can follow'.[38] In addition, Wanyan Linqing's *Illustrated Treatise on Hydraulic Tools* confirms that this instructional genre remained robust in the second half of the Qing dynasty.

Design drafts or models (*yang* 樣) constituted the other kind of documentary media that bannerman specialists made to provide an immediate reference in the framework of imperial projects. As mentioned above, humble bannerman bondservants were eligible for menial jobs in the Imperial Household Department, which included working as an apprentice in specialist institutions such as the Imperial Workshops, the Inner Department of Works (*Yingzaosi* 營造司) or the Imperial Armoury (*Wubeiyuan* 武備院). 'Those who have learned for one year are apprentices. Two years, half labourer. Three years, accomplished labourer. If one can still not make things after three years, one is fired'.[39] Designers were taught how to abstract formal patterns from specific materials so that they could coordinate workshops

37 Overlapping zones do exist in religious paintings and grandiose propagandistic visual projects.

38 '因事為圖，因圖系說，事惟其詳，辭惟其實。雖工匠之微皆得通曉，俾從事者有所守，而將來有所遵云'. See Jin Jian 金簡, *Wuying dian juzhen ban chengshi* 武英殿聚珍版程式 (*Siku quanshu* 四庫全書 edition. Reprint. Shanghai: Shanghai guji chubanshen, 2003), vol. 673, 34.

39 '一年者為學手，二年為半工，三年為整工。如有三年後不能造作，即革除'. See Wei Qingyuan 韋慶遠, '*Qingdai Neiwufu de jiangyi he yuyong shougongye* 清代內務府的匠役和御用手工業',

unlimited by material specificities. Hence, pattern drawers or designers (*huayang ren* 畫樣人) were able to translate between materials and were more suitable for taking on overarching managerial responsibility. Tang Ying, among other supervisor-designers, followed this career trajectory.[40]

The volume of products these visual specialists made designs for may even exceed the combined inventory of many modern corporations across industrial divisions, because the palace machine churned out artefacts ranging from monumental garden complexes, imposing cannons, to small candlesticks. The style and budget of every item had to be premeditated and the plans, in the form of design drafts, had to be submitted to the emperor for approval.

As we can observe from the extant drafts and models, the visual traits of every object – including their shape, colour, and decorative patterns – were always accompanied by textual directions which clarified the scale, measurements, and quantity of the commissioned items. To better capture the physical variables of myriad artefacts, design drafts were made as paper and wood models in both two-dimensional and three-dimensional formats. On rare occasions, the emperor encouraged novelty. For instance, in 1736, Qianlong ordered Tang Ying, the supervisor, to let local artisans freely draft designs for porcelain wares.[41] More often than not, though, designs and the strenuous rounds of modification usually made the process restrictive. Supervisor-designers across various fields struggled to meet the imperial preference summarized as the 'style respectfully made in the inner court (*neiting gongzao shiyang* 內廷恭造式樣)', which Emperor Yongzheng stated when reprimanding products made in a 'vulgar style outside the court (*waizao zhiqi* 外造之氣)'. The imagined edict and the very real cycles of commission gave the palace machine a regularity in its operation.

In summary, bannerman specialists' experience with instruction pictures and design drafts showed their skill with a system of visual symbols, primarily to transmit information and to ensure consistency and compliance with imperial directives and preferences. Their skill at this symbolic system complemented their capabilities in numerical and lingual systems and

in idem., *Mingqing shi xuxi* 明清史續析 (Guangzhou: Guangdong renmin chubanshe, 2006), 359-78.

40 Peter Lam, 'Tang Ying (1682-1756): The Imperial Factory Superintendent at Jingdezhen', *Transactions of the Oriental Ceramics Society*, vol.63 (1998-1999): 65-82. See also Zhao Bing, 'Administration, Savoir Techniques et Reconnaissance Impériale: Tang Ying 唐英 (1682-1756)', *Revue de Synthèse* 131, no. 3 (2010): 307-341.

41 Tang Ying 唐英, *Tang Ying dutao wendang* 唐英督陶文檔, ed. Zhang Faying 張發穎 (Beijing: Xueyuan chubanshe, 2012), 144.

were used to manage the flows of materials, labour, and funding. Individual bannermen's differing combinations of skillsets explains the absence of strictly hereditary professional skills in one lineage. Instead, we see bannerman specialists' recurrent professional service in military organizations, custom houses, manufactories, and river conservation efforts across different generations. Moreover, many of the figures mentioned in this chapter held positions consecutively or simultaneously in these bureaus, as it is evident from the enumeration of the many titles they held in archival documents. Some titles do not refer to positions reserved for bannermen, but to the civil government, where successful Han Chinese examinees dominated. This was because a sizeable number of bannermen also passed the civil service examination, so were appointed to civil service posts. Although emperors from the early- and mid-Qing had also appointed bannermen to high-ranking officials in civil service bureaus, the overflow of trained bannermen into the civil system was concurrent with the drastically increased population of bannermen, the decreasing opportunities at the inner court, and the general routinization of the inner court bureaus that occurred during Daoguang's reign (1821-1850).[42] That is where this chapter will end, leaving open the discussion of bannerman agents' role in the integration of the palace machine into the imperial machine.

Coda

Bannerman specialists' meticulous memorials reveal the striking variety of duties they performed, often glossed over in the official historiography, which only briefly lists the tasks. From Wanyan Linqing's *Memorials of Yunyin Hall* (Yunyin tang zougao 雲蔭堂奏稿), which includes his memorials from 1826-36 and 1839-41, we know that he conducted military manoeuvres, quelled rebellions, traded copper, sentenced criminals, and more, in addition to his famous (although sometimes unsuccessful) efforts to build dikes to stem floods.[43] An unmistakable rhythm emerges from his work, as he

42 For a definitive study of the tension between monarchical will and the power of bureaucratic routinization, see Philip Kuhn, *Soulstealers: The Chinese Sorcery Scare of 1768* (Cambridge, MA: Harvard University Press, 1990). A recent detailed study of the bureaucratic evolution of the Imperial Household Department is given in Huang Li-Chun's 黃麗君 dissertation, 'Huangdi jiqi baoyi nucai: lun Qingdai huangquan yu Neiwufu guanliao tizhi 皇帝及其包衣奴才：論清代皇權與內務府官僚體制' (PhD diss., National Taiwan University, 2014).

43 Wanyan Linqing 完顏麟慶 (Wanggiyan Linkin), *Yunyintang zougao* 雲蔭堂奏稿 (Mikrofilm. Beijing: Quanguo tushuguan wenxian suowei fuzhi zhongxin, 2005).

needed to respond to the rising and falling water levels dictated by nature. Similar cycles are also visible in porcelain supervisor Tang Ying's work – no matter how incessant the imperial thirst for porcelain was, there were only a limited number of weeks during spring and autumn when temperature and humidity were suitable for kiln firing in the porcelain town of Jingdezhen. Moreover, Tang Ying was simultaneously collecting rounds of taxes from the Jiujiang 九江 custom house in order to fund porcelain manufacture. Even though his seasonal trips took place in the furthest-flung reaches of the empire, the pulse of the palace machine is still vividly felt.

Bannerman specialists' vast diversity of work not only shows us the temporal regularity of the palace machine and the altering demand on their skills, but also delineates the changing nature of the ruling machine in three stages. At the beginning of the Qing dynasty, warriors were the driving engine of the war machine. Once the basic bureaucracy had been established, reliable bannermen with multilingual capacities, like Wanyan Ashitan and Hesu, were in high demand to aid cultural integration. In a bid to enhance his cultural capital among Confucian scholars, Kangxi even gathered elite Han scholars as an elite entourage into his 'Southern Study' (*Nanshufang* 南書房).[44] During the second phrase, between the 1680s and the early 19th century, as emperors continued to march to the west to expand Qing territories, the economic and cultural integration of Manchu, Mongolian, Tibetan, Han Chinese, and even a small population of Europeans was the priority for empire building. The palace machine, especially the bannerman managers in the Imperial Household Department and its annexed manufactories and custom houses, mobilized labour, materials and funding to literally build an expanding material empire. Bannermen like Cao Yin, Nian Xiyao, Gao Bin, and Tang Ying represent the generation of builders who laid the foundations of this imperial project in the first half of the 18th century.

There is one last visible generation from the Wanyan clan to mention. Linqing's son Chonghou 崇厚 (1826-1893) had a dramatic career which epitomizes the role of bannerman specialists in the third phase of the palace machine, during the second half of the 19th century. For lack of a proper name, we may call this the machine of damage control, given the rampant domestic disasters and frustrating encounters with Western powers at the time. When he was twenty-three years old, Chonghou bought the post of the magistrate of Zhili 直隸 prefecture, probably with the wealth passed

44 Zhu Jinfu 朱金甫, 'Lun Kangxi shiqi de Nanshufang 論康熙時期的南書房', *Gugong bowuyuan yuankan* 故宮博物院院刊, no. 2 (1990): 27-38.

down from Linqing.[45] He passed the provincial examination the next year.[46] Just like bannermen with that qualification, he was then able to serve as a vice commander (*fudutong* 副都統) in a couple of different banners and work for a variety of civil bureaus including the Boards of Rite (*Libu* 禮部), War (*Bingbu* 兵部) and Punishment (*Dalisi* 大理寺). What related his career trajectory all the way back to his ancestors Hesihen and Hesu were his diplomatic missions to European countries. His career ended in notoriety because of his epic failure at negotiating with Russia in 1878 and the consequent loss of sovereignty over a large share of territory in the northwest. Considering the unbalanced international power relations at the time, the bannerman specialist may appear to have been criminalized as a scapegoat. Yet the reason for the empire's downfall during this third phase seems to be incompetence at damage mitigation, such as Linqing's failure at flood reduction, in contrast to the risk of financial embezzlement that imperilled the early- and mid-Qing period.

In short, bannerman specialists constituted a leading force propelling the palace machine which lay at the heart of building the Qing material empire. It is no small irony that those technical experts who rose through the ranks by demonstrating their skills in a meritocratic system were recruited from a privileged hereditary genealogy, the banners. Ultimately, though, the community boundary of bannerman experts was fluid, shaped to a large extent by the priorities and agenda of the palace machine. The shifting configurations of the bannerman experts' skills illustrate the entanglement of the individual and the institutional that, together, made the palace machine work.

About the Author

Kai Jun Chen is Assistant Professor at Brown University. He specialises in the history of imperial institution, handicraft technology, and material culture studies. His monograph in progress examines the technocratic culture in the Imperial Ceramic Industry of the Qing dynasty. His other scholarly publications investigate the production of technological knowledge in the global circulation, collection, and replication of luxury artifacts, especially porcelains.

45 On the phenomenon of purchasing official positions, see Huang Li-Chun, 'Huangdi jiqi baoyi nucai', 167-186.

46 Wanyan Chonghou 完顏崇厚, *Hecha nianpu* 鶴槎年譜, 13 (rep. in *Beijing Tushuguan cang zhenben nianpu congkan* 北京圖書館藏珍本年譜叢刊. Beijing, Beijing tushuguan chubanshe, 1999, vol. 169, 128).

3 *Kupiao* and the Accounting System of the Imperial Household Workshops

Yijun Wang and Kyoungjin Bae

Abstract

Focusing on *kupiao*, a rudimentary document of accounting, this chapter explores the accounting system of the Imperial Workshops in Qing China. Spurred by a series of institutional reforms, a complex budgeting and auditing system developed at the Imperial Workshops during the eighteenth century. As the records of day-to-day transactions between various Works and other departments, *kupiao* instantiated the operation of production and finance as a correlated system. Tracing the paper trails of *kupiao*, therefore, we locate the manufacturing processes of the Workshops at the intersection of artisanal collaboration and the administrative cycles of budgeting and audits in which various bureaus participated. By comparing the accounting systems of the Imperial Workshops and the Qing state, moreover, we argue that the former modelled after the *zouxiao* system of the state. Both systems shared as their principles rigorous accountability and the pursuit of checks and balances.

Keywords: *kupiao,* accounting system, Imperial Workshops, budget, audit, *zouxiao* system

Maiban kupiao 買辦庫票 (hereafter *kupiao*), loosely translatable as requisition tickets, was a genre of accounting documents produced by the Imperial Workshops (*Zaobanchu* 造辦處) during the Qing. It first appeared in the 28th year of the Kangxi reign (1689) and lasted at least until the 25th year of the Jiaqing reign (1821).[1] *Kupiao* was the fundamental constituent of the

1 The last collection of *kupiao* in the First Archive appeared in December of the 25th year under Jiaqing's reign (1821). See 'Zaxiang maiban kupiao zanling piao 雜項買辦庫票暫領票

Siebert, Martina, Kai Jun Chen, and Dorothy Ko (eds), *Making the Palace Machine Work: Mobilizing People, Objects, and Nature in the Qing Empire.* Taylor & Francis Group, 2021
DOI: 10.5117/9789463720359_CH03

Imperial Workshop accounting system. For each activity in which cost and goods were transferred between different storage facilities, different specialized ateliers or works (*zuo* 作), and between the workshops and the market, one of these standardized tickets had to be filled out.[2] Similar to memoranda, these small 'paper tools' captured the day-to-day transactions in materials, money, and labour – the *fuel* that made the palace machine work. Every month, these tickets were bound into a semi-journal that became the basis for a higher level of accounting.[3] A study of *kupiao* thus provides an extraordinary opportunity to explore the multitiered accounting system of the Imperial Workshops.

This chapter examines the role of *kupiao* within the Imperial Workshops and investigates its multifaceted significance in the following sequence: first, it analyses the information held in individual *kupiao* within the context of the bureaucratic system of material and monetary transactions that produced it. *Kupiao* reflected an increasing complexity in the accounting system over the course of the 18th century with rising concerns about estimation, budgeting, and actual expenditure in the management of diverse projects. The chapter then discusses higher-level accounting processes in which such concerns were institutionalized into concepts of budget and actual expenditure, as well as regular cross-departmental audits achieved through mid-century reforms. Following the trajectory of a specific project recorded in *kupiao*, the third section takes an in-depth look at financial and manufacturing organs of the Imperial Workshops. The case study shows that cross-departmental collaboration happened not only within the manufacturing loops but also across accounting and administrative spheres. The last section compares the accounting system of the Imperial Workshops and the fiscal system of the Qing state, concluding that the former was linked seamlessly to the latter financially and administratively.

In so doing, this chapter weaves together two correlated spheres of the palace machine: the manufacturing of physical objects and the corresponding mechanism of accounting. Putting *kupiao* at the centre, it construes the

(Tickets of Budget Disbursement on Miscellaneous Items)', year 1821, *Gongzhong gechu dang'an* 宮中各處檔案 (Documents of the Palace Archives), no. 2252 (The First Historical Archives of China, Beijing).

2 *Kupiao* are approximately 17.94 centimetres wide and 21.79 centimetres long. For archiving they were bound into volumes whose covers are 26.41 centimetres wide and 28.9 centimetres long.

3 The order of tickets was not strictly chronological. For example, the *kupiao* of the 8th month of the 46th year of the Qianlong reign were not bound in date order. See the *Qing gong Neiwufu Zaobanchu dang'an zonghui* 清宮內務府造辦處檔案總匯, edited by Chinese University of Hong Kong and The First Historical Archives of China (Beijing: Renmin chubanshe, 2005) vol. 45, 193-5.

two spheres as two 'resistant bodies' operating by common motive forces toward efficient production.[4] Such a tendency was increasingly palpable during the 18th century despite the increasingly extravagant expenditure made at the Imperial Workshops. In the sphere of manufacture, *kupiao* inform us about not just financial transactions but also the division of labour among works, the mobilization of in-house and recruited artisans, and the outsourcing of tasks. They reveal an open system that connected the palace workshops with state warehouses and the market. In the accounting sphere, *kupiao* embody both the logic of accounting and its historicity within the overwrought processes of financial reasoning at the Imperial Workshops. It was the prospection, estimation, budgeting and compensation recorded in *kupiao* that made the complex and oft-prolonged production cycles accountable and sustainable at the court. The intersection between manufacture and accounting reveals a system of checks and balances through which many bureaus and works contributed to the goal of efficient and sustainable resource management.

Kupiao and the Financial Organ of the Imperial Workshops

In the 12th month of the 28th year of the Kangxi reign (1689), the Imperial Workshops established a system of 'red tickets' (*hongpiao* 紅票) in order to 'certify and verify the receipt of materials'.[5] These 'red tickets' were the beginning of *kupiao*. Although records show that *kupiao* were produced until at least the early 19th century, extant *kupiao* and related archival documents on palace accounting are fragmentary. Still, thousands of them (roughly from 1735 to 1741 as well as from 1782) are reproduced in facsimile in the *Qing gong Neiwufu Zaobanchu dang'an zonghui* 清宮內務府造辦處檔案總匯 (Archives of the Imperial Workshops of the Imperial Household Department), upon which this chapter is primarily based.[6]

4 Franz Reuleaux defines a machine as 'a combination of resistant bodies so arranged that by their means the mechanical forces of nature can be compelled to do work accompanied by certain determinate motions'. See Franz Reuleaux, *The Kinematics of Machinery: Outlines of a Theory of Machines*, trans. and ed. Alexander B.W. Kennedy (New York: Dover Publications, 1963), 35.

5 *Qinding Zongguan Neiwufu xianxing zeli er zhong* 欽定總管內務府現行則例二種, rep. in *Gugong zhenben congkan* 故宮珍本叢刊, vol. 306-310 (Haikou: Hainan chubanshe, 2000), vol. 4 (309), 291.

6 *Kupiao* first appeared in the 11th year of the Yongzheng era (1733) and appeared intermittently in 1735, 1736, 1739, 1741, 1743 and 1782. The records from 1735-36 are the most expansive and detailed of all of these. In 1736 alone, around 1,160 pieces of *kupiao* were stored in the published archives.

Figure 3.1 ***Kupiao*** **issued in 1736**

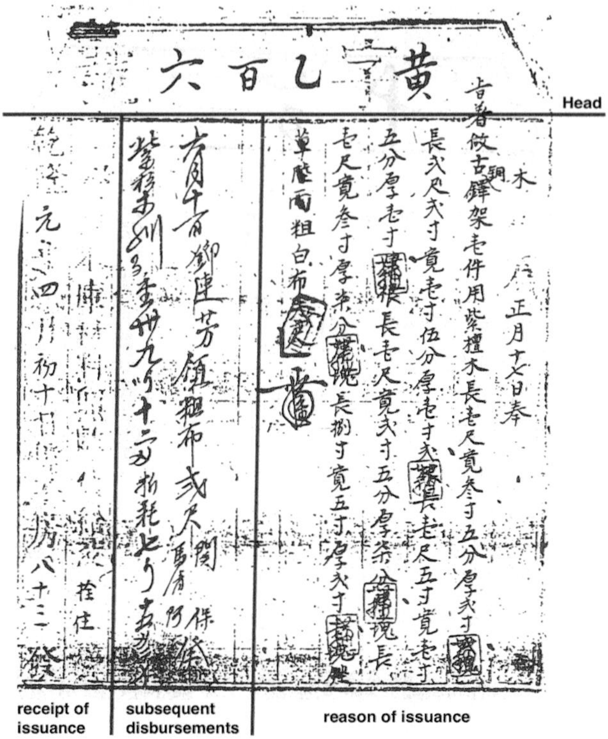

Source: *Qing gong Neiwufu Zaobanchu dang'an zonghui*, vol. 7, 295

The format of *kupiao* consisted of three parts: a preprinted form in red, handwritten contents in black ink, and stamps and other auditorial marks that were added later. A *kupiao* was issued by filling out the preprinted form with information such as date, occasion of issuance, amount of money or materials disbursed, and signature of the issuer. In the ticket, the information was visually divided into three sections and a heading (Figure 3.1).

The document begins in the far right columns with the name of the requisitioning works, along with an initial date of commission, occasion, and the amount of raw materials and/or silver requisitioned, together revealing the reason for issuance. In the far left, two columns of preprinted characters functioned as a receipt of issuance. Here the issuer wrote down the ticket's issuing date, the name of recipient, and his own signature. The middle section provided information about subsequent disbursement. Under another – often later – date, it would identify that the noted – or sometimes a different – recipient had actually 'received' (*ling* 領) the stated amount of supply. This section ended with yet another set of one or more clerical signatures. Finally, the heading, with the preprinted character '*zi* 字', assigned each ticket a filing number drawn from two numerical systems – Chinese numerals and that of the *Thousand Character Classic* (Qianziwen 千字文), a distinctive system in which every character was given a unique numeral value between 1 and 1,000. For instance, the number of the *kupiao* in Figure 3.1 is 106 of *huang* 黃 ('yellow', i.e. the fourth character in the *Thousand Character Classic*). In the sequence of time, this *kupiao* was issued to the Wood Works (*Muzuo* 木作) for a task commissioned on the 17th day of the 1st month of the 1st year of the Qianlong reign (28 February 1736). The works had requisitioned variously-sized pieces of red sandalwood and some coarse calico in order to make a stand for an ancient bronze bell. As a receipt, the filled-in text on the left reads: 'In accordance with the amount requested, our vault disbursed the material to Shuanzhu 拴住. Issued by Bashisan 八十三 of the Document House (*Dangzifang* 檔子房) on the 10th day of the 4th month in the 1st year of the Qianlong reign (20 May 1736)'.[7] Finally, in different handwriting, the middle section states: 'On the 11th day of the 6th month (19 July 1736), Deng Lianfang 鄧連芳 received two *chi* 尺of coarse calico, issued by Guanbao 官保 (Manchu: Guwanboo) and Maqing'a 馬清阿. The stock of red sandalwood at the beginning of this year was thirty-nine *jin* 斤 (catties) and two *liang* 兩 (taels). Of those, seven taels and five *qian* 錢 were consumed [in order to fulfil this request]'.[8]

Following the spatial order from right to left to centre, in other words, one can trace the sequence by which the *kupiao* activated a series of financial actions. A clerk named Bashisan at the Document House first wrote the requisition order and issued this *kupiao* to Shuanzhu. About two months later, Deng Lianfang took this ticket to Guanbao and Maqing'a to receive the stated materials. The sequence reveals an intricate bureaucratic network within and beyond the Imperial Workshops concerning financial administration. Although the office is not specified here, the material was probably provided by the Treasury of the Imperial Workshops (*Qianliangku* 錢糧庫).[9] Established in 1722 within the Imperial Workshops, the Treasury

7 *Qing gong Neiwufu Zaobanchu dang'an zonghui*, vol. 7, 295.

8 Only *kupiao* related to the precious timbers such as red sandalwood or *huali* 花梨 wood have had the additional information about balance after consumption. *Chi*尺 is a unit of length which is approximately 35.5 cm or 14 inches; *liang* 两 is a unit of weight. One *liang* (tael) was approximately 36.9 grams, and was equivalent to 10 *qian* 錢, 100 *fen* 分, and 1,000 *li* 釐. Above tael, there were units such as *jin* 斤 (catty; 16 *liang*) and *shi* 石 (100 *jin*).

9 *Qianliangku* literally means 'the storage of money and grain'. The term *qianliang* (food and grain) first emerged in the Tang dynasty (618-907) to refer to tax and military expenditure. See the imperial edict from Xianzong Emperor of the Tang Dynasty 唐宪宗 né Li Chun 李淳

supplied the workshops with silver cash and raw materials.[10] Administered by the Imperial Workshops, it was in effect funded by the Six Vaults (*Liuku* 六庫) of the Grand Storage Office (*Guangchusi* 廣儲司). When there were insufficient resources at the Six Vaults, it reached out to the vaults of the Board of Revenue (*Hubu* 戶部) and the Board of Work (*Gongbu* 工部) to refill its stock.[11] Taken from each work to the Treasury of the Imperial Workshops, *kupiao* were traded for the requested materials and likely archived there afterwards.

This shows that financial administration at the Imperial Workshops went through the joint activities of two bureaus. First, the Document House issued a *kupiao* for each transaction; then, the actual material was claimed using this ticket at the Treasury.[12] According to historian Wu Zhaoqing, the Document House was in charge of transmitting and archiving imperial memorials and the correspondence between the Imperial Workshops and other bureaus of the Imperial Household Department.[13] Yet, as Figure 3.1 demonstrates, the remit of the Document House was not limited to archiving administrative documents but extended to the operation of the Treasury through the issuing of *kupiao*.[14] The Document House prepared the preprinted ticket forms at the beginning of each month. A *kupiao* issued on the 1st day of the 3rd month of the 11th year under the Yongzheng reign

(778-820), 'Ping Liu Pi zhao' 平劉辟詔, in *Tang da zhao ling ji* 唐大詔令集, ed. Song Minqiu 宋敏求 (1019-1079), *juan* 124, 665 (Beijing: Zhonghua shuju, 2008). The usage of *qianliang* to refer to taxation might be due to the introduction of the double taxation system, in which the tax levied in money. In archives in the Yuan Dynasty (1271-1368), *qianliang* was used to refer to land taxes, tariffs, grain stocks and wages for soldiers and bureaucrats. See Pan Jie 潘潔, *Heishuicheng chutu qianliang wenshu zhuanti yanjiu* 黑水城出土錢糧文書專題研究 (Yinchuan: Ningxia renmin chubanshe, 2013). In the Ming and Qing dynasties, *qianliang* was used to refer to a wide range of things including taxes, wages, military and government expenditure. In the Imperial Workshops, the *Qianliangku* was used to store silver cash and raw materials for the workshop, therefore, we translate it as 'Treasury of the Imperial Workshops' or brief as 'Treasury'.

10 *Qinding Zongguan Neiwufu xianxing zeli er zhong*, vol. 4 (309), 296. Also, see Wu Zhaoqing 吳兆清, 'Qingdai zaobanchu de jigou he jiangyi 清代造辦處的機構和匠役', *Lishi dang'an* 歷史檔案, no. 4 (1991): 81.

11 *Zongguan Neiwufu xianxing tiaoli: Guangchusi* 總管內務府現行條例：廣儲司, in *Jindai Zhongguo shiliao congkan* 近代中國史料叢刊 (Taipei: Wenhai chubanshe, 1972), 43; *Da Qing huidian zeli* (Qianlong ban) 大清會典則例 (乾隆版), edited by Imperial Household Department (Beijing: Wuyingdian, 1764), *juan* 159, 25a.

12 The Document House was called Dangfang or Dangzifang. It appears in the *kupiao* of the Yongzheng and Qianlong periods.

13 Wu Zhaoqing, 'Qingdai zaobanchu de jigou he jiangyi', 82.

14 It is not clear which administrative tier '*benku* 本庫' in Figure 3.1 belonged to, but it is plausible that *benku*, which literally means 'this storage', refers to the Treasury of the Imperial Workshops.

(14 April 1733), for instance, states that the Document House purchased two kinds of papers 'in order to print and write notices and *kupiao*' within the bureau.[15] Such orders recurred regularly from the Yongzheng to the early Qianlong periods. Another *kupiao* issued on the 1st day of the 6th month (9 July) of 1736 shows that the Document House and the Treasury shared budgets for stationery items. Issued to the Document House, the ticket disbursed money to this bureau and the Treasury to purchase '*jinbuhuan* 金不換' brushes (literally, brushes 'more valuable than gold', a common name for scholarly stationery), fragrant ink and cinnabar in order to print and write *kupiao*, notices, monthly reports (*yuezhe* 月摺) and the base archives (*bendi dangan* 本底檔案).[16] This shows that the Treasury of the Imperial Workshops and the Document House, which were the makers, issuers and final recipients of *kupiao*, also issued and filed *kupiao* to themselves.

Kupiao reflected not just the financial bureaucracy of the Imperial Workshops but also an increasing complexity in their financial planning. In the first year of the Qianlong reign (1735), *kupiao* were differentiated into five categories identifiable by the heading written above the printed frame on each ticket. Tickets with no heading reflected an initial estimation of raw materials and silver cash to be disbursed from the Treasury in order to carry out a project (see Figure 3.1). Those labelled 'purchase' (*mai* 買) recorded an estimate of materials to be purchased from the market; those called 'preliminary budget' (*zan* 暫 or *zanling* 暫領) referred to the preliminary cost of a project in its early stage; those labelled 'balance disbursement' (*zhaoling* 找領) noted the reconciled cost of the project during its execution, and 'receipt of balance' tickets (*jiaohui can* 交回殘) indicated the balance of raw materials or cash returned to the vault after a project was completed.[17]

These five headings divided *kupiao* into two kinds of organization. On the one hand, the blank and 'purchase' tickets – which make up the majority of existing *kupiao* – concerned the source of materials, that is, either the Treasury of the Imperial Workshops or the market. As such, they differentiated the internal and external sources of requisition. In these tickets, while internal disbursements were calculated as stock, market acquisitions were calculated in silver cash. It is noticeable that, even though purchased goods were classified as 'purchase' tickets, the purchase of labour from the market was classified as blank tickets. Despite this difference, wages for hired

15 *Qing gong Neiwufu Zaobanchu dang'an zonghui*, vol. 6, 56.

16 Ibid., vol. 7, 445.

17 For examples of these type of tickets see *Qing gong Neiwufu Zaobanchu dang'an zonghui*, vol. 7, 339, 330, and 370 respectively.

workers were still calculated in silver, which suggests a dual status of silver in the accounting system as both stock and currency. In addition, *kupiao* of this kind reflected divergent channels of material circulation the palace was interlocked with. Materials obtained through the imperial tributes were released from the Treasury. Red sandalwood and *huali* wood 花梨木, Korean papers, textiles from the three Imperial Textile Manufactories (*Zhizaoju* 織造局), and precious metals were such tributary goods frequently found in *kupiao*. Yet a wider variety of materials and project-specific temporary labour were purchased from the market.

The other three headings, on the other hand, belonged to a different type of organization concerning work processes. They referred to a system of checks and balances in order to rationalize budgets and optimize the cost of production. Projects that produced such tickets often had complex procedures and took longer than others. The sequence of accounting for such projects included preliminary and subsidiary budgeting ('preliminary budget'), additional cash and resource disbursements to cover outstanding expenses that occurred during execution ('balance disbursement'), and a final cost reconciliation ('receipt of balance').

The concurrence of two organizing systems thus alludes to an embryonic stage of *kupiao* accounting in the first few years of the Qianlong period. In the 8th year of his reign (1743), the monthly filing of *kupiao* shifted its focus almost solely onto discerning the internal and external sources ('purchase' and blank) of disbursements. During this period, the presence of procedural *kupiao* diminished from 8.8% or 95 *kupiao* of the total of 1,074 in the first year to a meagre 1.15% or 6 out of 521. This does not mean that procedural *kupiao* were discarded; rather, it seems that they were bound into different ledgers. Although the lack of *kupiao* in the published archives after 1743 makes it difficult to reach a conclusive view, procedural *kupiao* from 1782 are found under the title 'Tickets of Actual Use' (*shiyong piao* 實用票).[18] Some unpublished *kupiao* of this sort issued during the Jiaqing reign (1796-1820) were bound under yet another title, 'Tickets of Budget Disbursements' (*zanling piao* 暫領票).[19] This shows that the *kupiao* headings functioned as practical indices – rather than a closed system of classification – for compiling different kinds of ledgers.

At the same time, the multiple steps of financial management revealed in the classification of *kupiao* indicate that these tickets were more than just

18 *Qing gong Neiwufu Zaobanchu dang'an zonghui*, vol. 45, 193-5.

19 See 'Zaxiang maiban kupiao', year 1821, *Gongzhong gechu dang'an*, no. 2252 (The First Historical Archives of China, Beijing).

ad-hoc memoranda. They show how the overarching accounting system of the Imperial Workshops was coterminous with that of other bureaus under the Imperial Household Department. According to the regulations of the Grand Storage Office issued in the 10th year of the Qianlong reign (1745), for instance, after receiving materials from the Six Vaults, the Imperial Workshops were required to submit to the Grand Storage Office an inventory of received items at the end of each month. The Grand Storage Office then compared the inventory from the Imperial Workshops with that from the Six Vaults, created its own inventory and sent it to the Imperial Workshops for reference. The Imperial Workshops returned the inventory upon confirming that there was no discrepancy with its own records.[20] Lateral checks of this kind were intended to ensure the proper management of resources. As the last section of the chapter discusses in depth, many of the accounting activities at the Imperial Workshops were incorporated into the accounting activities of the Grand Storage Office and its Six Vaults and, by extension, into the accounting activities of the Imperial Household Department itself.

The Accounting System of the Imperial Workshops and Its Reform

The Imperial Workshop's accounting system evolved with *kupiao*. During the early years of the Qianlong reign, several journals and ledgers existed in addition to *kupiao*. They included the 'Budgetary Ledgers of Gold in Each Work' (*Linian gezuo zanling jindang* 歷年各作暫領金檔), 'Budgetary Ledgers of Silver in Each Work' (*Linian gezuo zanling yindang* 歷年各作暫領銀檔), 'Annual Ledgers of Storage' (*Shouzhu qingce* 收貯清冊) and 'Annual Ledgers of Receipt' (*Xingqu qingce* 行取清冊).[21] The 'Annual Ledgers of Storage' and 'Annual Ledgers of Receipt' were inventory ledgers in the form of an end-of-year balance resulted from all transactions in cash and material. These annual ledgers used a four-column balancing system (*sizhu jiesuan* 四柱結算) – a longstanding accounting system since the Song dynasty[22] – which comprised a beginning balance (*jiucun* 舊存), new receipts (*xinjin* 新進), actual use (*shiyong* 實用), and ending balance (*xiacun* 下存).

20 *Zongguan Neiwufu xianxing tiaoli: Guangchusi*, 43.

21 For examples of *Linian gezuo zanling jindang* and *Linian gezuo zanling yindang*, see *Qing gong Neiwufu Zaobanchu dang'an zonghui*, vol. 12, 1-19. For examples of *Shouzhu qingce* and *Xingqu qingce*, see *Qing gong Neiwufu Zaobanchu dang'an zonghui*, vol. 7, 572-644.

22 Lin Jun, 'Chinese Double-Entry Bookkeeping before the Nineteenth Century', *The Accounting Historians Journal*, vol. 19, no. 2 (Dec. 1992): 108.

Figure 3.2 ***Budgetary Ledgers of Gold in Each Works*** **of 1740**

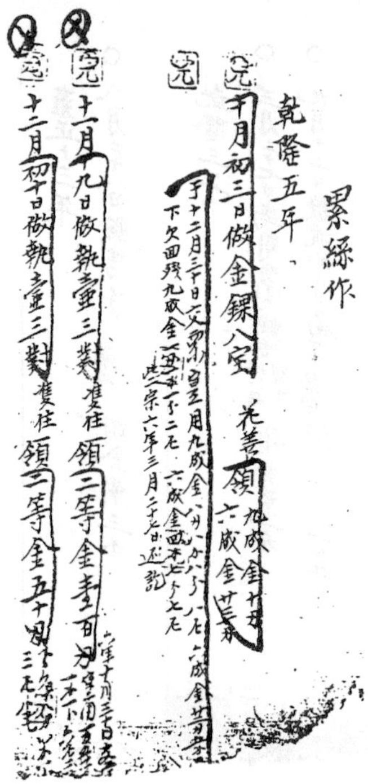

Source: *Qing gong Neiwufu Zaobanchu dang'an zonghui*, vol. 12, 3

The budgetary ledgers of gold and silver were special ledgers created to record the closely monitored transfer of precious metals into various works.[23] Each entry in these ledgers recorded a commission. The date of commission and the amount of gold or silver sent to each work were recorded in large characters (Figure 3.2).[24] To the left of these characters there was a note written in smaller characters that specified the date of completion, the amount of actual consumption in gold or silver, and the balance returned to the Treasury. In Figure 3.2, for instance, the note on a commission for the Filigree Works (*Leisi zuo* 累絲作) reads, 'on the 13th day of the 12th month (15 February 1741), [the Filigree Work] submitted a ticket that states the consumption of eight taels eight *qian* eight *fen* of gold of the ninth degree …'[25] The 'ticket (*piao* 票)' here refers to a *kupiao*,

23 On the influx into and the administration of precious metals at the Qing imperial court, see Chapter Five in this volume.

24 *Qing gong Neiwufu Zaobanchu dang'an zonghui*, vol. 12, 3.

25 Ibid.

by which one can infer that the ledgers of gold and silver were created based on *kupiao*. In addition, on top of each entry, there is a stamp mark indicating 'complete' (*wan* 完) or an 'X.' These marks might have been added during reviews or audits after a commission was completed. They hint at the intervention of auditing in the bookkeeping process: only after all the transactional and corresponding *kupiao* had been returned, collected and transcribed into ledgers could a commission be sealed as 'complete' in accounting terms.

In the 20th year of the Qianlong reign (1755), a series of reforms took place in the Imperial Workshops' accounting system that brought about a more coherent form of financial management. During the 3rd month, the emperor issued an edict that stated:

> Consolidate the 30-plus works at the Imperial Workshops into five based on the affinities between the vaults they use. Each of the [five] works should appoint a stock keeper, a project head and a deputy project manager to supervise tasks and the management of cash and materials. They should audit each other. ... Let the new works take over the ongoing tasks from the previous works and requisition materials such as gold, silver, copper, iron, tin, lead, gold and silver leaves, rims of satin and various kinds of silk, silk and woollen threads, felt blankets, mats, wood, paper, pigments, jade, and agate, ivory, walrus ivory (*qiujiao* 觩角), tortoiseshell, beeswax, chiselling sand (*baosha* 寶砂), borax, brocade ribbons, silk thread ribbons, yellow and white waxes, sandalwood sticks, rice glue, risk husks, coal, firewood and *hunao* 胡腦 [probably a type of herb similar to *longnao* 龍腦 or Borneo camphor] from the guarding officials of the Treasury of the Imperial Workshops, Six Vaults of the Grand Storage Office, the Inner Department of Works (*Yingzaosi* 營造司), the Imperial Armoury (*Wubeiyuan* 武備院), the Bureau of Weaving and Dying (*Zhiranju* 織染局), the Imperial Rice Growing Estates (*Daotianchang* 稻田廠), the Imperial Dispensary (*Yaofang* 藥房) and the Boards of Revenue and Work. As for the Three Textile Manufactories, send lateral communications and requisition by transferring the material.[26]

The five merged units of works were not a conclusive result of the reform, since sources show that three years after this order (1757) the works were

26 Although not specified in this passage, as shown earlier, it was the Treasury of the Imperial Workshops that administered and accounted for all requisitions directed outside the Imperial Workshops. *Qinding Zongguan Neiwufu xianxing zeli er zhong*, vol. 4 (309), 295-6.

regrouped into fifteen units altogether.[27] As can be gleaned from the order, it was financial accounting rather than actual production that was being merged here. By simplifying the accounting channels and consolidating the administration based on resources, the new workshop system intended to achieve more efficiency and accountability across multilateral flows of cash and materials. In light of this goal, the merged works became independent accounting units that produced individual ledgers. Sources show that, at least from the 1760s on, budget disbursement ledgers (*zanling dang* 暫領檔) were created for each merged unit of works. Inventorying each combined work's receipt of silver and materials for itemized tasks, these new ledgers stood between *kupiao* and the annual ledgers. The *Archives of the Imperial Workshops* include such work ledgers from the years of 1769 and 1775/6. The Works of Wood Varnish, Carving and Lathe-turning produced a budget disbursement ledger in 1761; the Works of Guangdong Woodwork, Casing and Mounting produced one in 1764; the Works of Gold, Jade, Inlay, Ivory and Inkstone made one in 1763; and the rest of the works, including Studio of the Ornamental Sceptre (*Ruyiguan* 如意館), Clock Works, Armoire Works and Glass Works also began to make budgetary ledgers from around 1760.[28] While it is unclear how long this bookkeeping practice lasted, the works undoubtedly remained as independent accounting units at least during the mid-Qianlong reign.

The structure of these ledgers was similar to the aforementioned four-column ledgers for gold and silver. Entries were made from right to left in the order of date, project name, and the amount of cash and materials received. Auditorial marks were added above each vertical line, with either a circle or the character *cha* 查 (meaning 'inspected'). The obvious difference between the pre-1755 gold and silver ledgers and these workshop ledgers is the focus of accounting: whereas the former focused on material, the latter focused on merged works as organizational units. By turning works into accountable units, the new accounting system strived to obtain a greater degree of financial transparency at every stage of production.

The accounting reform was not limited to reorganizing the works. In tandem with the bureaucratic reform, a novel and more sophisticated concept of accounting emerged to achieve, above all, a clear distinction between 'budget disbursement' (*zanling* 暫領) and 'actual use' (*shiyong* 實用). While

27 Although the regulation was created in 1755, it was not until 1757 that the newly merged works were reflected in the archives. See *Qing gong Neiwufu Zaobanchu dang'an zonghui*, vol. 23.

28 For more information, see *Qing gong Neiwufu Zaobanchu dang'an zonghui*, vol. 33, 10-117; vol. 39, 1-26; vol. 40, 1-15.

the concept of budget (*zanling*) had already appeared in the gold and silver ledgers, the juxtaposition of budget and actual use elevated the accounting system to a whole new level, since separating the estimated cost during planning from the actual expenditure incurred during execution meant that inventory management became more transparent and legible for inspection.

The separation of budget and actual use was carried out at every step in the accounting process. On a basic level, as mentioned in the previous section, this brought a change to the organization of *kupiao*. Whereas they had been chronologically bound into monthly reports in the early 18th century, they were now classified and archived into two separate journals, entitled 'Tickets of Budget Disbursements' and 'Tickets of Actual Use'.[29] Based on these journals, on a more strategic level, the new regulations mandated that the Treasury submit and archive an inventory that specified actual expenditures and returned balances from the budget at the end of each month. This 'Manuscript of Use and Balance' (*Shixiao gaojian* 實銷稿件) recorded the total expenditure of every project in cash, stock and wages, together with notes on the methods of work and the breakdowns of material estimates. At the level of execution, these budgets functioned as a guiding principle for planning and reviewing. For all big projects, the Audit Bureau (*Chahefang* 查核房) had to provide estimates (*yuegu* 約估) for market acquisitions and wages, based on the manufacturing method and measurements of specific products. Upon the completion of a project, the same bureau inspected the work period, expenditure and balance in order to 'avoid any abuse [of money and resources]' within the Imperial Workshops.[30]

Furthermore, the reconceptualization of accounting rationale and practice went hand-in-hand with the increasing bureaucratization of financial administration. The Audit Bureau, established in 1748, accumulated accounting documents and reviewed new budgets in reference to previous case histories. It functioned as a gatekeeper for the Imperial Workshops, for no work could initiate a project before it approved the budget. In addition, an Archiving Bureau (*Huigaochu* 彙稿處) was founded in 1755[31] and became

29 For examples of 'Tickets of Actual Use' see *Qing gong Neiwufu Zaobanchu dang'an zonghui*, vol. 45, 193-5; for 'Tickets of Budget Disbursements', see the one issued in the year 1821 and kept in the First Historical Archives of China, Beijing, under no. 2252 of *Gongzhong gechu dang'an*.

30 *Qinding Zongguan Neiwufu xianxing zeli er zhong*, vol. 4 (309), 296-7.

31 The Archiving Bureau was also called *Huizongfang* 匯總房 or *Huizongchu* 匯總處. Wu Zhaoqing has discussed the relationship between the *Huigaochu* (Archiving Bureau) and *Benfang* (Archiving House), pointing out that the Archiving House existed before the Archiving Bureau

responsible for collecting all documents of transactions in the Imperial Workshops after they had passed audits from the Audit Bureau.[32] Now the Treasury had to provide the Archiving Bureau with two types of ledgers each month – one classified by different works and the other by commissions.[33] After the Archiving Bureau received all the paperwork, a subordinate office called the Archiving House (*Benfang* 本房) produced monthly reports of actual expenditure. At the end of every year, the Treasury's annual registers were also archived at the Archive House.[34]

The increasing complexity of administrative routines and the increasing frequency of crosschecking between various bureaus enhanced the accountability of journals and ledgers. An example taken from an annual stock ledger called *Ledger of Actual Use, Budget Disbursement and Current Balance* (Shiyong zanling xiancundang 實用暫領現存檔) embodies the metrics of the reformed accounting system produced by the interaction between the 'budget' and the 'actual use' (Figure 3.3).[35]

The section shown in Figure 3.3 is taken from an account of the transactions between the Treasury and various works in Korean Summer Textiles (*gaoli xiabu* 高麗夏布) during the twelve months of the 59th year of Qianlong (1794). It comprises a beginning balance (*jiucun* 舊存), budget disbursement (*zanling* 暫領), additional budget disbursement (*tian* 添), actual use (*shiyong* 實用), resulting balance from budget (*xiao* 銷), current (monthly) balance (*xiancun* 現存), and actual remaining (monthly) balance (*xiacun* 下存). This breakdown allows the reader to grasp the estimated monthly outflow, current stock, as well as the entire volume of materials transferred to various works. As evident from the title, the account juxtaposes two separate yet correlated disbursements: the budgetary and the actual. These two categories were not just conceptually distinguished, but they were visually divided: labels related to budget ('budget disbursement', 'additional budget disbursement' and 'returned balance from budget') begin at the top of each column, while the labels indicating actual use ('actual use', 'beginning balance', 'current balance' and 'actual remaining balance') are written in indented lines.

was founded. When the Archiving Bureau was established in 1755, the Archiving House became a subordinate department of the Archiving Bureau. See Wu Zhaoqing, 'Qingdai zaobanchu de jigou he jiangyi', 80.

32 *Qinding Zongguan Neiwufu xianxing zeli er zhong*, vol. 4 (309), 297.

33 The first type of ledger recorded the budget disbursement, actual use and the returned balance of materials at each work at the end of each month. The second type contained the title of each commission and its actual cost.

34 Wu Zhaoqing, 'Qingdai zaobanchu de jigou he jiangyi', 81.

35 *Qing gong Neiwufu Zaobanchu dang'an zonghui*, vol. 55, 233.

Figure 3.3 Accounts for Korean Summer Textiles in the *Ledger of Actual Use, Budget Disbursement and Current Balance* in 1794

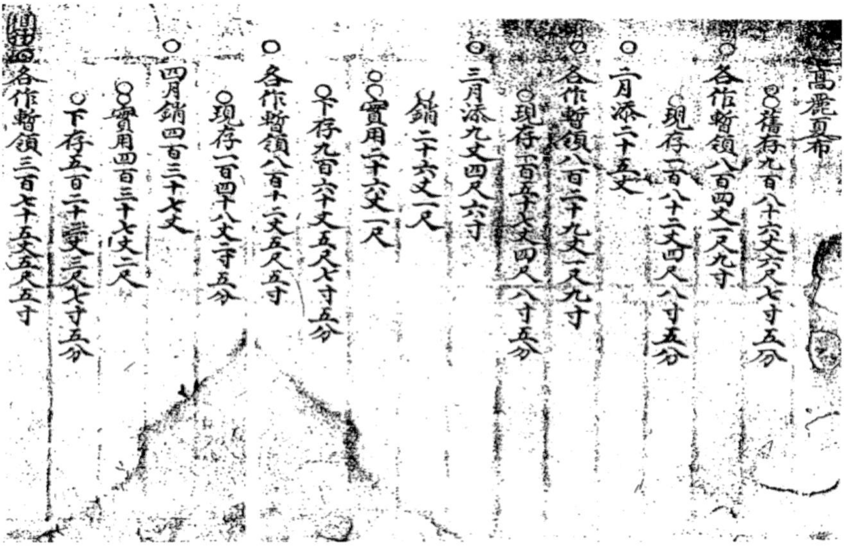

Source: *Qing gong Neiwufu Zaobanchu dang'an zonghui*, vol. 55, 235

The budget disbursement of each month (*zanling*) was calculated by adding the first and additional disbursements (first *zanling* and subsequent *tian*) and subtracting the resulting balances (*xiao*). In the first month, the current balance (182,485 *zhang*) resulted from simply deducting the budget disbursement (804,190 *zhang*) from the beginning balance (986,675 *zhang*). The subsequent current balances were calculated by subtracting additional budget disbursements (*tian*) and the actual amount of consumed materials (*shiyong*) from the previous month's current balance, and adding to it, if any, the amount of remaining materials from the previous total budget (*xiao*). For instance, the current balance of Korean summer textiles in the 3rd month (148,025 *zhang*) was obtained by subtracting from the current balance of the 2nd month (157,485 *zhang*) an additional budget disbursement (9,46 *zhang*) and the actual amount of use (26,1 *zhang*), then adding to it the returned balance (26,1 *zhang*). Finally, the actual remaining balance was obtained by subtracting actual uses from the beginning balance. This meant that the actual remaining balance did not appear every month, because an actual use figure could only be calculated once a project was completed and all the necessary documents had been returned and reviewed. Thus, in this example, the first actual remaining balance appeared in the 3rd month after the works that had requisitioned the textiles reported their actual use.

The formula in the Ledger of Actual Use, Budget Disbursement and Current Balance, therefore, was as follows:

'beginning balance' 舊存
– latest accumulated 'budget disbursement' 暫領 (increasing because of additional budget disbursements' 添 or reducing because of 'returned balances' 銷)
– accumulated 'actual use' 實用
= 'current balance' 現存

'beginning balance' 舊存
– 'actual use' 實用
= 'actual remaining balance' 下存

This accounting algorithm kept track of multiple expenditure channels across all of the workshops. Whereas the current balance reflected the amount of stock in the Treasury, the actual remaining balance indicated the total amount of materials available within the entire Imperial Workshops. Moreover, the difference between the budget disbursement and the actual use allowed auditors to map the lapse between a plan being submitted and a plan being executed, and to establish reasonable budgets by commanding a comprehensive overview of the stock available at the Treasury and all of the works. The *Ledger of Actual Use, Budget Disbursement and Current Balance* thus captured the dynamic state of resources across stock, budget and use, and it kept abreast of diverse interwoven trajectories of resource managements within the entire institution.

What was the ultimate goal of mid-century reforms that took place in both the financial and bureaucratic arenas? The dual assessment of budgetary estimation and actual expenditure was meant not only to keep track of present and future stock but also to rationalize plans and expenditures in the manufacturing process. The latter, to some extent, was a bigger concern. Sources show that there were lapses in the actual receipt of budgeted materials after the declared dates of commission. The policy announced in the 4th year of the Yongzheng reign (1727) stated that, if the requested materials were in stock, a work should collect from the vault the amount of preliminary budget within six days after a project started; metals and textiles should be collected within ten days.[36] This suggests that actual receipts might have taken longer than the stated period, causing difficulties

36 *Zongguan Neiwufu xianxing tiaoli: Guangchusi*, 36.

in stock management. The policy thus attempted to preclude discrepancies between a plan and its execution.

All of these reforms underscore an important attribute of the Imperial Workshop accounting system – its efforts to achieve an optimal inventory management. The goal of its accounting system was not profit maximization but cost minimization. While materials held in the Imperial Workshops could potentially become sellable assets – which many did after their initial lustre as imperial resources had faded – for accounting purposes they were considered non-capitalized assets for internal consumption. The purpose of accounting was therefore to closely monitor and regulate the organization of vault stock in order to fulfil both present and future productions successfully. Moreover, through auditing routines enforced across different accounting sectors, the system strived to reduce the risks of embezzlement and other corruption, while foregrounding an ideal of micromanagement over complex manufacturing processes.

The Paper Trails of Cross-Departmental Production: Processes and Characteristics

Having outlined the processes and purposes of an accounting system embodied by *kupiao*, this essay will turn to examine the mechanism of the interwoven processes of finance and manufacture at the Imperial Workshops by tracing an actual project recorded in *kupiao*. As mentioned above, *kupiao* show that the Imperial Workshops were not merely an assemblage of artisans but a complex system in which various components each played their role within an architecture of checks and balances. Tracing how a commission advanced through different works and produced various paperwork at different administrative and accounting units provides an insight into how artisans from different works and clerks at administrative bureaus collaborated to complete a project, as well as the way in which *kupiao* as paper tools enabled various bureaus and works to act in concert.

On the 2nd day of the 1st month in the 1st year of Qianlong's reign (13 February 1736), the warehousemen (*siku* 司庫) Changbao 常保 and Liu Shanjiu 劉山久 submitted a commission to the Project Management Bureau (*Huojifang* 活計房). The commission had come from the emperor and been orally transmitted to the Imperial Workshops by the eunuch supervisor-in-chief (*zonguan taijian* 總管太監) Liu Cangzhou 劉滄州, the supervisory eunuch (*shouling taijian* 首領太監) Wang Shougui 王守

貴, and a eunuch called Mao Tuan 毛團. The content of the commission was as follows:

> [The emperor] ordered [the Imperial Workshops] to make fifty pewter bowls painted with golden dragons in size one. Twenty of these bowls should be paired with stands and covers decorated with copper filigrees. [The emperor also ordered to make] ninety pewter bowls painted with golden dragons in size two. Sixty of them should be paired with stands and copper filigree covers.[37]

Since gold painting would have been done in the Southern Workshop (*Nanchang* 南廠), it appears that, after registering the commission at the Project Management Bureau, the tasks were immediately divided and transferred to the Southern Workshop and to the Copper Works (*Tongzuo* 銅作), which produced the actual pewter bodies for the bowls. These works would each have estimated the amount of materials and labour necessary to complete the tasks, and the preliminary budget made by them was probably reviewed by the Calculation and Archive Bureau (*Suandangfang* 算檔房).[38] For this particular project, it took 101 days for the budget to be assessed and approved. On the 14th day of the 4th month (24 May), Bashisan from the Archiving House issued two *kupiao* – one per work – that laid out all the budgetary details. On the *kupiao* for the Southern Workshop (Figure 3.4), Bashisan wrote:

> On the 2nd day of the 1st month (13 February 1736), in order to fulfil the imperial commission and to make 50 pewter bowls painted gold in size one, 90 bowls in size two, 20 stands in size one, 60 stands in size two, 120 railings (*bianlan* 邊欄) in size one, 300 decorated railings in size two, 100 copper wire covers in size one and 240 copper wire covers in size two, the Southern Workshop needs two taels and five *qian* of cottonseed oil, 160 *gong* 工 of oiling, 53 *gong* of lacquering, and 21.5 *gong* of gold painting. Every *gong* costs one tael and eight *qian* of silver, [so it amounts to] 42 taels five *qian* eight *fen* and eight *li* in total.

37 *Qing gong Neiwufu Zaobanchu dang'an zonghui*, vol. 7, 52-3.

38 'Suandangfang' appears in *kupiao* as the bureau which 'examines and audits projects' (*chadui huoji* 查對活計), see *Qing gong Neiwufu Zaobanchu dang'an zonghui,* vol. 7, 321, 340, 414. Wu Zhaoqing contends that it was the Audit Bureau which conducted the calculations, see Wu Zhaoqing, 'Qingdai zaobanchu de jigou he jiangyi', 80. However, as mentioned above, the Audit Bureau was not established until 1748, so it is possible that the Calculation and Archive Bureau performed a similar role prior to the creation of the Audit Bureau.

Figure 3.4 ***Kupiao* issued to the Southern Workshop in 1736**

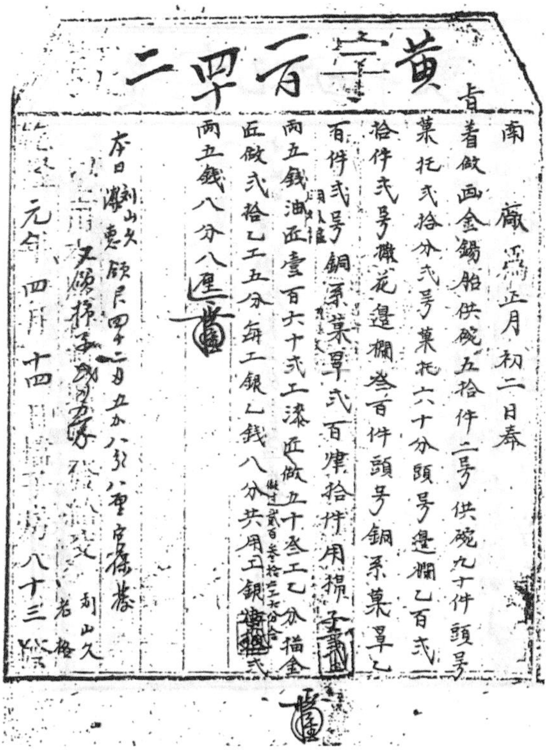

Source: *Qing gong Neiwufu Zaobanchu dang'an zonghui*, vol. 7, 303

Bashisan collated the estimated cost of materials and wages in order to lacquer the pewter bowls and issued it on a *kupiao* sent to the warehouseman Liu Shanjiu and a clerk at the Imperial Household Department named Laoge 老格.[39] Labour was measured by *gong* 工, a theoretical unit of the amount of work one skilled artisan could complete within a day.[40] Later on

39 Laoge was a bannerman, most likely a *booi*, serving the Imperial Household Department. He became Tang Ying's 唐英 assistant in the 6th year of the Qianlong reign (1741) and engaged in the imperial porcelain production. See Yiling'a's 伊齡阿 court memorial on QL 33/11/27 (4 Jan. 1769), *zou'an*, no. 03-0126-072: '*Zou ming xiezao Laoge jibing qing hui qi shi* 奏明協造老格疾病請回旗事' (The First Historical Archives of China, Beijing).

40 According to the regulations, a whole day's work by an artisan who had completed his apprenticeship could be counted as one *gong*. A whole day's work by an artisan who had received two years of apprenticeship was only half a *gong*. An artisan who had only undergone one year of training would have to work for three days to receive one *gong*. See *Zongguan Neiwufu xianxing tiaoli: Guangchusi*, 45.

that same day, Liu Shanjiu and Puhui 溥惠 took the *kupiao* to the Treasury in order to receive the cash and materials. At the Treasury, clerk Guanbao authorized the release and filled in the middle section of the *kupiao* in rather sloppy handwriting: 'Today Liu Shanjiu and Puhui received 42 taels, five *qian*, eight *fen*, and eight *li* of silver, issued by Guanbao. In addition, [they received] two taels and five *qian* of cottonseed oil'.[41] Guanbao might have recalculated the sum of required labour and appended the total number of '236 and three-fifths *gong*' in small characters next to the total amount of silver originally written by Bashisan. He then stamped the numbers to authenticate the amount of disbursement made to Liu Shanjiu and Puhui and signed the bottom of the ticket.[42] This recalculation of the total sum and Guanbao's signature show that the practice of checks and balances did not just occur in auditing but was already in place at the budgeting stage. Every transaction had to be examined and confirmed by a responsible person who then signed it for accountability and, later on, for archiving.

That same day, Bashisan wrote another *kupiao* for the Copper Works' requisition of 377 catties, four taels and seven *qian* of tin, and 251 catties, fifteen taels, and one *qian* of lead to produce 150 pewter offering bowls. The ticket was issued to Liudazi 六達子 and Zhang Si 張四. The middle section of the ticket remains empty, but the signature on the ticket indicates that it was also Guanbao who distributed the material (see Figure 3.5).[43]

According to the records kept in the *Huoji qingdang* 活計清檔 (Inventories of Commissions), this project ended on the 25th day of the 12th month (25 January 1737), about a year after the commission had originally been received.[44] Upon completing the project, the Archiving House collected all the transactional documents in order to produce monthly reports on actual expenditure, which it submitted to higher accounting units,

41 *Qing gong Neiwufu Zaobanchu dang'an zonghui*, vol. 7, 303.

42 The signature is neither Manchu or Chinese. It is likely to be a special mark called a *huaya* 花押.This kind of mark often appeared on reports in the Imperial Household Department next to the clerks' Manchu or Chinese signatures. For an example of this, see *zou'an*, no. 05-08-002-000174-0035: 'Wei xingqu zuocheng Yuanmingyuan deng chu tong xi huoji xuyong meitan muchai deng xiang shi 為行取成做圓明園等處銅錫活計需用煤炭木柴等項事' (The First Historical Archives of China, Beijing).

43 *Qing gong Neiwufu Zaobanchu dang'an zonghui,* vol. 7, 306. Although the two tickets related to one project were issued by the same person on the same day, they had different ticket numbers. This might be because the two *kupiao* were received and archived at different time.

44 *Huoji qingdang* was a genre of *huoji dang* 活計檔 which recorded the inventories of completed commissions.

Figure 3.5 *Kupiao* issued to the Copper Workshop in 1736

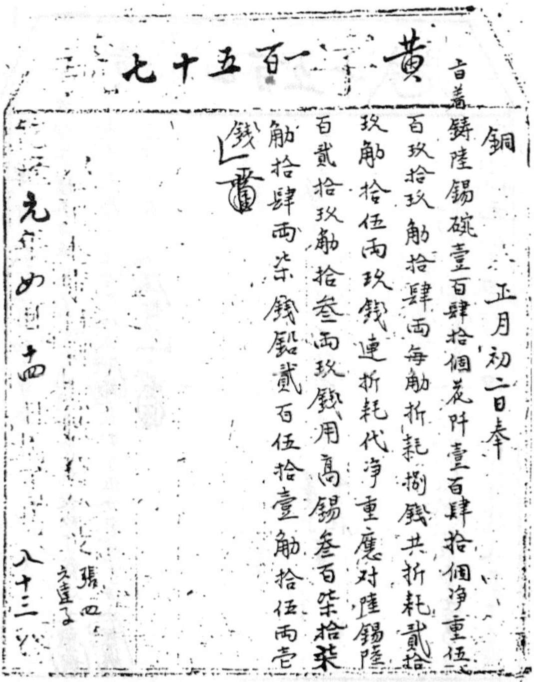
黄　一百五十七
銅
正月初二日奉
旨著鑄陸錫碗壹百肆拾個花件壹百肆拾個淨重伍
百玖拾玖觔拾肆兩每觔折耗捌錢共折耗貳拾
玖觔拾伍兩玖錢連折耗代淨重應对陸錫陸
百貳拾玖觔拾叁兩玖錢用高錫叁百柒拾柒
觔拾肆兩柒錢鉛貳百伍拾壹觔拾伍兩壹
錢

Source: *Qing gong Neiwufu Zaobanchu dang'an zonghui*, vol. 7, 306

likely the Headquarters of the Imperial Household Department (*Neiwufutang* 内務府堂).[45] Based on the documents examined above, the process through which the Imperial Workshops finished a commission can be broken down to the steps shown in Chart 3.1.

As Chart 3.1 shows, it was not just different works that collaborated in this process to complete a commission, but various administrative and accounting sectors – such as the Project Management Bureau, the Document House, the Calculation and Archive Bureau, the Archiving House, and the Treasury of the Imperial Workshops – also played an indispensable role in assigning tasks, dividing labour, creating *kupiao*, calculating budgets, and providing the funding and resources. If the works and artisans were the actual manufacturing components of the machine, the administrative and accounting sectors regulating and supplying them were analogous to the

45 The Archiving House was established no later than the 1st year of the Qianlong reign (1736). Until it was incorporated into the Archiving Bureau in around 1750 it operated as an independent office. See *Qing gong Neiwufu Zaobanchu dang'an zonghui*, vol. 7, 445.

Chart 3.1 Workflow of a commission at the Imperial Workshops from assignment to archiving

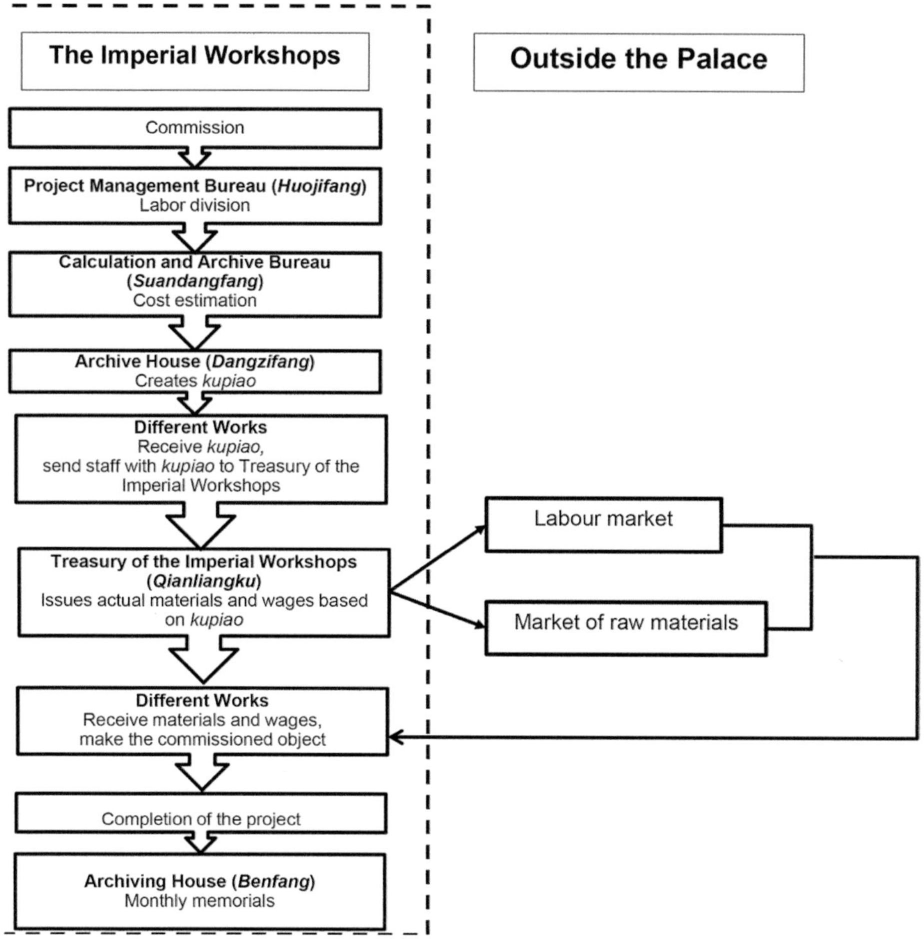

joints, hinges and screws that connected these components and enabled their operation. Through them, a pipeline of fuel – funding, materials, and human labour – was constructed for the operation of machine.

The content of *kupiao* was not limited to internal transactions in the palace. It also bridged the Imperial Workshops and the external market. As demonstrated by the *kupiao* with 'purchase' headings mentioned earlier, the Imperial Workshops procured a wide range of materials directly from the market. Frequently purchased items included various stationery goods such as brushes and ink, and raw materials such as pigments, wax, lacquer, glue, and nails.[46] The large volume of acquisition indicates that the Imperial Workshops relied heavily upon the local market for production, while the diversity of goods suggests that there was a mature market of raw materials in Beijing.

Furthermore, *kupiao* reveal that the palace often outsourced casual work to the labour market outside the palace. For instance, in the project discussed in Figure 3.4, the Southern Workshop hired oilers, gold painters, and lacquerers from outside.[47] In contrast to the 'provisioned artisans' (*shixiang jiang* 食餉匠) who received fixed monthly stipends, these temporary artisans were paid according to the amount of work (*gong*) they did.[48] *Kupiao* show that all casual workers (including ivory carvers, carpenters, lacquerers, masons and jade carvers) were paid at the same rate of one *qian* and eight *fen* of silver per *gong*, regardless of their specialities.[49]

Tapping the resources and workforce available outside the palace had certain benefits. Flexible outsourcing reduced the cost of production since the palace did not have to support a large number of artisans permanently. It also reduced the financial burden of the internal vault as the palace did not have to prepare every kind of raw material. The local material and labour markets were thus indispensable for the palace production. The Imperial Workshops were not a closed, autonomous system, but they were linked to the broader networks of commerce and labour in the capital city and beyond.

While Chart 3.1 represents the basic workflow that emerged during the first half of the eighteenth century, Chart 3.2 portrays the more complex and segmented trajectory of the paper trail that emerged after the 1755 accounting reforms mentioned in the previous section. Here, the newly added steps show how the accounting process continued even after a commission had been completed. The most noticeable change is that the Treasury now

46 For the purchase of *jinbuhuan* brushes and ink, see *Qing gong Neiwufu Zaobanchu dang'an zonghui*, vol. 7, 336; pigments: vol. 7, 337; different kinds of wax: vol. 7, 338; different kinds of glue such as *yujiao* 魚膠 (fish glue) and *guangjiao* 廣膠 (Cantonese glue): vol. 7, 338, 342; different-sized nails: vol. 7, 362.

47 Ibid, vol. 7, 303.

48 For a discussion of 'provision artisans' and other different types of artisans hired in the Imperial Workshop, see Chi Jo-hsin 嵇若昕, 'Qing zhong houqi (1821-1911) Neiwufu Zaobanchu nanjiang jiqi xianguan wenti 清中後期（1821-1911）內務府造辦處南匠及其相關問題', in *Gugong xueshu jikan* 故宮學術集刊, 32:2 (2015): 63-89. Wu Zhaoqing, 'Qingdai zaobanchu de jigou he jiangyi', 81. Wei Qingyuan 韋慶遠, 'Qingdai neiwufu jiangyi he yuyong shougongye 清代內務府匠役和手工業', in *Ming Qing shi xu xi* 明清史續析 (Guangzhou: Guangdong Renmin chubanshe, 2006), 359-379. On labour relations in the Qing palace, see Chapter One in this volume.

49 For information about outsourcing ivory artisans, see *Qing gong Neiwufu Zaobanchu dang'an zonghui*, vol. 7, 388, 447, 499, 502, 504; carpenters: vol. 7, 384, 473, 502; masons: vol. 7, 464; jade carvers: vol. 7, 464. Based on this, we speculate that 'one *qian* and eight *fen* of silver' was more or less the going rate for casual labour.

Chart 3.2 Workflow after the 1755 accounting system reform

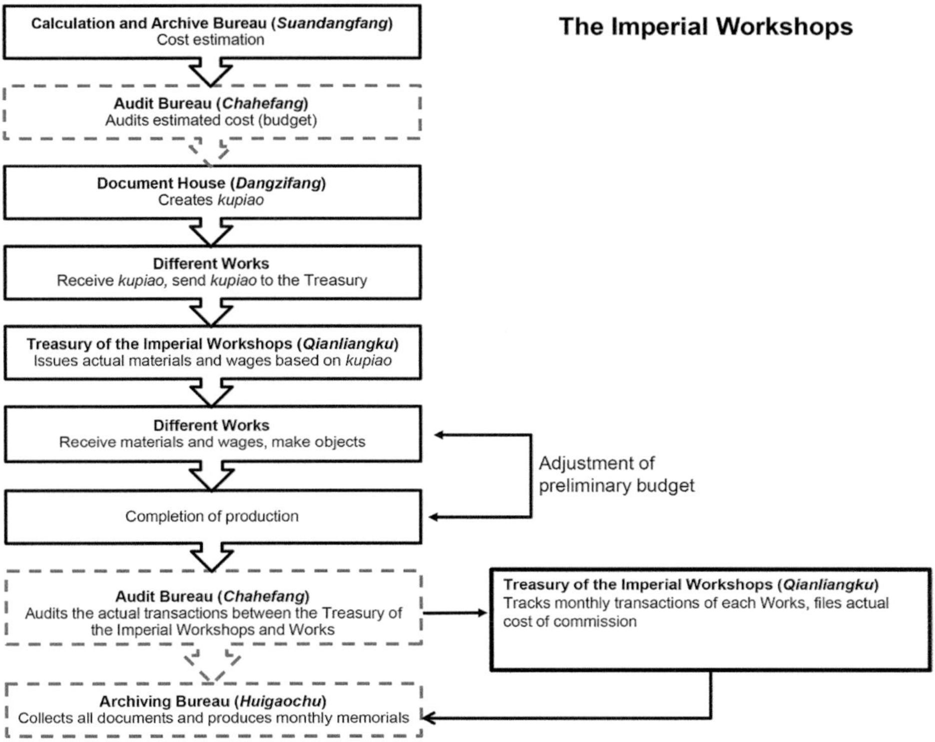

bore plural administrative roles. In addition to its original role as the central vault for the Imperial Workshops, it now performed as an accounting unit that produced various accounting documents. As an internal auditing bureau, in addition, the Audit Bureau became responsible for checking and approving the plausibility of requests. The ledgers and documents produced by the Treasury and the Audit Bureau were received and archived by the Archiving Bureau, which compiled the monthly reports.[50]

Alongside the actual production of objects, the Imperial Workshops thus produced, collected and archived an increasingly massive volume of accounting documents. As more rounds of cost estimation, audit and reassessment became mandatory, more accounting units were created to regulate the managerial process in addition to the production process. An extensive cross-departmental paper trail thus connected production and administration, artisans and materials at the Imperial Workshops.

50 *Qinding Zongguan Neiwufu xianxing zeli er zhong*, vol. 4 (309), 302.

The Palace and the State: The Accounting System of the Palace and the Fiscal System of the Qing State

Documents like *kupiao,* monthly reports, and annual ledgers represented different stages in the workflow of *zouxiao* 奏銷. The term *zouxiao*, which means to 'report financial matters to the imperial throne for approval', is better known among economic historians studying the Qing fiscal system. The documents in the *Archives of the Imperial Workshops* show that the process of *zouxiao* in the Imperial Workshops, which included many loops of bookkeeping and internal and external auditing on monthly and annual bases, was not only central to the Imperial Household Department's accounting system, but it was a microcosm of the fiscal accounting system of the Qing state.

The *zouxiao* process was widely employed in the administration of government revenue and expenditure. The regulations of the Board of Revenue stipulated that local administrators, governors and offices responsible for specific tasks had to report their tax revenues and expenses to the Board of Revenue on a regular basis, usually once a year.[51] If a report passed an audit by the Board of Revenue, it was sent to the emperor. If it did not pass the audit, it would be sent back to the submitters for correction. This *zouxiao* process was widely used to record fiscal items such as land revenues, salt taxes, custom duties, officials' salaries, government administration expenses and military expenditure. The receipt of taxation was called a ticket or *piao* 票, and – as in the accounting process of the Imperial Workshops – this served as the basis in the *zouxiao* process. During the Qing dynasty, different tickets were produced for different revenues and taxes: for salt, there were salt tickets or *yinpiao* 引票,[52] for tin mining, there were tin tickets or *xipiao* 錫票,[53] and for land revenue there were 'stringed' tickets or *chuanpiao* 串票. The tickets for land revenue were similar to *kupiao* in terms of format.[54] Both *chuanpiao* and *kupiao* were issued on preprinted forms and sequentially

51 Shi Zhihong, *Central Government Silver Treasury: Revenue, Expenditure and Inventory Statistics, Ca. 1667-1899* (Leiden; Boston: Brill, 2016), 14.

52 *Qinding Hubu zeli* 欽定戶部則例 (Regulations and Precedents of the Board of Revenue), vol. 1, 23 (rep. *Gugong zhenben congkan* 故宮珍本叢刊. Haikou: Hainan chubanshe, 2000, vol. 284-286).

53 Court memorials from the minister of revenue Su'erna 素爾訥 to Emperor Qianlong, 1771, *Huke tiben* 戶科題本 (Memorials of the Board of Revenue), no. 02-01-04-16407-006 (The First Historical Archives of China, Beijing).

54 *Chuanpiao* began their history in the Ming dynasty and were widely used in land revenues during the Qing dynasty. Guo Runtao 郭潤濤, 'Mingdai de chuanpiao 明代的串票' (Revenue Tickets in the Ming Dynasty), in *Yang Guozhen jiaoshou zhishi wushinian jinian wenji* 楊國楨

numbered for archiving purposes. Each *chuanpiao* had three copies and each copy recorded the same information: the name of the taxpayer, the date of payment, and the amount of revenue.[55] According to the *Collected Statutes of the Great Qing* (Da Qing huidian 大清會典), the taxpayer kept one copy as proof of payment, the tax farmer kept one as a receipt, and the third copy was sent to the local government to be archived for future audit.[56] Therefore it is possible to infer that *chuanpiao* and *kupiao* shared similar functions: both served as records of transaction, proof of payment, and records for future audit. In other words, both were basic elements of the *zouxiao* system. The similarities suggest the usage of *kupiao* in the Imperial Workshops was linked to the overarching *zouxiao* system of the state.

A better mirror-image of the *zouxiao* system of the Qing state in the Imperial Workshops was the production of monthly reports and annual ledgers. According to the regulations of the Board of Revenue, the Three Vaults of the Board of Revenue (*Hubu sanku* 戶部三庫) had to submit a report at the end of every month to report their monthly incomes and expenditures.[57] These monthly reports would be sent to the Court of Censors (*Duchayuan* 都察院) for audit within the first ten days of the following month.[58] In addition to monthly memorials, the Three Vaults had to produce and submit by the start of the following fiscal year two types of copies of its annual ledgers, respectively called yellow registers (*huangce* 黃冊) and blue registers (*qingce* 青冊). Both registers were written in the four-column format and contained the same information. Named after the colour of their silk covers, the yellow registers were presented to the imperial throne and stored in the Archives of the Grand Secretariat (*Neige daku* 內閣大庫) while the blue registers were archived in the relevant government departments.[59]

The production of monthly reports and annual ledgers in the Imperial Workshops thus resembled the accounting process of the Board of Revenue. As discussed earlier, the Archiving House was responsible for producing the Imperial Workshops' monthly reports.[60] The original monthly reports

教授治史五十年紀念文集 (Collected Essays for the Fiftieth Anniversary of Professor Yang Guozhen) (Nanchang: Jiangxi jiaoyu chubanshe, 2009), 32-55.

55 For the format of *chuanpiao,* see Huang Liuhong 黃六鴻, *Fuhui quanshu* 福惠全書, *juan* 6, 7b-8a. (1893 edition; rep. *Siku weishou jikan* 四庫未收輯刊, *shibu* 史部, vol. 19).

56 Guo Runtao, 'Mingdai de chuanpiao', 32.

57 The Three Vaults of the Board of Revenue were: the Bullion Vault (*Yinku* 銀庫), the Miscellany Vault (*Yanliao ku* 顏料庫), and the Textile Vault (*Duanpi ku* 緞疋庫). For monthly reports by the Three Vaults, see Shi Zhihong, *Central Government Silver Treasury*, 15.

58 Ibid., 15.

59 Ibid.

60 *Qing gong Neiwufu Zaobanchu dang'an zonghui*, vol. 7, 445.

Figure 3.6 Monthly summary (*yuezong* 月總) of 1793

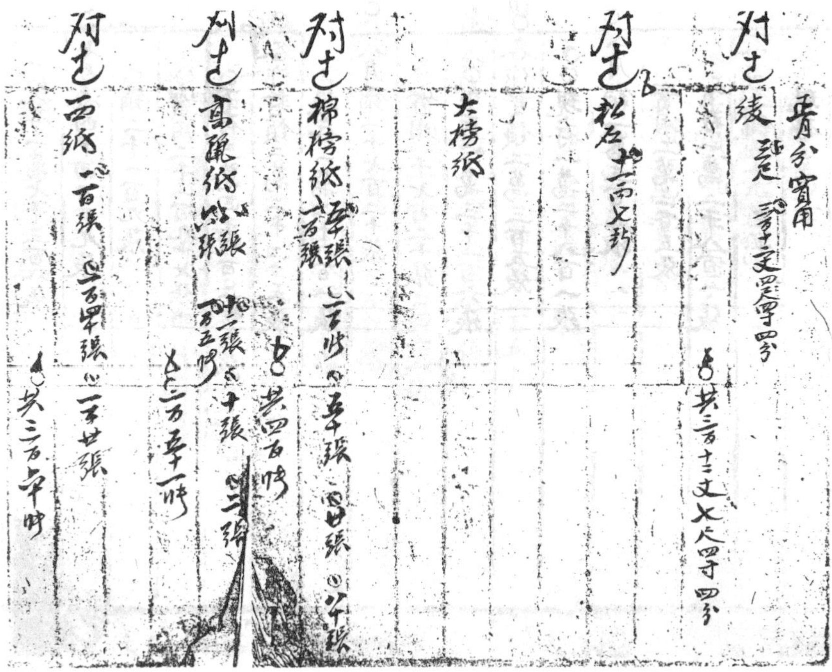

Source: *Qing gong Neiwufu Zaobanchu dang'an zonghui*, vol. 54, 162

produced by the Imperial Workshops no longer exist in the published collection, probably because they were submitted to higher accounting units. Some of the drafts that the Imperial Workshops crafted in preparing the monthly reports still remain, however. The regulations of the Imperial Household Department show that, in order to produce monthly reports at the Archiving Bureau, the Treasury re-examined all the material and cash transactions before submitting its own monthly ledgers.[61] The 'monthly summary' (*yuezong* 月總) of the 58th year of the Qianlong reign (1793) appears to be a draft for such a ledger recording monthly expenditures at the Treasury.[62] The mark '*duiguo* 對過' (meaning 'compared') added to the upper margin of the monthly summary demonstrates that an internal audit had taken place (see Figure 3.6). Following the internal audit was an external one by the Headquarters of the Imperial Household Department.[63]

61 *Qinding Zongguan Neiwufu xianxing zeli er zhong*, vol. 4 (309), 297.

62 *Qing gong Neiwufu Zaobanchu dang'an zonghui*, vol. 54, 162-194.

63 *Qinding Zongguan Neiwufu xianxing zeli er zhong*, vol. 4 (309), 299.

Within this overarching structure, the monthly reports seem to have served two major functions: first, they provided the documents necessary for auditing; second, they collated datasets to prepare annual ledgers. The annual ledgers marked the final stage of the *zouxiao* process, in which all elements – beginning balance, new receipt, actual disbursement, and ending balance – were audited and settled. The extant annual ledgers of the Imperial Workshops, of which the earliest dates to 1733,[64] are divided to yellow registers (mentioned above), inventory registers (*qingce* 清冊), and navy registers (*lance* 藍冊). Few yellow registers appear in the *Archives of the Imperial Workshops*, likely due to the fact that they were stored in the Archives of the Grand Secretariat after being presented to the emperor.[65] The inventory registers and navy registers belonged to the category of 'blue registers'; both were copies of yellow registers, but they were submitted to, and stored in, individual departments.[66] Inventory registers were copies submitted to the Headquarters of the Imperial Household Department, while navy registers were archived at the Treasury. The contents of the two were slightly different, however. Inventory registers contained a complete list of all materials, whereas the navy registers only recorded a narrower range of materials, such as textiles, pigments, firewood, paper and wax.[67] In addition, the inventory registers appeared much earlier than navy registers. The earliest copies of the former included in the published collection are from 1733, while the latter only started to appear after 1758.[68] After 1758, the two types of ledgers coexisted until 1794.[69]

64 The earliest existing *Shouzhu qingce* and *Xingqu qingce* in the collection are from 1733. *Qing gong Neiwufu Zaobanchu dang'an zonghui*, vol. 2, 231-206.

65 One example of the few remaining yellow registers in the published collection is the 1795 *Yellow Register of Gold and Silver Materials in Stock* (Huangce jinyin cailiao xiancun dang 黄册金銀材料現存檔), in the *Qing gong Neiwufu Zaobanchu dang'an zonghui*, vol. 55, 508-538. This yellow register was stored in the Treasury of the Imperial Workshops, so it might be one of a few that was not submitted to the higher accounting units. See Shi Zhihong, *Central Government Silver Treasury*, 15.

66 Shi Zhihong claims that there were three types of blue registers in the Board of Revenue: inventory registers (*qingce* 清冊), blue registers (*qingce* 青册), and navy registers (*lance* 蓝册). Ibid.

67 For example, the '*Original Records of the Blue Register of the 23rd Year under Qianlong's Reign* (Qianlong ershisan nianfen lance xiaosuan didang 乾隆二十三年份藍冊銷算底檔)' were divided into two register books, one of which recorded silk textile and pigment transactions, while the other recorded transactions of wood, paper and wax. See *Qing gong Neiwufu Zaobanchu dang'an zonghui*, vol. 24, 1-7. The navy registers in the same pattern can be found from 1758 to 1794.

68 For *Shouzhu qingce* and *Xingqu qingce* in 1733, see *Qing gong Neiwufu Zaobanchu dang'an zonghui*, vol. 2, 231-206. For navy registers from 1758, see ibid., vol. 24, 1-7.

69 For deep blue registers from 1794, see *Qing gong Neiwufu Zaobanchu dang'an zonghui*, vol. 55, 470-478. For inventory registers from 1794, see ibid., vol. 54, 732-752.

The emergence of navy registers was largely concomitant with the abovementioned reforms of the workshop accounting system during the 1750s. It thus reflects a growing attention paid to specific materials and different sources and channels of circulation. It was no coincidence that textiles, pigments, firewood, paper and wax all happened to fall into the category of materials supplied by warehouses outside the Imperial Workshops: textiles and pigments came from Three Vaults of the Board of Revenue, firewood came from the Office of Palace Construction, while paper and wax came from the Grand Storage Office.[70] In accordance with the regulations of the Imperial Household Department, the Imperial Workshops had to report 'all the amount of materials requisitioned from external departments' to the Office of Inspection and Censors of Affairs at the Imperial Household Department (*Jicha neiwufu shiwu jiancha yushi yamen* 稽查內務府事務監察御史衙門) for external audit.[71] It is therefore likely that the Treasury of the Imperial Workshops kept an extra copy of the ledgers that recorded transactions between the Imperial Workshops and external vaults for auditing purposes. The navy registers also embodied the practice of checks and balances in the cross-departmental *zouxiao* process: transactions between the Treasury, the Grand Storage Office and the Vaults of the Board of Revenue were subject to reviews by the Office of Inspection and Censors.

The *zouxiao* procedure was not limited to the Imperial Workshops, as it was central to the accounting system of other departments in the Imperial Household Department. For instance, at least from the 4th year of the Yongzheng reign (1726), the Six Vaults of the Grand Storage Office had to routinely submit monthly reports, blue registers and yellow registers to the Headquarters of the Imperial Household Department.[72] In addition, from 1729 onwards, a regular external audit was introduced to the Grand Storage Office's *zouxiao* process: an official inspector was sent to the Six Vaults to audit their financial records. Beginning in 1730, all bureaus and workshops subordinate to the Grand Storage Office had to annually declare their receipts and balances and have their accounts examined by an external official. In the 30th year of the Qianlong reign (1748), the significance of the five-year audit increased to include several chief supervisors of the Imperial Household Department (*Neiwufu dachen* 內務府大臣) within the body of

70 *Qinding Zongguan Neiwufu xianxing zeli er zhong*, vol. 4 (309), 291 and 303.

71 Ibid, vol. 4 (309), 299.

72 *Zongguan Neiwufu xianxing tiaoli: Guangchusi*, 35-6. Similarly, the Office of the Old Summer Palace (*Yuanmingyuan* 圓明園) had to submit monthly reports and yellow registers to the Headquarters. *Qinding Zongguan Neiwufu xianxing zeli er zhong*, vol. 2 (307), 108-9.

inspectors, and the Grand Storage Office was inspected as well as several other bureaus subordinate to the Imperial Household Department, such as the Imperial Estates (*Guanfang* 官房), Leased Vaults (*Zuku* 租庫) and the Imperial Bureau of Weaving and Dying (*Zhiranju* 織染局).[73]

This shows that the Imperial Workshops and bureaus in the Imperial Household Department shared the same *zouxiao* system and produced similar documents. By auditing divergent monthly reports and annual ledgers, the Headquarters of the Imperial Household Department wove the *zouxiao* processes of different accounting units into the financial system of the Imperial Household Department as a whole. Furthermore, higher financing bureaus such as the Office of Inspection and Censors and the grand ministers of the Imperial Household Department were asked to audit and inspect not only internal transactions within the Imperial Household Department but also transactions between the Imperial Household Department and the Boards of Revenue and Work. Documents such as *kupiao*, monthly reports, and annual ledgers thus indicate that the *zouxiao* system of the Imperial Workshops, embedded in the Imperial Household Department's accounting system, was modelled on the fiscal system of the state. In other words, the accounting system of the Imperial Workshops was a microcosm of the accounting system of the Imperial Household Department as a whole, and the latter mirrored the Qing state's overarching fiscal system. The multilateral flow of resources between the Board of Revenue and the Imperial Workshops' vaults thus instantiated the interweaving of the state's revenues and that of the Imperial Workshops.

Conclusion

Whereas the institution of the Imperial Workshops and its accounting system had been set up exclusively for the emperor and his family, its operation was connected to that of the state and the market outside the palace. The Imperial Workshops' accounting system coordinated a complex flow of human and material resources as well as the information network between the Imperial Household Department, the external market and the Boards of Revenue and Work. Although the Imperial Workshops were not a constituent of the fiscal body of the Qing state, it operated in the same accounting mechanism that aimed to make heterogeneous economic and financial sectors function like cogwheels revolving on a gear.

73 *Zongguan Neiwufu xianxing tiaoli: Guangchusi*, 35-6.

From the 1720s to the 1750s, the accounting system experienced several changes: stricter regulations were instated regarding deadlines for paperwork, several rounds of internal audits were demanded, more and more departments started to engage in the verification of accounting books, and more attention was paid to certain materials over others. These changes instantiated the logic and goal underlying both the emperor's private purse and the empire's treasury: the need to construct a system of checks and balances in order to prevent corruption and ensure an efficient use of resources. The increasing number of bureaus established at the Imperial Workshops during the 18th century put into action numerous regulations for internal audits, cross-departmental inspections, and the archiving and memorializing of various information. These changes reflected increasing concerns about, and demands for, control over funds and resources. Through them, the Imperial Workshops pursued the goal of rational planning and cost economization at various stages of its work process. In this regard, the accounting system of the palace was itself a machine run by 'a combination of resistant bodies'.[74] *Kupiao*, tickets of the Treasury as the smallest documentary unit, embodied both the principle and the praxis of accounting. On the one hand, they recorded the day-to-day minutiae of the Imperial Workshops' financial activities in all their abundance, heterogeneity, and complexity. Abstracting such information into concise ledgers for financial appraisal and imperial perusal, on the other hand, epitomized the principles of accounting that emerged as a composite mechanism within the Imperial Workshops.

About the Authors

Yijun Wang is Assistant Professor of History at New York University. Her current project examines the transmissions of tin mining technology and changes in statecraft in China from 1700 to 1850s. She has recently contributed an article to *The Silver Age: Origins and Trade of Chinese Export Silver.*

Kyoungjin Bae is James P. Storer Assistant Professor of Asian history at Kenyon College. She is a historian of late imperial China, specializing in the history of craft, everyday technology, and Sino-European interaction.

74 Reuleaux, *The Kinematics of Machinery*, 35.

Part II

Producing the Court: Materials and Artefacts

Part Two explores three material artefacts that were pivotal products of the Imperial Workshops due to their high technical demands and politico-ideological importance to the Qing multiethnic empire: porcelain (Chapter Four), gilded metals (Chapter Five), and jade (Chapter Six). The process of sourcing and making these items was so onerous or exacting that carrying out these projects was, in itself, a veritable display of imperial power and the court's managerial and technical expertise. Once completed, the translucent porcelain vases, jade sculptures, or gilded temple roofs and statues further conveyed the magnificence of the Qing through their materiality.

Imperial time and space were brought together in these projects of making and knowing imperial objects. The logistical challenge was such that they made the Imperial Household Department's arm extend to such remote areas as jade quarries in the newly-subjugated northwestern territory, while new technology and craftsmanship was drawn in from Tibet, Nepal, and Europe to build the imperial summer residence and religious centre at Chengde. Porcelain, an iconic Chinese invention, had been produced in Jingdezhen, near the economic and cultural heartland of Jiangnan, about 1,400 kilometres from the capital, since medieval times. But during the Qing the products of the Imperial Porcelain Manufactory were injected with Qing characteristics – not only in design and the chemistry of the glazes, but also in less conspicuous aspects such as fiscal management and labour organization. Thus, in its logistical, technological, and symbolic aspects, the fabrication of imperial material culture was coextensive with empire-making.

The vignette essay by Qiong Zhang that opens Part Two illuminates the manipulation of time and space required in both artefact-making and empire-making, by analysing the trajectory of a Buddhist court painting and its textile reproductions. The first reproduction was ordered by Emperor Qianlong on the occasion of his mother's 70th birthday and more were commissioned twenty years later, reusing the same image, when Qianlong himself turned seventy. The precisely-scheduled conveyance of the painting, as well as its embroidered and woven copies, between Beijing and the court's textile centre in Suzhou served to integrate imperial space over a span of time. The management of time was important to the functioning of the extensive palace machine for another reason as well: economizing and cost-control. In the making of porcelain, gilded metals and jade objects, the emperor held the operators in the machine accountable not only for the material expenses but also for the use of time: the time from commission to final delivery was carefully tracked, as was the time spent on each of the small steps in between.

The accounting of time found its codified expression in the 'Regulations and Precedents' (*zeli* 則例), which specified the required quantities of materials and labour calculated in temporal 'work' units (*gong* 工), or the number of a craftsman's working days needed for a certain procedure.[1] Late delivery was punished, being seen as an indication of the operators' incompetence or delinquency at various hierarchical levels, although time-lags or delays were built into the long-distance nature of operating the palace machine. Since many officials and workers involved in a project were responsible for various elements of it, and had to commute between an office in the capital city and another in remote regions, they were likely to fail in one way or another.

The chapters in Part Two also illustrate the extent to which place was an important factor in the performance of the machine. Precious materials and high-ranking officials moved between the central hub in Beijing and places that were either as geographically and culturally remote as the 'new territories' of Xinjiang, or as dense as the narrow lanes occupied by skilful artisans in Suzhou. The material products served to manifest and integrate the various locales involved in their creation. Thus, the large-scale gilding project analysed by Su and Lai in Chapter Five collapsed the space of Qinghai, where the gold came from, Tibet, where the gilding technique (and the religion of Tibetan Buddhism) originated from, with the imperial sites in Beijing and Chengde, where the objects were on perpetual display.

The ability of the palace machine to fabricate imperial objects by assembling resources across time and space attests to the complexity of its operations and the dexterity of its operators. This ability can be described as *jiqiao* 機巧 (literally 'machine artistry'). Taken together, the manufacture of imperial objects examined in Part Two reveals the high demand for technical artistry at the heart of the Qing empire-making project, namely the art of control. The supervisors' and artisans' choice of materials, styles, and production procedures were not simply aesthetic decisions; concerns with technological innovation, political efficacy, and budgetary constraints were also at play. The recycling of misfired porcelain objects or broken shards discussed by Guangyao Wang in Chapter Four suggests that the pursuit of technical perfection and cost control were both priorities in Qing imperial porcelain production. But the very fact that misfired pieces existed suggests that even such tightly-controlled procedures as kiln firing remained a volatile operation, just as uncertainty and the spectre of failure remained an enduring challenge for the palace machine.

1 Song Jianze and Christine Moll-Murata, 'Notes on Qing Dynasty "Handicraft Regulations and Precedents" (*jiangzuo zeli*), with a Special Focus on Regulations on Materials, Working Time, Prices, and Wages', *Late Imperial China*, vol. 23, no. 2 (December 2002): 117-18.

Vignette essay II

The Story of An Image

Ding Guanpeng's 'Ultimate Bliss' and the Auspiciousness of Reproduction

Qiong Zhang

Abstract
Focusing on the uniqueness of tapestries and embroideries based on paintings, this vignette retrieves the production process of one set of 'embroidered paintings' based on the same painting. These objects are outstanding pieces of artwork, often marking important historical events. The essay investigates the sequence of commissioning the painting in different media and the diverse patterns of production procedures, management, and finance in the process.

Keywords: Qing dynasty, court, tapestries, embroidery based on painting, *kesi*

The Qing court produced hundreds of thousands of silk textiles for various purposes over the course of its existence. A very special type of these textiles is *zhixiu hua* 織繡畫, embroideries and tapestries modelled on paintings. These textiles were generally made from silk but occasionally also from wool. Distinct from most other silks, they were individualized pieces and were almost always commissioned by the emperor.

The following example is representative in many ways. It exemplifies how the palace machine, with its power and artistry, fabricated one design motif across several pictorial media to glorify the religious aura of the dynasty. It reveals the command process whereby manifold textile renditions of the same model painting were actualized, with the production technique changing from woven tapestry to embroidery and back; finally the addition

Siebert, Martina, Kai Jun Chen, and Dorothy Ko (eds), *Making the Palace Machine Work: Mobilizing People, Objects, and Nature in the Qing Empire*. Taylor & Francis Group, 2021
DOI: 10.5117/9789463720359_ZHANGQ

of a dedication text allowed the textiles to be used at multiple auspicious events. Moreover, the example shows the coloured threads of one series of these exceptional pieces, linking painting with textiles and images with text. These elaborate craft objects connected the Beijing court to the Chinese south; the production and financing also brought various actors together.

It all began with a painting. The sequel to the catalogue of the imperial collection of religious paintings, the *Midian zhulin xubian* 秘殿珠林續編, lists a coloured painting by the famous court painter Ding Guanpeng 丁觀鵬 (1737-1768) with the title 'Realm of Ultimate Bliss' (*Jile shijie tu* 极樂世界圖) which depicts 'Buddha on a lotus seat with his ten main disciples, encircled by the heavens with heavenly ladies playing music'.[1] The catalogue also records the text of the calligraphy and stamps on the image, as is customary. Of interest to us here is firstly, a stamp that identifies the painting as Ding's work that was completed in the second month of the Qianlong year 24, i.e. March 1759; secondly, a dedication by Emperor Qianlong in four scripts – Manchu, Chinese, Mongolian, and Tibetan – dated to the year *renyin* 壬寅, i.e. 1782, that adorns the top part or 'jade pond' (*yu chi* 玉池) of the huge scroll, which is about 1.4 metres wide and 2.8 metres high.[2] Both dates mentioned on the painting relate to personal events in Qianlong's life and both triggered a series of textile productions.

The year of Ding's painting suggests that its production was already part of preparations for festivities to celebrate the 70th birthday of Qianlong's biological mother, Empress Dowager Chongqing 崇慶皇太后 (1693-1777). It was not unusual, especially for someone like the filial pietistic Qianlong emperor, to start planning and preparing for such substantial events three years in advance.

Immediately after Ding had finished the painting in March 1759, the eunuch Hu Shijie 胡世傑 (fl. 1758-1770) transmitted the following imperial order: 'Ding's painting "Realm of Ultimate Bliss" is to be sent to the south (i.e. Suzhou) and Anning 安寧 (n.d.) is ordered to produce a tapestry following its model'. In the fifth month of Qianlong 26 (June 1761), that is two years and three months later, secretary Bai Shixiu 白世秀 (n.d.) handed the tapestry sent by Anning to Hu Shijie for perusal. It seems to have not fully met the emperor's expectation, as it was ordered that Ding's model should be resent

1 On Ding Guanpeng and one of his even more oversized paintings see the study by Wang Ching-ling, *Praying for Myriad Virtues* (Berlin: Staatliche Museen zu Berlin & Kunsthistorisches Institut in Florenz – Max-Planck-Institute, 2017).

2 Today the painting is housed in the National Palace Museum in Taipei (http://www.npm.gov.tw/exh98/oversize9804/html/pic_tw_01.html; accessed Feb. 2020).

to Anning to produce one more tapestry scroll, this time in brighter colours.[3] It is not clear when this second order was finished but it would have taken a long time because of its enormous size, the same as the original painting, and the labour-intensive, sophisticated *kesi*-weaving (緙絲) technique used to make it. The Empress Dowager Chongqing was born on the 25th day of the 11th month in the 31st year of Kangxi reign (i.e. 1 January 1693) so, counting age the Chinese way, her 70th birthday party was to be held on 1 January 1762. Thus, the first version that had been completed in June 1761, six months before that event, was right on time.

Today, the Palace Museum in Beijing owns four *kesi*-tapestry scrolls reproducing Ding's 'Realm of Ultimate Bliss'; three of which have Qianlong's 1782 dedication in four scripts, as mentioned in the painting catalogue *Midian zhulin xubian*. Thus, they seem to be based on Ding's painting in the form it exists today, namely having Qianlong's dedication in four scripts glued onto it, which was written 23 years after the painting was completed. The fourth tapestry in possession of the Beijing Palace Museum is different (see Figure II.1). Here, the space of the 'jade pond' (or 'poetry hall', *shi tang* 詩堂, as it is also called) in the scroll is taken up by a duckling-yellow damask with the woven pattern of two dragons offering up an ornamental version of the character *shou* 壽 ('longevity') floating among *ruyi* 如意 ('happiness') clouds. The tapestry shows no trace of the 1782 dedication and – extremely rare for image objects of the palace – no collection stamp. So this fourth piece is certainly the one presented to Empress Dowager Chongqing on the occasion of her 70th birthday. But when and why were these other tapestries produced?

Among the official communications of the Imperial Workshops of the Palace (*Zaobanchu* 造辦處) we find the following document:

> In the 47th year of Qianlong reign, on the 29th day of the second month (i.e. 11 April 1782) eunuch Eluli 鄂魯里 (n.d.) was handed a scroll of the painting 'Ultimate Bliss' by Ding Guanpeng and one sheet of white paper with a dedication by Qianlong to the 'Ultimate Bliss' painting in four scripts and given the following order: 'The emperor has already acknowledged sending the 'Ultimate Bliss' scroll to the official Side 四德

3 Qianlong 24/2/9 (7 March 1759), 'Qianlong ershisi nian ge zuo chengzuo huoji dang'an, xingwen chu 乾隆二十四年各作成做活計清檔, 行文處', in *Qing gong Neiwufu Zaobanchu dang'an zonghui* 清宮內務府造辦處檔案總匯, vol. 24, 615; Qianlong 26/5/1 (3 June 1761), 'Qianlong ershiliu nian ge zuo chengzuo huoji dang'an, xingwen chu 乾隆二十六年各作成做活計清檔, 行文處', in ibid., vol. 26, 656.

> (n.d) at the Suzhou Imperial Textile Manufacture to use the image as a template just like before but this time produce two embroideries based on it. Above the image, at the place of the 'poetry hall' embroider the dedication to 'Ultimate Bliss' in blue characters on a yellow background with the imperial stamp in red.[4]

About a year later, in March 1783, vice director Wude 五德 and foreman Dadaseshudian 大達色舒典 suggested manufacturing also two *kesi*-tapestries in addition to the two embroidered scrolls, because 'the incomparable exquisiteness of Ding's painting and the imperial dedication which should be transmitted to thousands of generations to come will be even more lustrous and beautiful when produced with the *kesi* method'. Donation money for this task had already been collected from officials, but one detail in Qianlong's text was no longer accurate. The two officials therefore further asked for approval to change the two Chinese characters for 'embroidery' (*xiuxian* 繡線) into 'work of tapestry' (*kegong* 緙工), to match the text with the object. This change was approved and the dedication text sent to the Hanlin academy located at Maoqin Hall (*Maoqin dian* 懋勤殿). The academy was instructed to exchange the two characters; no changes were considered necessary for the other three languages. The updated text would then be handed back to Side to produce the two *kesi* tapestries.

About two years later, at the end of the 49th year of Qianlong reign (i.e. early 1785) the two tapestry scrolls and Ding's original painting scroll were sent back. The tapestries were stored at Qixiang Palace (*Qixiang gong* 啟祥宮) (where the original ordered embroidered scrolls were already housed), and the painting scroll was returned to the Yuhubing 玉壺冰 Buddha Hall in the gardens of Jianfu Palace (*Jianfu gong* 建福宮).[5]

Several times in his life Qianlong invested in large Buddhist projects. Every time they were connected to either his or his mother's auspicious festivities. The painting and embroidery initiated in 1759 was connected to Empress Dowager Chongqing's 70th birthday in 1762; the personal dedication to, and reproduction of, image and dedication as embroidery and *kesi*-tapestry launched in 1782 were an aftermath to Qianlong's 70th birthday in 1780. This example of the 'Ultimate Bliss' image in its different technical and event-related realizations exposes the workings of the Imperial Workshop,

4 Qianlong 47/2/29 (20 March 1784), 'Qianlong sishiqi nian ge zuo chengban huoji qingce, xingwen 乾隆四十七年各作承辦活計清冊, 行文', in *Qing gong Neiwufu Zaobanchu dang'an zonghui*, vol. 45, 395-6.

5 Ibid., 396.

Figure II.1 ***Kesi*** **tapestry of Ding Guanpeng's 'Ultimate Bliss' without Qianlong's 1782 dedication**

Source: The Palace Museum, Beijing (故72704)

with its connections to other units inside and outside the palace. Especially in the case of imperial birthdays and for financing special craft works related to these events, not only expertise but also money from various sources – donated by officials, from the imperial coffers, etc. – was collected to realize these projects.

There are even more physical and textual traces of textile reproductions of Ding's 'Ultimate Bliss': the National Palace Museum in Taiwan owns another tapestry of this image which also has the dedication; moreover, the Palace Museum in Beijing has a third embroidery of 'Ultimate Bliss' as well as the two ordered by Qianlong in 1782, and a brocade (*zhijin* 織錦) version of the image that is also listed in the sequel to the *Midian zhulin*. One of the remaining questions is why aren't there two *kesi*-tapestries without Qianlong's dedication? One hypothesis could be that the *kesi* housed in Taiwan is actually the second one to which the dedication was added later – just as it was added to Ding's painting. This still leaves one of the four Beijing *kesi*, one of the three embroidered versions housed in Beijing and the brocade version for which the relevant archival documents have not yet surfaced. Today, the Palace Museum in Beijing houses a total of 1,149 embroideries and tapestries based on paintings; most, but not all, are objects produced in the context of the palace.

Translated from Chinese by Martina Siebert in close consultation with the author

About the Author

QIONG ZHANG 張瓊 is a Researcher of the Palace Museum in Beijing. She is a specialist of pre-modern textiles and the history of dress and costumes of the Ming and Qing dynasties. Her publications on the subject include *Guxiu yuanliu bianxi* 顧繡源流辨析 (Investigation of the Origin and Development of the Gu Family Embroidery) and the edited volume *Qingdai gongting fushi* 清代宮廷服飾 (Costumes and Accessories of the Qing Court).

4 Piecing Shards Together

The Uses and Manufacturing of Imperial Porcelain

Guangyao Wang

Abstract

The imperial manufacture of porcelain in the Qing dynasty was always related to the broad state economy and politics, especially ritual. As has been revealed in administrative regulations and raw material supply, porcelain production seemed to be an independent operation like other court art production. However, as evidence regarding managerial personnel, finance, quality control, and design shows, the technology for producing porcelain was integral to other material production at court. Thus, multiple productive processes were interdependent and they influenced each other. The productive processes and products of court art showcase the character of monarchical industry.

Keywords: Imperial Manufactory of Porcelain, ritual vessel, porcelain supervision, labour

Tang Ying 唐英 (1682-1756), the chief supervisor of the Imperial Ceramic Manufactory (*Yuyaochang* 御窯廠), once remarked that, 'Although ceramic production is a minor concern, it is of relevance to the affairs of the state' (*tao sui xishi, guan hu guozheng* 陶雖細事，關乎國政).[1] Referring to this statement, many researchers have exaggerated the importance of imperial porcelain,[2] or focused only on its political significance, at the expense

1 'Ciwu shiyi shiyu gao xu 瓷務事宜示諭稿序 (Tang Ying's Preface to the Draft Advise on Ceramic Affairs)', in Tang Ying 唐英, *Tang Ying dutao wendang* 唐英督陶文檔, ed. Zhang Faying 張發穎 (Beijing: Xueyuan chubanshe, 2012), 13. Tang Ying uses the word 'ceramics' (*tao* 陶) here to refer broadly to any fired clay. His quote indirectly refers to a quotation from the *Chunqiu Zuo zhuan zhushu* 春秋左傳注疏. See footnote 3.

2 In modern terminology, porcelain refers to ceramics fired beyond 1300 degrees. In pre-modern Chinese terminology, many so-called porcelains would actually count as stoneware

Siebert, Martina, Kai Jun Chen, and Dorothy Ko (eds), *Making the Palace Machine Work: Mobilizing People, Objects, and Nature in the Qing Empire*. Taylor & Francis Group, 2021
DOI: 10.5117/9789463720359_CH04

of the technical aspects. Few have examined its actual position in the political life of the Qing empire or viewed its production sequence within the broader operating mechanism of the state. This chapter investigates the role of imperial porcelain production in the palace machine and analyses the material and other factors that restricted its production. Specifically, utilizing some common historical sources as a point of entry, the chapter examines porcelain production in relation to other goods and materials of the court, the responsibilities of the officials involved, and the relationship between porcelain production and state finance.

Porcelain as Ritual Vessels and Diplomatic Gifts

The relevance of porcelain production to state affairs that Tang Ying spoke of firstly pertained to the performance of state rituals. According to an ancient account in the *Zuo chronicles to the Spring and Autumn Annals* (Chunqiu Zuo zhuan 春秋左傳), the most important state affairs were military campaigns (*rong* 戎) and sacrifices (*si* 祀).[3] Whereas the military used force to defend the country against invaders and keep order within the realm, sacrifices used rituals – and ritual objects – to establish and maintain decorum and the moral regimen, mediate between humans and the divine, the high and the low, and keep all polities (*bang* 邦) in harmonious accord with each other. Thus, sanctioned violence and ritual decorum constituted the basis of stately practice in imperial history. To understand the role and significance of the ceramic sector in Qing political rituals and how it fitted into the complex workings of the palace machine, this chapter begins by contextualizing ceramic manufacture in the Qing against the background of the long history of imperial ceramic production which had begun in the Northern Song dynasty (960-1127). This is indispensable for understanding the larger pattern of the development and specific changes to this sector that took place in the palace machine of the Qing dynasty.

The extensive use of ceramics as ritual vessels only began in the Southern Song dynasty.[4] An imperial decision of 1143 permitted 'the use [of objects

by modern standards. However, we choose to translate '*ci*' 瓷/磁or '*ciqi*' 瓷器 literally as porcelain.

3 See for example the commented version of this canonical work by Kong Yingda 孔穎達 (574-648), *Chunqiu zuozhuan zhengyi* 春秋左傳正義, *juan* 14, 12 ('13th year of King Chenggong 成公') (Beijing: Wuying dian 武英殿 edition, 1739).

4 Modern scholars consider celadon wares and various other categories of ceramics produced in the middle period as being stoneware rather than porcelain, because of their level of vitrification

made from] ceramic and wood out of expedience' as substitutes for bronze and jade.[5] This was originally meant to be a provisional arrangement in response to the fiscal strain on the country after the dynasty's loss of its northern territory to the Jurchens and relocating the court to Lin'an 臨安 (present Hangzhou), which was so named in 1129 and established as the capital of the Southern Song in 1138. Archaeological excavations in the city have shown that the majority of ritual objects unearthed in the site of the royal palace's residential areas and major religious places such as the Imperial Ancestral Temple (*Taimiao* 太廟)[6] were actually earthen wares (*tao* 陶), even though the two kilns at Tiger Grotto site (*Laohudong* 老虎洞)[7] and Turtle Hill site (*Wuguishan* 烏龜山) did produce beakers (*gu* 觚), wine goblets (*zun* 尊), and other imitations of ritual vessels in high-quality porcelain.

When the Mongols established rule over China in the late-13th century, they eclectically adopted various rituals of the Song and the Jin dynasties and used their remaining ritual vessels. It was not until the 4th year of the Zhizhi 至治 reign (1324) that the Yuan government ordered the various kilns in Jiangzhe province (*Jiangzhe xingsheng* 江浙行省) to start producing ceramics for ritual usage, thereby launching large-scale production of these kinds of objects. At about the same time, celadon basins (*pan* 盤), a famous product from the Chuzhou 處州 kilns of the same area, were deployed in the 'Hair and Blood' (*maoxue* 毛血) sacrifice.[8] This further epitomized the new official status of ceramics as ritual vessels of the state.

When the Han-ethnic Ming took rule over China back from the Mongolian Yuan they claimed to establish a ritual system that was modelled on the ancient dynasties of Xia, Shang, and Zhou. But they continued some practices from the recently-ousted Yuan dynasty, such as using ceramic vessels. The

and the kiln temperature in production. Chinese historical documents, despite the many different grades of vitrification, differentiate only between the two categories of *tao* (ceramics) and *ci* (porcelain).

5 'In the 1st month of the 13th year of Shaoxing 紹興 reign (1142), the Court of Imperial Sacrifice (*Taichang si* 太常寺) and the Board of Rites (*Libu* 禮部) request to follow the ritual statute (*lizhi* 禮制) of our dynasty... In place of ritual vessels that should be bronze or jade wares we use ceramics and wood out of expedience (*jiqi yingyong tong yu zhe quan yi tao mu* 祭器應用銅玉者權以陶木)', see Qian Shuoyou 潛說友 (1216-1277), *Xianchun Lin'an zhi* 咸淳臨安志, *juan* 3: '*jiaomiao* 郊廟 (Sacrifice)', in *Song-Yuan fangzhi congkan* 宋元方志叢刊 (Beijing: Zhonghua shuju, 1990), vol. 4, 3373-1.

6 Hangzhou shi wenwu kaogu suo 杭州市文物考古所, *Nansong taimiao yizhi* 南宋太廟遺址 (Beijing: Wenwu chubanshe, 2007).

7 Du Zhengxian 杜正賢, 'Laohudong nansong guanyao zhi 老虎洞南宋官窑址', *Wenwu* 文物, no. 10 (2002): 4.

8 Song Lian 宋濂, *Yuan shi* 元史 (History of the Yuan Dynasty), *juan* 72 (Beijing: Zhonghua shuju, 1976), 1798-1799, 1845-1846.

Ming government's use of ceramics as ritual and sacrificial vessels had already started in the 2nd year after the dynasty's inception, i.e. in 1369. The *Collected Statutes of the Great Ming* (Da Ming huidian 大明會典) recorded this among numerous other material changes, as follows: 'It was ordered in the 1st year of the Hongwu reign (1368) that utensils in the Imperial Ancestral Shrine should be exchanged with those made from gold; as for the miscellaneous items used by the emperor, such as the decorations on his attire and carriages, wherever the use of gold was originally required it should be substituted with brass. In the second year, it was determined that all sacrificial vessels should be made of porcelain'.[9] One year later, in 1370, a memorial by the Board of Rites reconfirmed the issue, declaring that, even though the porcelain food plates and water jars used as sacrificial vessels did 'not conform with the ways of the ancients (*buhe guzhi* 不合古制)', their use nonetheless accorded with the principle of 'advocating simplicity' (*shangzhi* 尚質), which had governed the use of pottery ritual vessels in the ancient dynasties; hence, the ministry petitioned to revise the ritual regulations so that the porcelain plates and jars used as ritual vessels would be named according to the ancient system. This received the Hongwu emperor's approval.[10] Thus, the institute sanctioned the use of ceramic vessels in imperial rituals and these humble ceramic wares officially acquired archaic names which carried much political weight.

Exchanges in material not only happened from bronze to porcelain, i.e. from a seemingly more precious to a more 'humble' material, but also in the other direction. In the 4th year of Hongwu (1371), the rules were adjusted further, meaning that the wooden sacrificial vessels in the Temple of Confucius were all to be replaced with porcelain ones.[11] The *Veritable Records of the Ming* (Ming shilu 明實錄) notes more generally that, 'in the twelfth month [of the same year], it was decided to substitute all [bronze] *dou* 豆, *fu* 簠, *gui* 簋, *deng* 登, *xing* 鉶 with porcelain vessels'.[12] In the

9 Both editions of the *Da Ming hui dian* contain this phrase. For the edition compiled by Xu Pu 徐溥 et al. during the Zhengde 正德 reign (1506-1522) in 180 *juan*, see *Siku quanshu* 四庫全書 (Shanghai: Shanghai guji chubanshe, 2003), vol. 618, 588; for the edition compiled by Shen Shixing 申時行 et al. during the Wanli 萬曆 reign (1587) in 228 *juan*, see *Xuxiu Siku quanshu* 續修四庫全書 (Shanghai: Shanghai guji chubanshe, 1995-2002), *juan* 201 (vol. 792), 1014.

10 Zhang Tingyu 張廷玉 et al., *Mingshi* 明史, *juan* 47 (*zhi* 志 (treatise) 23: '*li* 禮 (rites)'), 1237 (Wuying dian edition, 1739; rep. Taipei: Dingwen shuju, 1980).

11 Zhang Tingyu et al., *Mingshi*, *juan* 50, 1296; *Da Ming huidian* (Wanli edition), *juan* 91 (chapter 49: 'Libu 禮部 (Board of Rites)'), 523.

12 *Ming shilu* 明實錄, Taizu huangdi 太祖高皇帝 (Emperor Taizu), *juan* 70, 1297 (12th month of the 4th year of the Hongwu 洪武 reign) (rep. Taipei: Zhongyang yanjiuyuan Lishi yuyan yanjiusuo, 1966).

subsequent years of Hongwu's reign (1368-1402) ritual regulations were amended so that eventually all official sacrificial and ritual vessels – from the Temple of Heaven to the Imperial Ancestral Shrine and the Temple of Culture, from central to local governments, from the inner court to the princes' residence – would be replaced by porcelains.

What, then, was the actual shape of these porcelain pieces that replaced the ritual bronzes? The 1370 memorial had already noted the interesting point that Ming ritual vessels had not only changed in material from bronze to porcelain, but an even more drastic change was that the archaic, ritually-charged shapes of the Three Ancient Dynasties' vessels were being replaced by everyday objects such as plates and bowls. The *Collected Statutes of the Great Ming* addressed this replacement even more clearly when it chronicled the demand that, from the 26th year of the Hongwu reign (1393) on, annual sacrifices to a wide range of deities – from the gods of wind, thunder and rain to the city god and Confucius – must be performed to a strict schedule by all prefectures, provinces, and counties throughout the country. As an authoritative reference, the *Statutes* quoted the *Ritual Code of the Hongwu Reign* (Hongwu lizhi 洪武禮制), which stated that, as far as the sacrificial vessels were concerned:

> for all [bronze] *bian* 籩, *dou* 豆, *fu* 簠, and *gui* 簋, use porcelain plates (*cidie* 瓷碟); as substitutes for the *fu* and *gui* vessels slightly larger [porcelain plates] should be used; for the three wine goblets of the *zun*-type 尊, use porcelain *zun* (*cizun* 瓷尊), each equipped with one covering towel (*gaibujin* 盖布巾) and one wine-ladling spoon (*diao jiushao* 酌酒勺); for the six spouted tripod goblets of the *jue*-type 爵, use porcelain ones; to substitute for the one cauldron of the *xing*-type 鉶, use a porcelain bowl (*ciwan* 瓷碗).[13]

In the entry on 'washing utensils' (*guanju* 盥具) directly following this we find a similar directive: 'for the one *zun* [needed in the ceremony], use a porcelain urn (*ciweng* 瓷瓮)... for the one basin (*pen* 盆), use one from any material like tin, copper or porcelain'.[14] Moreover: '[for sacrificial rites] at

13 *Da Ming huidian* (Wanli edition), *juan* 94 (chapter 52: 'Libu'), 533. No original copy of the *Ritual Code of the Hongwu Reign* is existent today. But chapter 7 of the *Huang-Ming zhishu* 皇明制書 (System of the August Ming) – a collectanea from the late-16th century containing various sources on law, regulations and jurisdiction of the Ming – reproduces one part of this work where the above quotation is found. See Yang Yifan 楊一凡, ed., *Huang-Ming Zhishu* 皇明制書 (Beijing: Shehui kexue wenxian chubanshe, 2013).

14 Ibid.

any prince's estate… the production of porcelain wine *zun*, *bian*, *dou*, *fu*, and *gui* is assigned to Fuliang county 浮梁縣 in Jiangxi…' (The famous ceramic production centre of Jingdezhen was under the jurisdiction of Fuliang.)[15]

Not only in prince's fiefdoms and local capitals, these replacements in material and shape were also made at ceremonial sites of central importance, as suggested by the *Collected Statutes of the Great Ming* (Wanli edition). In *juan* 82, the caption to the diagram on the 'Display Plan at the Round Altar, Frontal Position (facing south)' – that is, in front of the spirit tablet of the Emperor of the Heaven – explains, that '*deng* 登-type covered dishes and *xing* cauldrons are to be substituted with porcelain bowls, and vessels of the type *fu*, *gui*, *bian* and *dou* with porcelain plates. The same applies to all other altars and temples'.[16] This note explicitly pointed out that the *fu*, *gui*, *bian*, *dou*, *deng* and *xing* among the Ming ritual vessels were actually common porcelain plates and bowls. Only the porcelain *jue*, the spouted tripod goblet, retained the shape of the corresponding ritual [bronze] vessel – the others only carried the 'name' of the ancient object, even though both their material and shape had changed. This fact might also explain the massive bowls and plates from the early Ming which are often found in excavated Jingdezhen and Longquan kilns (*Longquan yao* 龍泉窯): they are actually these replacement vessels mentioned in the ritual books.

Setting up altars to offer sacrifices to heaven was considered the most important sacrificial rite in all Chinese dynasties, and rituals at the Imperial Ancestral Shrine to commemorate and offer sacrifices to the ruler's ancestors was the most important family rite of the imperial states. In using porcelains, the Ming dynasty had simply continued the practice of the Yuan, which in turn had perpetuated an expedient practice that emerged out of the national distress of the Southern Song. The fact that porcelain bowls and plates were systematically replacing bronze ritual vessels in the Ming altars and temples indicates that, by that time, porcelain had acquired the status of a top-level ritual object and, more importantly, the substitution was promulgated by state law. The dynasty's founding emperor Ming Taizu, being the son of a commoner, viewed ritual vessels with both a reverence for state institutions and with a peasant's frugality. Unsurprisingly, his policies on ritual vessels were closer to real life and he considered vessels of common

15 '凡親王之國，瓷酒尊、籩、豆、簠、簋行江西浮梁縣燒造'. See *Da Ming huidian* (Wanli edition), *juan* 208 (chapter 28: 'Nanjing Gongbu dushui qingli si 南京工部都水清吏司 (Board of Works in Nanjing, Directorate for Waterways)'), 2773-2.

16 '登、鉶以磁碗代， 簠、簋、籩、豆以磁盤代，凡壇廟同'. See the diagram 'Yuan qiu diyi chenshe tu zhengwei (nan xiang)' 圜丘第一陳設圖正位 (南向), in *Da Ming huidian* (Wanli edition), *juan* 82 ('Lisi er 郊祀二 (Ancestral Sacrifice, section 2)'), 1298-1.

shapes and types to be suitable to serve state sacrificial rites. He thought that most of them should be porcelain but, to comply with ancient rules, they should be named in accordance with the practice inherited from the Xia, Shang, and Zhou dynasties.

Even though these regulations on ritual vessels convey the impression that the main purpose of the imperial kilns was to produce these porcelain simulacra, in fact, scrutinizing the surviving ceramic works from the Ming imperial kilns reveals that the number of ceramics made for everyday use or decorative display there far exceeded those produced for ritual and sacrificial purposes. Ritual vessels certainly did not comprise the majority of the hundreds of thousands of imperial porcelains that were mentioned in historical records. Lu Wan'gai 陸萬垓 (1533-?) made this observation in his *Great Gazetteer of Jiangxi Province* (Jiangxi sheng dazhi 江西省大志), claiming that most of the porcelains produced during the Wanli period were everyday vessels for food and beverage.[17] Archaeological excavations at the imperial kilns at Jingdezhen suggest that, as well as these kinds of utensils for daily use, bird feeding cups and playthings like cricket bowls also constituted an important line of products made in the imperial porcelain factory. But, of course, the excavations also provide indisputable material evidence of the manufacture of ritual and sacrificial vessels, including, for example, 'sacrificial vessels for official use' (*guanyong gongqi* 官用供器) that date back as far as the Hongwu reign, ritual and sacrificial vessels inscribed with the character 'altar' (*tan* 壇) from the Xuande reign (1426-1435), and ritual vessels for ancestral sacrifices from the Jiajing period (1522-1566). This indicates that, despite the small number of ritual objects created, the Ming imperial kilns were indeed in charge of producing them. Furthermore, the rules of proprieties stated that the manufacture of ritual vessels should take priority over that of banquet vessels. Although imperial kilns did not fully comply with these rules, ritual objects were unquestionably among their products. Thus, no matter how small their number might have been in comparison to 'banquet vessels', the imperial needs for – and the actual production of – these ritual vessels justified the operation of imperial kilns in the Ming dynasty, as they were needed to comply with the demands of the traditional ritual system.

The Qing imperial kilns inherited and basically continued the manufacturing system and managerial structure from the Ming. The *Precedent Cases*

17 Wang Zongmu 王宗沐 et al., *Jiangxi sheng dazhi* 江西省大志, *juan* 7 ('*Taoshu* 陶書: *Gongyu* 供御 (Treatise on Ceramics: Imperial supplies)'), 32ff (edition of 1597; reprinted in *Nanjing tushuguan cang xijian fangzhi congkan* 南京圖書館藏稀見方志叢刊) (Beijing: Guojia tushuguan chubanshe 2012, vol. 107-108).

of Statutes of the Great Qing (Da Qing huidian shili 大清會典事例) explicitly state that 'porcelains for the emperor's use (*shang yong* 上用) should be produced according to the designs and in the quantities promulgated by the inner court [which probably meant the Imperial Household] and transported to Beijing under the responsibilities of Raozhou prefecture in Jiangxi (*Raozhou fu* 饒州府) [the prefecture where the porcelain town Jingdezhen is located]; sacrificial vessels for use at altars, temples and imperial tombs should be manufactured following the designs, colours, and quantities designated by the Court of Imperial Sacrifices (*Taichangsi* 太常寺) of the central government and the manufacture orders should be sent to Jiangxi to be completed and then delivered to the departments concerned'.[18]

These regulations demonstrate that, during the Qing, porcelains for imperial use and for sacrifices at imperial tombs were all made by the Imperial Ceramic Manufactory, based on designs issued from the Imperial Household or the Court of Imperial Sacrifices. The *Qing Statutes* clearly juxtapose ritual vessels for altars, temples, and imperial tombs alongside everyday porcelains used by the emperor. On most occasions the Qing dynasty continued the practice of the Ming and used everyday porcelain items as vessels for rituals and sacrifices. It was not until the Qianlong reign (1735-1795) that the emperor ordered the imperial kilns to produce porcelain vessels that mimicked the shapes of ancient ritual bronzes such as *fu*, *gui*, *bian*, *dou*, *deng*, *xing* and the like. Moreover, the bodies of these porcelains were manufactured in five distinct colours – yellow, red, white, blue, and moon white (*yuebai* 月白) – and decorated with 'Five coloured' patterns (*wucai wenyang* 五彩紋様), in order to distinguish their different spheres of use.[19]

Apart from their use in imperial rituals, the other important role porcelains played in state affairs was comprising diplomatic gifts for dignitaries from foreign countries. This became a prominent practice during the Ming and Qing dynasties.[20] In the early Ming, close to ten thousand items of

18 '凡上用瓷器, 照内頒式樣數目, 行江西饒州府燒造解送; 凡壇、廟、陵寢需用祭器，照太常寺圖式、顏色、數目，頒發江西燒造解部'. See Yunlu 允祿 et al., *Da Qing huidian shili* (Yongzheng ban) 大清會典事例 (雍正版)), *juan* 201 ('*Gongbu wu, Dushui qinglisi: Qiyong* 工部 五, 都水清吏司: 器用 (Board of Work 5, Directorate for Waterways: Utensils)'), 13a (rep. in *Jindai Zhongguo shiliao congkan sanbian* 近代中國史料叢刊三編. Taipei: Wenhai chubanshe, 1985). This phrase is repeated from the equivalent section of the Kangxi edition of the *Precedent Cases, Da Qing huidian shili* (Kangxi ban), *juan* 138, 13a-b (rep. in Wang Guangyue 王光越 et al., eds., *Qingdai wubu huidian* 大清五部會典. Beijing: Shutongwen Guiji shujuku 書同文古籍數據庫, 2007; database accessed via the CrossAsia platform, Feb. 2020).

19 Wang Guangyao 王光堯, '*Qingdai cizhi jiliqi luelun* 清代瓷質祭禮器略論', *Gugong bowuyuan yuankan* 故宮博物院院刊, no. 2 (2003): 70-79.

20 Wang Guangyao, *Mingdai gongting taoci shi* 明代宮廷陶瓷史, 196-207.

porcelain gifts had already been lavished on other countries. Assuming that all those items were manufactured by official kilns, it is clear that the imperially-sponsored production of porcelain was considered of high political significance. The Qing dynasty continued this system from the Ming and ordered the imperial kilns to produced ceramic items that were presented to the upper classes of Tibetan and Mongolian societies – such as ewers in the shape of a Tibetan monk's cap (*sengmao hu* 僧帽壺), Tibetan nectar vases (*zangcao ping* 藏草瓶), and Tibetan-style altar bottles (*benba hu* 賁巴壺). These vessels, from their beginning in the imperial kilns to their conclusion as a gift in the hands of Mongolian or Tibetan recipients, constituted a vital part of Qing political life and state operation. This kind of material diplomacy can be traced back to the Yuan, when large numbers of ceramics were sent by the Yuan government to the Ilkhanate and other states. These gifts were also of a political nature, meant to foster friendly ties with the khans.

This brief historical overview of state porcelain in the Song, Yuan, Ming, and Qing Dynasties has shown that the manufacture of porcelain ritual vessels was regulated by official codes and records. The relatively modest quantity of these objects in relation to other luxury or everyday ceramics reveals a gap between the ostensible claim of the ritual necessity of porcelain production and the de facto desire for porcelain wares for daily use in the imperial and princely courts. In other words, it is likely that the assertion of the necessity of state-run ceramic production to the state ritual system was just an excuse that made the establishment of such an industry morally defensible to begin with. This discrepancy between words and deeds was characteristic of the palace machine, as outlined in the other chapters of this volume.

Unfortunately, although ceramics were manufactured in the name of ritual obligation or political diplomacy, they did not bring peace and prosperity. Even worse, during the Ming dynasty, the imperial kilns production ended up being the biggest loophole in state finance – the onerous financial and labour demands even led to popular uprisings in Jingdezhen. The Qing tried to control the production costs better by professionalizing the administration and management of the kilns as well as improving the porcelain's quality through technological advances.

Porcelain Supervision in the Qing Administrative Machine

To better understand the Qing's administrative move to establish the position of supervisor of the Imperial Ceramic Manufactory (*dutao guan* 督陶官), as the central official in charge of porcelain production, a brief review of how the production was managed and financed in the preceding dynasties seems necessary. In the Song dynasty, two kinds of officials were concurrently related to, and in charge of, the manufacture of ceramics. This was, firstly, local officials at the various production sites who received and implemented the orders to produce ceramics for the emperor, and secondly, the officials who supervised tax matters regarding the ceramic kilns.[21] Archaeological evidence of the responsibility of the former can be illustrated, for example, by an inscription found on numerous ceramics unearthed in the kiln site at Hutian 湖田 near Jingdezhen: 'Manufactured under the supervision of Zhang Ang, Magistrate of Fuliang County'.[22] The main duty of tax inspectors, on the other hand, was, obviously, to supervise the collection of the taxes. Tax inspection was standard practice in all trades during the Song dynasty and performed mainly as a sideline job. The same held true for the officials in charge of overseeing ceramic tax matters. The mix of responsibilities was even more varied for the magistrates of counties with a kiln that produced porcelain for the emperor. To them, the supervision of ceramics production was a temporary and mostly, out of the ordinary, task as well. Thus, at that time, there was no such thing as an expert in managing ceramic production from a state perspective.

This more or less continued in the Yuan dynasty. The town supervisor (*jianzhen* 監鎮) of Jingdezhen still only managed tax matters related to porcelain manufacture and, even after local magistrates were entrusted with the task of overseeing the production of imperial porcelain in the Taiding 泰定 reign (1324-1328), it remained an ad hoc and short-term duty. The *Treatise of Ceramics* (Tao shu 陶書) contained in the *Great Gazetteer of Jiangxi Province* expresses this as follows: 'When there were imperial orders, porcelains were [manufactured and] supplied to the court; when there was none, production ceased'.[23] Moreover, although the Yuan established a special 'Porcelain Bureau of Fuliang' (*Fuliang ciju* 浮梁瓷局) in 1278, its duties were not limited to manufacturing porcelain, as it was also required to produce

21 Wang Guanyao, '"Jian ciyaowu" guan kaobian "監瓷窯務"官考辨 (Clarification of the 'Supervisor of Porcelain Kilns')', in idem, *Zhongguo gudai guanyao zhidu* 中國古代官窯制度 (Beijing: Zijincheng chubanshe, 2004), 50-56.

22 '*Fuliang xiancheng chen Zhang Ang cuozhi jianzao* 浮梁縣丞臣張昂措置監造'. See Specimen 97F·T3③:39, in Jianxi sheng kaogu yanjiu suo 江西省考古研究所 et al., eds., *Jingdezhen Hutian yao zhi 1988-1999 nian kaogu fajue baogao* (shang) 景德鎮湖田窯址 1988-1999年考古發掘報告（上）(Beijing: Wenwu chubanshe, 2007), 436.

23 'You ming ze gong, fou ze zhi 有命則供，否則止', see Wang Zhongmu et al., *Jiangxi sheng dazhi, juan* 7 ('Tao shu'), 3a.

other artefacts, such as lacquer wares, hats made from horse-tails, hemp, and conical rattan (*qizao* 漆造, *mawei* 馬尾, *zong* 棕, *teng* 藤, *limao* 笠帽). The Porcelain Bureau was staffed with a chief and a vice director (*dashi* 大使 and *fushi* 副使), and the former could be as low as the sub-ninth rank (*cong jiupin* 從九品), the lowest in the bureaucratic hierarchy. Even though the chief holding such a low-level post, he was charged with an array of demanding tasks.

Almost nothing had changed in the early Ming dynasty regarding local magistrates' management of ceramic production. The magistrate (*cheng* 丞) of Fuliang county was still in charge of porcelain production; hence, the management of official kilns in the early Ming dynasty very likely continued the practice of the Song and the Yuan. Supplying the inner court with porcelains was only one of the many responsibilities of the Fuliang local government and it was often temporary and sporadic, since porcelain manufacture was not yet a regular or important task.

The crucial transition of the Jingdezhen locality happened between the Yongle (1403-1424) and Xuande (1426-1435) eras of the Ming dynasty. During that time, the official kilns at Jingdezhen underwent a management reform which finally brought about a true imperial system of porcelain manufacture.[24] To knit the production more strictly into state politics and the court's luxury needs, officials from the Board of Work began taking part in the administration of the imperial kilns. In Xuande reign, Zhang Shan 張善 (n.d.) was the first eunuch dispatched to supervise porcelain production in the whole of Raozhou region – Fuliang was one of its counties. At that time both forms of porcelain management – both by officials from the Board of Work and by local magistrates – still persisted in parallel. But, from Zhang Shan's tenure, it became normal practice for a trusted eunuch to supervise ceramic manufacture. Ceramic production was important enough for Ming emperors to dispatch eunuchs to oversee the production, but not enough for them to create the post of a dedicated supervisor.

In the late Ming, after the Jiajing and Wanli reigns, popular resentment against eunuchs led to a slight retrenchment in the policy of dispatching these men to oversee imperial kilns, although eunuch supervisors

24 See my chapter 'Zailun Mingdai yuyaochang de jianli shijia – Mingdai yuyao yizhi de kaoguxue fenqi 再論明代御窯廠的建立時間——明代御窯遺址的考古學分期 (Re-examining the Founding Time of the Ming Dynasty's Imperial Ceramic Manufactory – Archaeological Periodization the of Ming Imperial Kiln Sites)', in idem, *Zhongguo gudai guanyao zhidu* (er) 中國古代官窯制度 (二) (Beijing: Gugong chubanshe, 2017), 122-31.

were still a common sight. On several occasions the vice magistrates of Raozhou and its seven neighbouring prefectures needed to assist one another in all sorts of affairs related to imperial porcelain production. This is proved, for example, by the relocation of the Bureau of the Police Office (*Xunyansi yamen* 巡檢司衙門) to Jingdezhen. It was also the local magistrates from Raozhou who had to discharge the eunuchs, sending them back to the capital, and during that time of vacancies they had to preside over production at imperial kilns for a considerable period. In the late Ming, a routine was set up whereby local magistrates took it in turn to manage the imperial porcelain manufacture. This meant that, once again, no personnel were specifically dedicated to porcelain affairs. This no doubt partially accounted for the inefficiencies in production and the decrease in the quality of that era's porcelains.

The turnover of dynasties from Ming to Qing did not immediately bring about many changes to porcelain production, as the imperial kilns still generally followed Ming practice.[25] The system of eunuch supervisors, however, was replaced early on by assigning that role to Manchu officials from the Imperial Household Department. Other management practices – from the allocation of funding to the appointment of officials – remained intact, and the local Raozhou officials still featured among those mainly responsible for imperial porcelain production. This pattern of close cooperation between the 'private' bureaucracy serving the royal house and the 'public' bureaucracy of the state was, as we see elsewhere in this book, an important feature of the Qing palace machine.

It was only about eighty years after the inception of the Qing dynasty that the administrative and managerial responsibilities for porcelain production changed drastically. Starting from the Yongzheng reign (1723-1735), the imperial kilns in Jingdezhen were placed under the direct jurisdiction of Huai'an Customs House (*Huai'an guan* 淮安關) over 500 kilometres away. The customs' income from the Grand Canal funded porcelain production costs and customs officials were charged with supervising the production remotely. At the site itself, it was the assistant director (*yuanwailang* 員外郎) from the Imperial Household Department, who served as the de facto director of porcelain manufacture. Thus, the dedicated role of ceramic production supervisor was born. Tang Ying, whose quote opened this chapter,

25 See my chapters 'Wanli zhi Yongzheng shiqi de yuyao 萬歷至雍正時期的御窯 (Imperial Kilns from the Wanli to Yongzheng Reigns)' and 'Qingdai yuyaochang shulue 清代御窯廠述略 (An Overview of the Imperial Ceramic Manufactory in the Qing Dynasty)', in idem, *Zhongguo gudai guanyao zhidu* (er), 151-64.

was the most famous and successful individual to hold this post, serving in that capacity for 26 years.[26]

In the 8th year of the Qianlong reign (1743), the administrative duty of the kiln was transferred to the supervisor of Jiujiang Customs House (*Jiujiang guan* 九江關), while officials from the Imperial Household Department continued to assist in the kilns. That arrangement was rescinded in the 51st year of the Qianlong reign (1786), when the supervisor of the Jiujiang Customs House was appointed as general manager of the imperial kilns, and was required to perform annual inspections of all the kilns. At the production site itself it was the sub-prefectural magistrate (*tongzhi* 同知) of Raozhou stationed in Jingdezhen, along with the Bureau of the Police Office of Jingdezhen, who were jointly in charge of overseeing the manufacture and transportation of porcelains – but the latter focused mainly on its safety.

Although the task of the officials from the Imperial Household Department was ostensibly to supervise imperial porcelain production, just as with those Imperial Household officials supervising the Jiangnan Weaving Bureau or tax inspections in the Jiangnan area, that was only one of their many concurrent duties. Their primary mission, to some extent, was to serve as emperors' watchdogs in this financially prosperous region. As in the case of the production of porcelain ritual vessels, a conspicuous gap between de jure proclamations and de facto practice was a consistent feature of the operation of the palace machine.

Managing the Labour and Workflow of Imperial Porcelain Production

The management of the actual production of imperial porcelains can be explored from two perspectives: the artisans and labourers who carried out the hands-on work, and the procedures into which the whole process was structured. Before the late Northern Song, the Song court did not maintain its own production sites but instead procured ceramics through local tributes and taxes in kind. Hence, the actual producers of imperial porcelains were artisans with private kilns. Even in cases where a local government agency was commissioned to manufacture ceramics for the court, the production itself took place in selected high-quality private kilns in the area, without

26 See Tang Ying's biography in Zhao Erxun 趙爾巽 et al., *Qingshi gao* 清史稿, *juan* 505 ('*Liezhuan* 列傳 (Biographies)' section 292, '*Yishu* 藝術 (Art and Crafts)', subsection 4), 13926 (rep. Beijing: Zhonghua shuju, 1977).

a fixed body of artisanal staff. It was common practice during the Song to recruit independent artisans to undertake the production and that was how most kilns operated, from the Ru kilns (*Ruyao* 汝窯) in the Northern Song to the official kilns of Xiuneisi 修內司 that were built in the Southern Song.[27] People with technical expertise were officially registered as craftsmen (*jiangji* 匠籍), and were recruited and paid by the hour when the state needed their labour.

During the Yuan dynasty, a system of government craftsmen (*guanjiang zhi* 官匠制) was established and the kilns started to function as state-run enterprises. Craftsmen trades were made hereditary, thus the descendants of registered craftsmen were not allowed to change profession or de-register themselves from the official list. Specifically, ceramic artisans were employed at the kilns and they were paid in the form of daily food rations.[28] The official kilns in places like Cizhou 磁州, Junzhou 鈞州, Raozhou 饒州 and Chuzhou 處州 probably all operated according to this system of state-run enterprises.

Even though the names of the specific kiln workers are nowhere to be found, they were without question registered craftsmen and presumably had to work under the local government's direction.[29] In addition, some foreign artisans worked for the Yuan government. For instance, material evidence has corroborated textual records about Arabic 'guest artisans' (*kejiang* 客匠) joining imperial porcelain production in China: the emergence of blue and white porcelain in the Yuan dynasty can be directly associated with these craftsmen from the Middle East – mainly Persian speakers who were experts in the decorative arts of underglaze blue and underglaze red.[30] The system of government craftsmen highlights the Yuan government's

27 According to archaeological findings, the production history of the Ru kiln can be divided into different periods. The period of imperial monopoly was marked by the uniform destruction of inferior products. At that time, the Ru kiln existed in the form of an official kiln, just like the Laohudong kiln in the Southern Song. The Laohudong and Wuguishan kilns from the Southern Song are now commonly referred to as the Xiuneisi kiln and the Jiaotanxia 郊壇下 kiln respectively. I have pointed out the problem this different naming of kilns creates many times elsewhere, so I will not elaborate on it here.

28 See the entry on 'artisans' (*zhujiang* 諸匠) in Su Tianjue 蘇天爵, *Guochao wenlei* 國朝文類, *juan* 42 ('Gongdian zongxu 工典總敘 (General Preface of the Statutes on Work))', in *Sibu congkan chubian* 四部叢刊初編 (Shanghai: Shanghai shudian, 1989), 19.

29 For the production of lacquerware in Yuan official workshops, see Yuan Quan 袁泉, 'Lue lun Song-Yuan shiqi shougongye de jiaoliu yu hudong xianxiang — yi qiqi wei zhongxin 略論宋元時期手工業的交流與互動現象——以漆器為中心', *Wenwu* 文物, no. 11 (2013): 63.

30 Huang Wei 黃薇 and Huang Qinghua 黃清華, 'Yuan qinghua ciqi zaoqi leixing de xin faxian — cong shizheng jiaodu lun Yuan qinghua ciqi de qiyuan 元青花瓷器早期類型的新發現——從實證角度論元青花瓷器的起源', *Wenwu* 文物, no. 11 (2012): 79-88.

command over artisans across its empire, while the arrival of guest artisans signifies its endeavour to produce outstanding and exceptional porcelains, by enlisting masterful craftsmen from thousands of miles away. These artisans, Chinese and foreign, were engaged in handiwork that might appear small in number from a national perspective, but their work and products were closely related to the imperial politics and embodied state control over porcelain production.

The imperial kilns of the Ming and Qing dynasties retained, to a certain degree, the organizational and economic structure established by the Mongol Yuan, while reforming some parts of it. One reform which had a substantial influence on the labour structure in the kilns was the 'artisans shift payoffs' (*banjiang yin* 班匠銀) that began in the Jiajing period (1522-1566), when social changes wreaked havoc on the system of hereditary status. The reform allowed registered artisans to escape from their corvée duties by commuting their shifts into a payment in silver. Whereas before this reform workers in the imperial porcelain factory mostly consisted of registered official craftsmen; after the *banjiang yin* system came into practice hired artisans (*guyong jiang* 雇佣匠) with no special status became the majority.

This substitution was possible because, by the mid- to late-Ming, porcelain production in Jingdezhen was already being organized as a production line comprising individual specializations, with each artisan attending to just one part of the process. As the saying goes, 'engage the hand in one activity so that the mind will not be distracted' (*yi qi shou er bu fen qi xin* 一其手而不分其心). Under the technological influence of the imperial kilns, private kilns after the mid-Ming also brought in fine divisions of labour in the form of an assembly line, where one finished product might go through the hands of seventy-two craftsmen. Each worker was therefore reduced to a small cog in a big machine, interdependent and mutually restraining. From a longue durée perspective, this change marked the last stage in the development of porcelain production in pre-modern China: from household sideline production, to family-based handicraft business, and then to specialized workshops.

With a production line entailing many steps that was robustly designed and meticulously controlled, Qing imperial porcelain production strived to guarantee a uniformity in quality and meet demands for enormous amounts of porcelain to tight deadlines. As a result, the porcelains manufactured by the imperial kilns differed from those produced in the popular kilns, as well as from those submitted as tributes by local governments. One of their main characteristics was that they were constantly under strict imperial control – from design to production to disposal.[31] The complete life cycle of a porcelain object designed, produced, and disposed of within this state monopoly can be broken down into the following steps:

1 The emperor required certain ceramic items and issued an order for their production.
2 Decisions were made on their designs and quantities.
3 Models for the items were drafted.
4 The drafts were sent to the emperor for approval and then issued to the imperial porcelain factory.
5 The various workshops in the porcelain factory undertook their specific manufacturing tasks.
6 The finished products received a preliminary assessment.
7 Unwanted and defective items were disposed of, while quality items were delivered to the capital.
8 The items were submitted to the Imperial Warehouse and re-appraised.
9 Finally, the items were distributed to the different palaces and divisions of the court.
10 If items were damaged, they were either repaired or disposed of.

Examining this operation sequence from the perspective of history of technology, the first six steps of imperial porcelain production can be viewed as being concerned with ceramic technology – from designing to shaping the bodies, glazing, and controlling the firing temperature. But, whereas only the fifth operation, i.e. the various workshops' participation in the actual manufacture, belonged to technology in the narrow sense, the whole life cycle can be considered technologies in a broader sense, as they pertain to relevant knowledge that is not strictly speaking ceramic in nature.

Certainly, for imperial porcelain production, technology in the narrow sense was extremely important. Yet, throughout Ming and Qing China, many kilns mastered these techniques – the necessary expertise was not limited to artisans in Jingdezhen kilns alone. It was the patronage of the court that made Jingdezhen into the official provider of imperial porcelains. Presumably, had the blueprints and design drafts been sent to other kilns, the same kind of high-quality imperial porcelains would have been produced there. Indeed, this was the case for the Cizhou, Junzhou and Longquan kilns in

31 Li Ji 李季 et al., 'Gugong bowuyuan Nan Daku cipian maicang keng fajue jianbao' 故宮博物院南大庫瓷片埋藏坑發掘簡報', *Gugong bowuyuan yuankan* 故宮博物院院刊, no. 5 (2016): 24.

Chuzhou during the Yuan and early Ming.[32] The imperial kilns at Jingdezhen precisely followed the instructions on shape, pattern, colour, and number that were conveyed by official drafts and orders, and thereby carried out the emperor's intent. They were the hands that performed tasks triggered by a remote mind. Thus, officials and artisans alike followed the blueprints and mechanically completed the tasks given, without any opportunity to add their own creativity or initiative into the process.[33]

This point, that workers on Qing imperial porcelain were merely tiny yet indispensable component elements linked together into a production line, can be seen in the decoration of bowls with the design of a reign mark and dragon among clouds. Not only were the dragons and clouds painted by different artisans, but the circles at the rim and the base were also each completed by different workers. Even the task of making the reign mark at the bottom of the bowl was divided into two steps: Certain artisans wrote the characters of the reign name while others painted the circular or square frame around it.[34] This extreme division of labour was also evident in several other production steps – such as selecting, purchasing and preparing the raw materials, modelling the clay bodies, applying the glaze, decorating, placing the clay bodies into protective saggars, and firing the kilns.

If one looks at the technologies of porcelain production in the broader sense, from design to transportation, however, the situation appears to be somewhat different. Even though the design drafts were completed in specialized departments, the designers needed to understand closely and intimately the intentions and the preferences of the emperor because he had the final say over the designs and dominated the whole process. Thus, it was not possible for this part of the production line to function as a purely mechanical and unchanging workflow. Transporting the porcelains back to the imperial palace – a step one might not necessarily consider an important technology relevant for the production of imperial porcelain – was also strictly controlled, to avoid logistical and technological difficulties as far as possible. But, of course, a smooth transport of the finished goods was

32 See the section 'Gongbu: Yaoye, ciqi 工部: 窯冶, 陶器 (Board of Work: Kiln Firing, Ceramics)', in *Da Ming huidian* (Wanli edition), *juan* 194, 2631-1.

33 See my chapter 'Qianlong shiqi yuyaochang de guanli tizhi yu guanyang zhidu 乾隆時期御窯廠的管理體制與官樣制度 (Administration and the *Yang* Design System of the Imperial Ceramic Manufactory during the Qianlong Period)', in idem, *Zhongguo gudai guanyao zhidu (er)*, 199-200.

34 This is discussed in more detail in chapter 4: 'Mingdai gongting yongci de shengchan yaochang 明代宫廷用瓷的生产窑场 (The Kiln Sites of Production of Court Ceramics in the Ming Dynasty)' in my *Mingdai gongting taoci shi*, 136-38.

not always achieved. For example, in the Xianfeng reign (1851-1861), when the Taiping army occupying Nanjing blocked the waterways in the Yangzi, the porcelains could not be transported to Beijing, so had to be stored in the government repository (*fanku* 藩庫) at Nanchang 南昌. In another case, in the 28th year of the Guangxu reign (1902), new means of transport were employed to save time, reduce costs and adapt to modern needs and standards. The porcelains were carried by steamers from Jiujiang to Tanggu 塘沽 near Tianjin and then onwards to Beijing by train. However, the train caught fire near Yangcun 揚村, resulting in an accident.[35]

The design of the production procedure fostered the consolidation of state control over the imperial porcelain production, but also promoted technological advances and increases in product quality. Even more importantly, the evolving workflows transformed not only the production process, but the whole economic modes and social structure of the city of Jingdezhen and the Chinese porcelain industry in general. Meanwhile, porcelain production in private kilns came under the control of commercial capital and similarly changed its character in late Ming and Qing times. Viewed in this light, porcelain production was only one part of the broader changing economic system of the time, where different historical forces acted upon each other.

Financing Imperial Porcelain Production

So far this chapter has demonstrated that imperial porcelain production in the Ming and the Qing was a vast business run by an orderly administrative system. However, putting its political significance – discussed in the first part of this chapter – aside, ceramics were merely one among the many goods that the palace machine needed and acquired. To gain a better view of porcelain's role in the court's mobilization of resources, it is instructive to investigate the cost of manufacturing porcelain.

During the Song and the Yuan dynasties, when Jingdezhen or other kilns received official commissions to produce ceramics for the court, the production outlays were very likely paid for out of local tax revenues. This was probably a continuation of regulations inherited from the Tang and the Song, and were actually substitutes for tax payments to the central government which could be contributed in the form of local product tributes. Since ceramics as local products could cancel out part of the taxes owed in silver,

35 Wang Guangyao, *Zhongguo gudai guanyao zhidu* (er), 199-200.

they were produced with local funds to satisfy some of the tax debt, so can be thought of as a tax in kind. The ceramic tributes were, in other words, no different from local ceramics that remained in the area or were sold to other customers. However, for the period from the early monopoly of the Ru kiln to the time of the official Xiuneisi and Jiaotanxia kilns in the Southern Song dynasty, evidence about how they funded their production is yet to be found.

In the Ming dynasty, porcelain production incurred tremendous expenses. Together with textile production, it was the most exploitative state enterprise of the time and faced constant opposition from government officials. No cost estimate or budget was set for the production outlays, and whatever resources were needed were just taken out of Jiangxi's ordinary tax revenues. For large-scale bulk productions, additional taxes were collected throughout the province to make up for insufficient funds.[36] The severity of this burden of additional taxes led to the outbreak of riots, with people demanding the suspension of porcelain production and the reform of relevant policies.

When the Qing government resumed porcelain production at Jingdezhen, they initially adopted the policies of the Ming. During the Kangxi reign (1662-1722), the production of imperial porcelains was still funded by Jiangxi's ordinary tax revenues. In the 5th year of Yongzheng reign (1727), however, the responsibility for production outlays and costs moved into the hands of the Imperial Household Department. From then on, funds to pay for porcelain production were allocated from the Huai'an Customs House surpluses.[37] Consequently, the nature of manufacturing imperial porcelain fundamentally changed, since it was no longer part of the state finance, but was funded entirely out of the emperor's private pocket. After this change to the funding structure, the Yongzheng emperor spent roughly 8,000 taels of silver per year to cover the costs of porcelain production.[38] About ten years later, in the 2nd year of the Qianlong reign (1736), that amount had increased to 20,000 taels per year,[39] although this trend was halted and, in the fourth year (1738), the annual amount was reduced to 10,000 taels. Moreover, the source of the money changed from being the surplus from

36 See the entry on 'Material Prices' (*liaojia* 料價), in Wang Zongmu et al., *Jiangxi sheng dazhi*, *juan* 7 ('Tao shu'), 17.

37 See the entry on 'Producing porcelain vessels' (*shaozao ciqi* 燒造磁器) in *Zongguan Neiwufu xianxing tiaoli: Guangchusi* 總管內務府現行條例: 廣儲司 (Current Regulations of the Imperial Household Department: The Grand Storage Office), in *Jindai Zhongguo shiliao congkan* 近代中國史料叢刊 (Taipei: Wenhai chubanshe, 1972), *juan* 1.

38 Fu Zhenlun 傅振倫 et al., 'Tang Ying ciwu nianpu changbian 唐英瓷務年譜長編', *Jingdezhen taoci*, no. 2 (1982): 18-67.

39 Ibid.

the customs house at Huai'an to that of Jiujiang, where it remained until the end of the dynasty. The expenditure on porcelain production further shrunk drastically from the 4th year of the Jiaqing reign (1799) on.[40]

In addition to imposing a strict control on porcelain expenditure, Emperor Qianlong took up an idea first proposed in the late-Ming, issuing an edict to sell off second-rate porcelains in the 7th year of his reign (1742).[41] Prince Yi 怡親王 (Yinxiang 胤祥 1686-1730) and Haiwang 海望 (?-1755) further set down the rules that second-rate products should be sold off at no less than 50% of their production costs and demanded that the damage rate in imperial kilns should not exceed 30%. The supervisor would have to pay for the loss of any damaged goods above that rate. These rules were further elaborated in the 12th year of the Qianlong reign (1747). 'The Memorial of Provisional Regulations for his Majesty's Approval Regarding the Production of Porcelain Vessels' (*Zouwei yiding shaozao ciqi zhangcheng* 奏為議定燒造瓷器章程) revised the rules so that the cost of every article, 'which was converted into the volume of clay slabs in the size of one *cun* in width and one *chi* in length, should then be computed by calculating the itemized costs of clay, glaze and labour in making the clay bodies and in decoration, etc. one by one'.[42] It further decreed that the production failure rate should not exceed 20% and that the discount on second-rate porcelains 'should not exceed 30%' of the original production costs.[43]

The applicability of this rule was facilitated by supervisor Tang Ying, who renewed the list of the necessary costs for all 25 working steps in porcelain production by adjusting the respective items and amounts based on actual situations. This list provided a functional guideline for cost computation. In practice it became clear that the sales prices calculated by that method were relatively high and unappealing to customers; as a result, the expected sales revenues were always in arrears. Therefore, in the 36th year of the Qianlong reign (1771), it was decided that the price calculated from the production costs should be reduced by 30%,[44] in the hope that this would recover at least some of the costs by selling more rejects on the open market.

40 Ibid.

41 Tie Yuan 鐵源 et al., eds., *Qing gong ciqi dang'an quanji* 清宮瓷器檔案全集, vol. 2 (Beijing: Zhongguo huabao chubanshe, 2008), 209-230.

42 '經過折寬一寸、長一尺，應用泥土、釉料、做坯、做細等項銀兩，逐一分析', see ibid., vol. 3, 73.

43 Fu Yuhong 傅育紅, 'Shaozao ciqi zhangcheng zhiding hou yuyao shaozao jingfei de hexiao 燒造瓷器章程制定後御窯燒造經費的核銷', *Gugong bowuyuan yuankan* 故宮博物院院刊 = Palace Museum Journal, no. 2 (2006): 73-80.

44 Tie Yuan et al., eds., *Qing gong ciqi dang'an quanji*, vol. 11, 126-127.

Conclusion

Initially, in the Ming and early Qing, there was neither a fixed budget for porcelain production nor a statutory specification about the sources of finance for it. Particularly during the Ming era, porcelain production was financed in an extremely arbitrary manner. Even after the Yongzheng emperor had changed the source of funding from regional treasury to the superfluous revenue from customs duties, the scale and requirements of production in any specific year were entirely at the emperor's whim. Even under Qing rule, with their elaborate reporting routines and controlling workflow procedures, porcelain production could never become a constant, indispensable project in state budgeting and expenditure, because it remained tied to the emperors' changing desires and preferences. In this sense, imperial porcelain was different from commercial operations by its very nature, even though both sectors shared workers and techniques. At the same time, the sales of second-rate products – sanctioned in the Qianlong reign – indicate that even the so-called imperial porcelains were treated as simply one among the many goods produced and managed by the court. Porcelain production was not only subject to constraints caused by the economic conditions of the state, but its organization was also reformed to better fit into the government's financial organization and lessen the burden on the state's expenditures.

In the dynastic histories of pre-modern China, peace, prosperity, and longevity resulted from maintaining the balance between numerous interacting and interlocking forces that were simultaneously interdependent and mutually restrictive. The production of imperial porcelains was only one tiny thread amongst the tens of thousands of strands which wove the huge web of the state. This chapter has demonstrated how marginal imperial porcelain was on various levels: in the role porcelain actually played in state politics and the ritual system, in the minor hierarchical position held by the officials involved, and their lack of a recognized ceramic supervision profession and, last but not least, in production expenditures and management. Hence, even though Tang Ying's remark quoted at the beginning of this chapter does have a point, to stay faithful to the historical reality of imperial porcelain production we should shift its emphasis by reversing the two clauses to create the amended assertion that: 'Ceramics, although of concern to state affairs, was nevertheless a trivial matter' (*sui guan hu guozheng, tao reng xishi* 雖關乎國政，陶仍細事).

Translated from Chinese by Yanqing Shen

About the Author

Guangyao Wang 王光堯 is researcher (equivalent to curator) at the Palace Museum in Beijing. He specialises in ceramic archaeology and the institutional history of imperial courts. He has published *Zhongguo gudai guanyao zhidu* 中國古代官窯制度 (The Systems of Official Ceramic Manufacture in Premodern China), book I and II.

5 Resplendent Innovations

Fire Gilding Techniques at the Qing Court

Te-cheng Su and Hui-min Lai

Abstract

This chapter investigates how the innovation of fire-gilding techniques by the Imperial Household Department helped integrate political and religious power within the empire. During the Qianlong reign (1736-1795), Tibetan style *sisa* forging, rubia cordifolia burnishing, and multi-layered gilding techniques were introduced while equipping temples with resplendent brass roof tiles and gilded artefacts. The sophistication of manufacturing gilded copper objects relied on the multicultural exchange with Nepalese, Tibetan, and Xinjiang handicrafts. It also required careful management of the flow of such precious materials as gold and copper as well as the necessary labour resources in the form of skilled court and contract artisans.

Keywords: Qianlong emperor, fire gilding, metal manufacturing, multiculturalism, Tibetan style

Introduction

The technology of gold plating, which enjoyed a long history in China, witnessed significant breakthroughs in both materials and techniques in the mid-18th century, during the reign of the Qianlong emperor. The context of these innovations was politico-religious in nature and stood in close connection with the conquest of the former Mongolian region of Dzungaria in Central Asia in 1758. This conquest not only expanded the territory of the Qing empire considerably but also furthered the Qianlong emperor's (r. 1736-1795) attempt to find ways to integrate political and religious power within the empire.

Siebert, Martina, Kai Jun Chen, and Dorothy Ko (eds), *Making the Palace Machine Work: Mobilizing People, Objects, and Nature in the Qing Empire*. Taylor & Francis Group, 2021
DOI: 10.5117/9789463720359_CH05

How the empire ruled different peoples across its large territory has aroused much interest amongst historians in recent years.[1] This chapter contributes to this debate by investigating a nexus of technology, religion, resources, and knowledge management, whose reins lay in the hands of the Imperial Household Department at the core of the 'palace machine'. These activities aimed to culturally bind the new Mongol subjects to China by re-enacting the religious culture of the imperial periphery in the power centres of the empire. A prominent example of this is the copy of the Potala Palace in Lhasa, Tibet, residence of the Dalai Lama and religious centre for the Mongols and Tibetans, which was built at the imperial summer residence in Chengde. In an effort to attract the Mongols to China on pilgrimage, Qianlong built temples outside Tibet in the Tibetan style, equipped with resplendent brass roof tiles and gilded artefacts.[2] This chapter examines the sophisticated process of manufacturing fire-gilded artefacts needed for this endeavour at the Qing Imperial Workshops, control of the material flows of the precious metals, and the European and Tibetan influences on this technology. As the discussion below makes clear, this workflow, knitted together from old and new parts of the palace machine, came into fruition during the Qianlong reign.

Gold, the key ingredient of fire gilding, was a precious commodity whose sourcing and movement through the palace machine were strictly controlled and meticulously logged. The Bullion Vaults of the Grand Storage Office (*Guangchusi Yinku* 廣儲司銀庫) of the Imperial Household Department (*Neiwufu* 內務府) was responsible for handling imperial finances and stocking all the treasures collected by the emperors. The stockpile of gold held by the Imperial Household Department was also stored in the Bullion Vaults. The deposit was regularly augmented by annual tributes of gold from the peripheral areas which amounted to tens of thousands of taels, far exceeding those received by individual European kings in the 18th century. The Bullion Vaults' directors (*tangguan* 堂官) had to present 'monthly reports' (*yuezhe dang* 月摺檔), also known as 'four-column ledgers' (*sizhu*

1 See, for example, the works of Patricia Berger, *Empire of Emptiness Buddhist Art and Political Authority in Qing China* (Honolulu: University of Hawai'i Press, 2003); Philippe Forêt, *Mapping Chengde: The Qing Landscape Enterprise* (Honolulu: University of Hawai'i Press, 2000); Pamela Crossley, *A Translucent Mirror: History and Identity in Qing Imperial Ideology* (Berkeley: University of California Press, 1999); and Peter Perdue, *China Marches West: The Qing Conquest of Central Eurasia* (Cambridge, Mass: Harvard University Press, 2005).

2 Lai Hui-min 賴惠敏, *Qianlong huangdi de hebao* 乾隆皇帝的荷包 (Taipei: Zhongyang yanjiuyuan jindaishi yanjiusuo, 2014). See especially Chapters Five and Seven on the costs of the Qing Court in constructing Tibetan Buddhist temples.

qingzhe 四柱清摺) detailing the vaults' opening balance (*jiucun* 舊存), new receipts (*xinshou* 新收), actual use (*chuyong* 除用), and closing balance (shizai 實在).[3] (See Chapter Three for the Imperial Household Department's accounting system.)

The flow of materials, labour resources, and their monetary equivalents was managed with care and in accordance with constantly-evolving precedents. When constructing and repairing imperial palaces and gardens, Qianlong set up specific construction offices for each project, such as the Long River Construction Office (*Changhe gongcheng chu* 長河工程處) and the Old Summer Palace Construction Office (*Yuanmingyuan gongcheng chu* 圓明園工程處). Each construction office would have to issue its own 'Current Regulations and Precedents' (*xianxing zeli* 現行則例), detailing the amount and price of raw materials needed for the project, as well as the number of artisans and their salaries. Artefacts were ordered to be manufactured 'according to the precedents on manufacture' (*an chengzao zhi li* 按成造之例). Most of these construction offices' 'Regulations and Precedents' (*zeli* 則例) were compiled during the Qianlong reign, turning these larger building projects into standardizing models and financial orientation for future projects.[4] Qianlong pressed the Imperial Household Department to follow the 'Current Regulations and Precedents'. The *Current Regulations and Precedents on Fire Gilding in the Yangxin Palace* (Yangxin dian dujin zuo zeli 養心殿鍍金作則例) and the *Current Regulations and Precedents on Fire Gilding in the Old Summer Palace* (Yuanming yuan dujin zuo zeli 園明園鍍金作則例) are examples of these kind of documents containing information on the raw materials and artisans required; while the *Current Regulations and Precedents on the Price of Miscellaneous Items* (Zaxiang jiazhi zeli 雜項價值則例) and the *Precedents on Compositions of*

3 *Qianlong chao Neiwufu yinku jinxiang yuezhe dang* 乾隆朝內務府銀庫進項月摺檔 (Microfilm. Beijing: The First Historical Archives of China, 2001).

4 These documents are not datable exactly but were, according to Wang Shixiang's analysis, compiled after finishing the project that gave the regulation its title. Most of the manuscript copies we have today, such as the *Neiting Yuanmingyuan neigong zhuzuo xianxing zeli* 內庭圓明園內工諸作現行則例 (Current Regulations and Precedents on the interior handicrafts in the Inner courts of the Yuanmingyuan) or those concerning the building of the palaces at Rehe date from Qianlong's time; the regulations excerpted from the building project of the Yuanmingyuan and Wanshou shan 萬壽山 palace gardens and the Inner Palaces (*neiting* 內庭) date from after that time. Before Qianlong reign the only existing orientation for projects was the *Gongcheng zuofa* 工程做法 compiled in the 22nd year of Yongzheng (1734) by an editorial team headed by Yunli 允禮 (1697-1738). Wang Puzi 王璞子 annotated it and published a modern edition. See also Wang Shixiang 王世襄, ed., *Qingdai jiangzuo zeli* 清代匠作則例 (Zhengzhou: Daxiang chubanshe, 2000).

Materials (Wuliao jinliang li 物料觔兩例) provide valuable information concerning the price of raw materials and the formulas of various mixtures.[5] Taking these regulatory and other archival sources together, supplemented by actual extant artefacts, this chapter scrutinizes fire gilding as a case study to illustrate the complexity and meticulousness of managing manufacture and technical processes in the palace workshops, as well as the multicultural exchange between Chinese, Tibetan, and European craft knowledge that fuelled the respective technical innovation.

Fire gilding, known in historical Chinese as *liujin* 鎏金, was a chemical gold plating technique developed in China as early as 300 B.C.E. It entailed first combining gold and mercury to make an amalgam paste, which was spread onto the surface of a copper or silver alloy object. The object was subsequently heated to volatilize the mercury, producing a firmly-bonded gilding layer. The descriptions of the process found in historical sources are very similar to how the process is explained by craftsmen today.[6] The *Complete Dictionary of Chinese Crafts* summarized the process into five stages: (i) preparing a gilding-stick (*zuo jingun* 做金棍) to stir the paste and apply the amalgam onto the objects; (ii) melting the gold together with mercury to produce the amalgam (*shajin* 煞金); (iii) applying the gold [amalgam] onto the object (*mojin* 抹金); (iv) 'unclosing the gold', i.e., heating the object to evaporate the mercury so that a gilded layer is formed (*kaijin* 開金); (v) press polishing the object (*yaguang* 壓光).[7] Conservation chemist Kilian Anheuser suggested that a possible variation of this historical technique was to spread gold foil onto a surface coated with mercury, then heat the surface directly.[8]

Traditional fire gilding is a technique that offers several advantages: first, a relatively cheap form of gold, namely gold leaf, and mercury were

5 Jiang Yasha 姜亞沙 et al., eds., *Qingdai gongyuan zeli huibian* 清代宮園則例彙編 (Beijing: Quanguo tushu guan wenxian suowei fuzhi zhongxi, 2011). For *Yangxin dian dujin zuo zeli* see ibid., vol. 5, 327-329; for *Yuanming yuan dujin zuo zeli* see ibid., vol. 18, 355-356.

6 Liang Xudong gave an extended introduction to the different gilding techniques, relying partly on oral reports by craftsmen (Liang Xudong 梁旭東, 'Zhongguo chuantong de liujin jishu 中國傳統的鎏金技術', *Cailiao baohu* 材料保護, vol. 23, no. 1/2 (1990): 83-86. For a short analysis of the history of Chinese gilding technique see also Joseph Needham et al., *Science and Civilization in China*, vol. V: Chemistry and Chemical Technology, part IV, 'Spagyrical Discovery and Invention: Apparatus, Theories and Gifts', (Cambridge: Cambridge University Press, Cambridge, 1980), 80.

7 Wu Shan 吳山, ed., *Zhongguo gongyi meishu dacidian* 中國工藝美術大辭典 (Nanjing: Jiangsu meishu chuban she, 1989), 272.

8 Kilian Anheuser, 'Archaeotechnology: The practice and characterisation of historic fire gilding techniques', *JOM-Journal of the Minerals Metals & Materials Society*, vol. 49 (1997): 58-62.

used as raw materials and, as both can form gilding layers, it was possible to minimize waste; second, the task can be performed with simple equipment and tools and thus only required low capital investment beyond the raw material; third, any plating flaws in the first attempt can be repaired by a second or even third attempt. Gold amalgam can be easily spread on hollow parts or the interior of blind holes, and this technique is therefore still widely used on silver and hollow copper-alloy jewellery and sculptures today.[9]

The following sections of this chapter will investigate the sources of two essential raw materials used in fire gilding, namely gold and copper,[10] before examining the management of fire gilding artisans. The chapter will then explore two major fire-gilding projects of the Qianlong-era court: first, the gold roofs of the Chengde Potala and second, a set of gilded Buddhist statues for a Beijing temple. Closely connected to the large Chengde project was a change in fire gilding to use copper alloy bodies. Prior to the Qing, fire gilding had largely been applied onto silver objects, which was a relatively simple procedure. The more involved and complex procedure of fire-gilding copper alloys was, as will be shown later, a Tibetan technology introduced to the Qing court by the lama envoys residing in Beijing.

Sourcing Raw Materials used in Fire Gilding for the Qing Court

The Qianlong court went to great lengths to secure a consistent and adequate supply of gold from multiple sources for its gilded politico-religious projects. Gold was the fuel for the sub-unit in the palace machine in this case. An important and steady source of gold during the Qianlong reign was the yearly

9 Electroplating engineers Wu Yuankang 吳元康 and Chu Rongbang 儲榮邦 compared traditional fire gilding to electroplating – the common technique in modern times – and pointed out several benefits of the traditional techniques, see their 'Liudu – zhongguo gudai faming de yizhong zai cailiao biaomian shang du jinshu de jishu 鎏鍍—中國古代發明的一種在材料表面上鍍金屬的技術', *Tuzhuang yu diandu* 塗裝與電鍍, no. 1 (2011): 3-9 and its sequel 'xu wan 續完', ibid., no. 3 (2011): 25-28.

10 China's indigenous resources of mercury are comparatively small and thus, in the 18th century, it was imported via the port of Guangzhou. As *The Chronicles of the East India Company Trading to China, 1635-1834* shows, mercury was one of the major import products of the time. Starting with 64 *dan* 擔 (picul) worth 2,864 *liang* (tael) of silver in 1700 the amount grew yearly by up to 100 *dan*. Several countries were involved in the mercury business with Qing China. By the first half of the 19th century the largest amounts of that metal came from the US – 8,210 *dan* in 1823 and 10,154 *dan* in 1832 worth 492,600 and 629,548 US dollars respectively. See Hosea Ballou Morse, *The Chronicles of the East India Company, Trading to China 1635-1834*, vol. 1 (chapter 6), 69; vol. 4 (chapters 80 and 91), 84 and 339.

tributes coming from the newly incorporated peripheries of the Qing empire; but ad hoc sourcing via other administrative channels also played a part in feeding the gold-hungry court. In addition, when copperware replaced silverware during the Qing in the fire gilding process, copper of a higher purity had to be imported from Japan to ensure optimal results. Sourcing the materials for fire gilding thus involved new itineraries of materials and new logistical arrangements.

(I) Gold in the court

The Qianlong court's appetite for gold to use in fire gilding was insatiable, and the acquisition of this precious material was enmeshed with the conduct of politics and diplomacy of the empire. During the Kangxi reign the court's main source of gold was tribute gold from Annan 安南, modern Vietnam. The steep rise in the volume of gold flowing into the court began under Qianlong's rule as new sources were tapped, such as regular deliveries of gold from the new Qing territories to the West and gold tributes arriving from vassals and other foreign countries. Generally speaking, gold in the 18th-century Qing court can be distinguished as coming from two sources: inner-Chinese or foreign sources. From the inner-Chinese sources, the largest amount of gold in the period under review here was not obtained from any gold mines but from the pockets of salt merchants, who had to pay fines of 'indulgence gold' (*shuzui jin* 贖罪金) to the Lianghuai Salt Administration, and from the fines of the famous 'Lianghuai Salt Certificate Case' (*Lianghuai yanyin an* 兩淮鹽引案, 1768). Both went to the Imperial Household Department. In 1768, the Lianghuai Salt Supervisor (*yanzheng* 鹽政) Youbashi 尤拔世 handed in 2,988 taels of 'indulgence gold', a fine paid by merchant Hong Zhenyuan 洪箴遠 and others. In the following year, Youbashi sent another ten boxes of gold leaf, amounting to 5,000 pieces and 490 taels in weight in total, this time directly to the Imperial Workshops (*Zaobanchu* 造辦處).[11] The practice of salt supervisors collecting salt merchants' debts and sending them to the Imperial Household Department as gold continued in the years after the Salt Certificate Case. The monthly report for October 1770, for example, reports that 'the touring salt censor (*xunshi changlu yanzheng* 巡視長蘆鹽政) Li Zhiying 李質穎 is obligated to deliver the debt of one million taels of silver in total. From this instalment, 5,000 taels of gold were bought'.[12]

11 *Junjichudang zhejian* 軍機處檔摺件 (Archives of the Grand Council), no. 010831 (QL34/10 = Oct. 1769) (Taiwan National Palace Museum).

12 *Qianlong chao Neiwufu yinku jinxiang yuezhe dang*, QL35/10 = Nov.-Dec. 1770).

By 1786, the Lianghuai Salt Administration had submitted a total of about 50,000 taels of gold.[13]

Another domestic source of gold was the compulsory annual deliveries by provincial gold refineries to the Board of Revenue (*Hubu* 戶部) in amounts set by the *Precedent Cases of Statutes of the Great Qing* (Da Qing huidian shili 大清會典事例). As a matter of fact, part of this gold was given as a 'tribute' (*gongjin* 貢金) to the Imperial Household Department. Archival documents show that, from 1759 on, the Imperial Household Department received such 'tribute gold' from Yunnan, Guizhou, Shanxi, and Gansu.[14] For instance, in 1759 the provincial governor (*xunfu* 巡撫) of Yunnan Liu Zao 劉藻 and the provincial governor of Guizhou Zhou Renji 周人驥 sent 131.4 and 111.36 taels of gold respectively;[15] thereafter they contributed more than 100 taels annually.[16] Qianlong also ordered the provincial governor of Yunnan to buy gold locally. For example, in 1753 Aibida 愛必達, following an imperial order, bought 1,000 taels of gold in different qualities: 200 taels of 80% gold, 200 taels of 75% gold, and 600 taels of 70% gold. At that time gold was cheaper in the gold-producing province of Yunnan by over 2,000 silver taels per catty than it was in the capital, Beijing.[17]

The Imperial Household Department also procured gold directly from production sites on an ad hoc basis in the form of official gold tax. The sites for gold sand production located in the mountains north and south of Shazhou in Dunhuang, Gansu, are one example. Traces of this gold and how it made its way into the Bullion Vaults (*Yinku* 銀庫) of the Imperial Household Department can be found in the monthly report dated January 1781. There the director of the Bullion Vaults (*yinku langzhong* 銀庫郎中) Bandaersha 班達爾沙 and others registered the following extractions: 19 gold ingots of *zhengkejin* 正課金 (standard gold tax) quality,

13 Lai Hui-min, *Qianlong huangdi de hebao*, 366-367.

14 For a list of the yearly taxes in gold and silver (*jinyin kuangke* 金銀礦課) received from the various gold refineries in Yunnan etc. see *Da Qing huidian shili* (Guangxu ban) 大清會典事例 (光緒版), eds. Kungang 崑岡 et al., *juan* 243 (chapter '*Hubu · zafu: Jin yin kuangke* 戶部 · 雜賦: 金銀礦課 (Board of Revenue · Miscellaneous Tax: Levy from Gold and Silver Mines)' (rep. Beijing: Zhonghua shuju, 1991, vol. 3, 871-879). The fact that part of this gold went to the vaults of the Imperial Household Department can be detected only from the archival documents.

15 *Qianlong chao Neiwufu zouxiao dang* 乾隆朝內務府奏銷檔, vol. 248, 137-138 (QL24/11/19= 6 Jan. 1760) (The First Historical Archives of China, Beijing). These volume numbers refer to the *ce* 冊 of the original accounting reports. A reproduction of all existent accounting reports of The First Historical Archive was published in 2014 (*Qing gong Neiwufu zouxiao dang* 清宮內務府奏銷檔. Beijing: Gugong chubanshe). The 300 volumes cover the years 1723 to 1912.

16 *Qianlong chao Neiwufu zouxiao dang*, vol. 310, 8 (QL37/3/2 = 4 April 1772).

17 Ibid., vol. 228, 60-62 (QL18/12/20 = 12 Jan. 1754).

each weighing 10 taels; one gold ingot of *weijin* 尾金 ('tail' gold), weighing 2.5 taels; and one ingot of *sasanjin* 撒散金 ('scattered' gold), weighing 5.78 taels. The ingots had been submitted by the governor-general of Shaanxi and Gansu (*Shaan-Gan zongdu* 陝甘總督), Leierjin 勒爾謹, as a product of the local gold refinery. On their standard route, they should have been delivered to the Board of Revenue as ordinary gold tax. But Leierjin had labelled them as 'tribute' so that they were sent directly to the Imperial Household Department.[18]

After Qianlong's conquest of Dzungaria, the 'new territories' of Xinjiang also became a source of gold. In 1766, the Grand Council (*Junjichu* 軍機處) noted down the arrival of 137.5 taels of tribute gold from the areas of Yarkand, Hotan (Hedian 和闐, present Hetian shi 和田市), Shaerhuer 沙爾胡爾 and Kashgar.[19] An additional source was confiscated 'private' gold. According to the regulations in the Qianlong reign, gold, silver and jade were state-owned; any private trade in these materials was prohibited. Between 1784 and 1795, 1,595.24 taels of private gold were confiscated in Urumqi in Xinjiang and sent to the Imperial Household Department.[20]

Gold also came pouring into the Qing court from foreign sources, mainly as tribute gold sent by tributary states and envoys. Korea had been among these tributary states up until the 32nd year of the Kangxi reign (1693), when the Korean king was granted exemption from the donation of 100 taels per year, because he argued that 'gold was not a product of his country'.[21] Among the other tributary states, Annan delivered comparatively regular tributes of gold – 533 taels in 1743, 418 in 1748, 475.5 in 1754, and 418 in 1760. In 1762 and 1765, Annan offered a total of 42 gold bullions weighing 418 taels, and 12 golden ritual artefacts, weighing 115 taels. In 1766, it sent another 533 taels.[22] In 1785, Xianluo 暹羅 (modern Thailand) sent several gold objects as tribute, including a ceremonial document made from one sheet of gold leaf (*jinye biaowen* 金葉表文) and 16 small gold rings.[23]

18 *Qianlong chao Neiwufu yinku jinxiang yuezhe dang*, QL46/1 = Jan. or Feb. 1781. On Qianlong's use of the Board of Revenue as his 'own purse', see Lai, *Qianlong huangdi de hebao*.

19 *Qianlong chao Neiwufu zouxiao dang*, vol. 284, 10-26 (QL31/10/6 = 7 Nov. 1766).

20 *Qianlong chao Neiwufu yinku jinxiang yuezhe dang*, QL46/1 to QL60/12 = Jan. or Feb. 1781 to Jan. or Feb. 1796.

21 He Xinhua 何新華, *Qingdai gongwu zhidu yanjiu* 清代貢物制度研究 (Beijing: Shehui kexue wenxian chubanshe, 2012), 231.

22 *Qianlong chao Neiwufu zouxiao dang*, vol. 284, 26-28 (QL31/10/6 = 7 Nov. 1766).

23 *Qing gong Neiwufu Zaobanchu dang'an zonghui* 清宮內務府造辦處檔案總匯, eds. The First Historical Archives of China, Beijing, and Hong Kong Chinese University Art Museum (Beijing: Renmin chubanshe, 2005), vol. 48, 286-287 (*jishi lu* 計事錄 'record notes'; QL50/2 = March or April 1785).

The Qing court also maintained tribute relations with several European countries. Some of the items presented contained gold, but there was no direct transfer of 'tribute gold' of any kind. For example, in 1752, King Joseph I of Portugal sent envoys and 28 items as tribute goods, including flintlock rifles and pistols, enamel knives, silver candleholders, pure gold stationery and snuff boxes, gold-threaded brocade, and gold-embroidered textiles.[24]

Even with all these channels of ad hoc and routine sources of gold, the demand ultimately outstripped supply. According to the available 'Monthly receipts and expenditure reports of the Bullion Vaults of the Imperial Household Department' (*Neiwufu yinku jinxiang yongxiang yuezhe dang* 內務府銀庫進項用項月摺檔), a total of 428,909.7 taels of pure gold (*chun jin* 純金) were withdrawn between 1740 and 1795, but only 373,824 taels were deposited, as shown in Chart 5.1.[25]

Chart 5.1 below shows a number of exceptionally high new deposits or withdrawals, especially for the years Qianlong 18 (1753) and Qianlong 44 (1779). In 1751 the high-ranking Manchu official Fuheng 傅恆 (1720-1770) suggested melting down decommissioned gold and copperware stored in the vaults of the Qing Court, including the Ming artefacts and gold books (*jince* 金冊) and seals (*jinyin* 金印) of deceased empresses and consorts. This was carried out, adding 10,919 taels of gold in 86% purity to the Bullion Vaults' storage registers. In 1773 officials of the Imperial Household Department melted down objects in the Eastern and Western Qing Tombs, amounting altogether to 28,992.75 taels of goldware, 31,955.82 taels of silverware, 1,751 catties of copperware, and two gold artefacts from the Imperial Princes of He and Ding (*He Ding qinwang* 和定親王), weighing an additional 443.4 taels.[26] The peak in gold withdrawal from the vault in 1779 was mainly caused by demands in relation to constructions in preparation for the visit of the Sixth Panchen Lama to Rehe 熱河 (modern Chengde) and Beijing, in the wake of the celebration of Qianlong's 70th birthday. To decorate the various buildings of the newly-built Zongjing Dazhao Temple 宗鏡大昭廟 at Xiangshan near Beijing, 13,300.69 taels of first-grade gold were requested from the Bullion Vaults. Because the vaults could not provide the full amount, a total of 13,176.55 taels of second- and third-grade gold, as well as 90%, 80%, and 70% gold was melted and purified, producing 11,820.86

24 *Da Qing huidian shili* (Guangxu ban), *juan* 503 (vol. 6, 827.2).

25 According to the Bullion Vault's 'Monthly receipts reports', all the various gold qualities were exchanged into values of 'pure gold'.

26 *Qianlong chao Neiwufu zouxiao dang*, vol. 319, 175-179 (QL38/4/27 = 16 June 1773). Also see *Qianlong chao Neiwufu yinku jinxiang yuezhe dang*, QL18/7 = July or Aug. 1753, and QL38/6 = July or Aug. 1773.

Chart 5.1 Withdrawals and deposits of 'pure gold' in the Bullion Vaults of the Grand Storage Office (*Guangchusi*) in the Qianlong reign

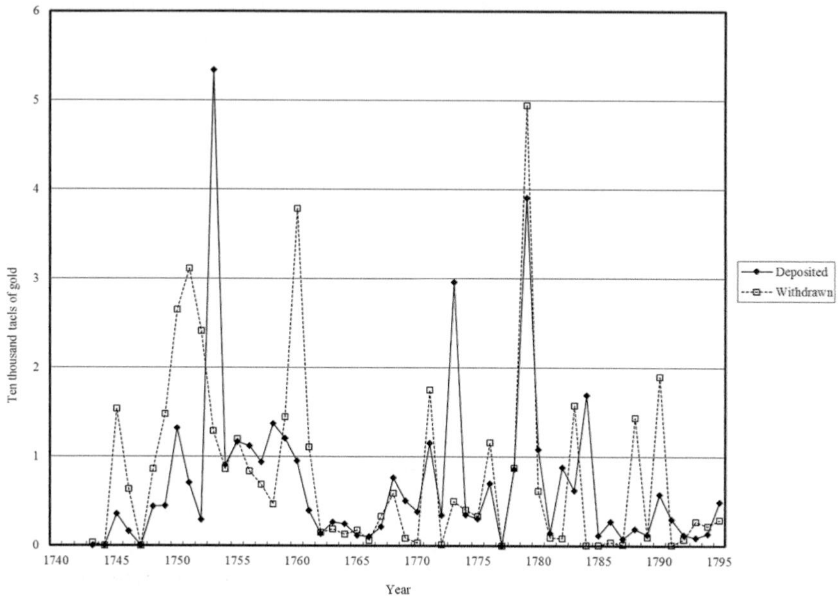

Data collected from the various monthly reports on receipts and expenditure of the Imperial Household Department's Bullion Vaults of the Qianlong period (*Qianlong chao Neiwufu yinku jinxiang yuezhe dang* and *Qianlong chao neiwufu yinku yongxiang yuezhe dang*)

taels of first-grade gold. An additional 4,266.94 taels of tribute gold were collected for this project from the gold refineries in Yunnan or donated from governors, governor-generals and provincial treasures of Guizhou, Sichuan, Shandong, Guangxi, Zhili, and Liang Guang.[27] To prepare for the visit of the Sixth Panchen Lama in Rehe, Qianlong started to build the Xumi Fushou Temple 須彌福壽之廟 in 1779. To fire gild the copper roof tiles and other items, 15,315.35 taels of gold leaf were used for the first gilding, and another 15,315.35 taels for the second gilding.[28] The objects in the Zongjing Dazhao Temple were also gilded two times, thus the figure of 13,300.69 taels of gold mentioned above had to be budgeted and consumed twice. In 1780 a total of 57,232.08 taels were used in these two projects.

The Jesuit missionary Father Giovanni Laureati (1666-1727) suggested the enormous volume of transactions in gold. He stated that 'Chinese gold is not as pure as Brazilian, but its price is very low. When shipped to Europe one

27 *Qianlong chao Neiwufu yinku jinxiang yuezhe dang*, QL44/11 to QL44/12 = Jan. to Feb. 1780.
28 Ibid., QL44/11 = Jan. or Feb. 1780.

can make 70% profit'. This was not because the Chinese lacked experience in handling and assaying gold. On the contrary, as Laureati further explained to his European colleagues, Chinese artisans were highly proficient in identifying pure gold, silver, and their alloys. But, as gold was only sometimes used as payment and was primarily a commodity rather than a currency, purity was more an aesthetic than a financial issue.[29]

(II) Copper used in the Qing court

The new gilding technology preferred by the Qianlong court required ample supplies of not only gold, but also copper. Before the Qing, most gold-plated objects in China had been silverware; however, silver was much more expensive than copper. According to a guide to prices compiled by the Board of Work (*Gongbu* 工部) in 1736, one catty of copper cost 0.5 tael of silver; one catty of silver could, in turn, buy 32 catties of copper.[30] In the Kangxi reign copper still had to be imported from Japan, but during the Qianlong reign, large copper reserves were discovered in Yunnan, the Chinese province in the far south bordering on what is now Vietnam. This made copper more easily accessible for the court, therefore it is not surprising that there were more gold-plated copper alloys in the Qing than the Ming.[31] However, the Japanese possessed better copper refining technology, so the Imperial Household Department still preferred to use Japanese copper (called 'foreign copper', *yang tong* 洋銅), which was labelled 'water basin copper rods' (*shuicao hongtong tiao* 水槽紅銅條) in its accounting system. The more impure the copper, the more difficult it was to achieve good quality in the fire-gilding process, as the mercury in the gold paste would not evaporate completely, resulting in a surface with an uneven colour and low stability.[32]

The Qianlong emperor himself was aware of the perils of low-grade copper. The record notes (*jishi lu* 記事錄) of August 1744 noted that the eunuch Hu Shijie 胡世傑 had delivered an imperial order: 'Due to the impure

29 Du Hede 杜赫德 (J.-B. du Halde), *Yesuhui shi zhongguo shujian ji* 耶穌會士中國書簡集, trans. Zheng Dedi 郑德弟, Lu Yimin 吕一民 and Shen Jian 沈坚 (Zhengzhou: Daxiang chubanshe, 2005), vol. 2, 117-118.

30 *Jiuqing yiding wuliao jiazhi* 九卿議定物料價值, ed. Board of Work of Qing Court, Beijing, 1736 (rep. in *Gugong zhenben congkan* 故宮珍本叢刊. Haikou: Hainan chubanshe, 2000, vol. 317).

31 A search in the Scripta Sinica Database for gilded copper alloys in the Ming only returned two results, one in *Ming shilu* 明實錄 and the other in *Qixiu leigao* 七修類稿, http://hanchi.ihp.sinica.edu.tw (accessed via Institute of History and Philology, Academia Sinica, on 11 March 2015).

32 Liu Wanhang 劉萬航, *Jinyin zhuangshi yishu* 金銀裝飾藝術 (Taipei: Xingzhenyuan wenhua jianshe weiyuanhui, 1989), 26.

copper used, the artefacts made by Deng Bage are coarse and lack rigidity. The issue is to be handed over to Prince Yi 怡親王 and Haiwang 海望 for admonishment. In the future, pure copper must be used'.[33] In a similar case in 1779 the chief supervisor of the Imperial Workshop reported to the emperor that 'the copper used for the *vajra* heads (*chutou* 杵頭) were made from reclaimed copper, which contains glaze and other residues rich in lead, preventing strong bonding between the gilding and the substrate'.[34] Thus, the resource-minded practice of reclaiming or recycling used materials was thwarted by the technical demands of fire gilding.

Qianlong was not the first Qing emperor to look to Japan for a reliable supply of pure copper. As early as 1725, a Copper Refinery (銅吹所, *tongchui suo* in Chinese; *dōfukiya* in Japanese) was established in Osaka, Japan, to process the copper ore into copper bars (棹銅 Chinese *zhaotong*; Japanese *saodō*) for the export market. A bar measured two centimetres in diameter and weighed 300 grams. Each box contained 200 copper bars, weighing a total of 60 kilograms, and bore the lettering of 'Copper Bars for Imperial Use' (*Yuyong zhaotong* 御用棹銅) on the cover (see Figure 5.1). As with gold, the court's appetite for high-grade copper appears to have been insatiable.

Once conveyed to Beijing, copper was stocked together with other less precious metals, such as tin and mercury, in the Ceramic Vaults (*Ciku* 磁庫) of the Grand Storage Office in the Forbidden City to be used for gilding, casting cannons, and other purposes. Between 1743 and 1795 (no record is available for 1777), a total of 1,263,718 catties of copper entered the vaults; during the same period 1,274,725 catties of copper were disbursed.[35] In addition, the Imperial Workshop stocked its own inventory of copper. The annual 'Inventory register of the Imperial Workshops of the Yangxin Palace' (*Yangxin dian zaobanchu shouzhu qingce* 養心殿造辦處收貯清冊) was structured in the same way as those of the Ceramic Vaults, in the form of 'four-column ledgers'.[36] Research into the Imperial Workshops' storage during Qianlong reign reveals that storage data is missing for 24 of the 60 years, as is withdrawal data for 20 years. For the remaining years a total of 265,978 catties in copper deposits and 233,705 in withdrawals were made. Over the years, the Imperial Workshops and the Ceramic Vaults received a

33 *Qing gong Neiwufu Zaobanchu dang'an zonghui*, vol. 12, 301 (QL9/8 = Sep. or Oct. 1744).

34 *Qing gong Neiwufu Zaobanchu dang'an zonghui*, vol. 42, 738-739 ('Zhuluchu 鑄爐處, Casting Workshop'; QL44/10 = Nov. or Dec. 1779).

35 *Neiwufu Guangchusi Liuku yuezhe dang* 內務府廣儲司六庫月摺檔 (Monthly reports of the Six Vaults of the Imperial Household's Grand Storage Office), *Ciqi ku* 瓷器庫 (Ceramic Vaults) (The First Historical Archives of China, Beijing).

36 For an analysis of their logic and character, see Chapter Three in this volume.

Figure 5.1 Japanese copper bars (*saodō* 棹銅) and a replica of the box used to send the bars to the Qing empire

Collection of the Osaka Museum (photograph by the authors)

total of 1,529,696 catties of copper and dispensed 1,508,430 catties. A third storehouse for copper was the Bullion Vaults of the Old Summer Palace (*Yuanmingyuan yinku* 圓明園銀庫), but its records were destroyed during the Anglo-French expedition to China in 1860. Chart 5.2 illustrates the amount of copper deposited into and withdrawn from the Ceramic Vaults and the Yangxin Palace. It shows that during the early and late years of Qianlong reign, the amount of copper handled by the Imperial Household Department was quite low. From the 9th year of the Qianlong reign (1744) on, however, the court began to build Buddhist temples in a Tibetan style which were covered with gilded copper tiles, so the use of copper rose steeply. The 1779 peak in copper use was caused by constructing the new Dugang Hall (*Dugang dian* 都罡殿) of the Xumi Fushou Temple – which required 120,163 catties of copper for the roof tiles.[37] As mentioned above, in the same year the Zongjing Dazhao Temple was built and roofed with gilded copper tiles. We do not know the exact number of tiles used, but

37 *Junjichudang zhejian* 軍機處檔摺件, no. 028524 (QL45/10/16 = 2 Nov. 1780) (National Palace Museum, Taipei).

Chart 5.2 The aggregated amount of copper stocked in the Ceramic Vaults and the Yangxin Palace 養心殿 of the Imperial Household Department during Qianlong reign

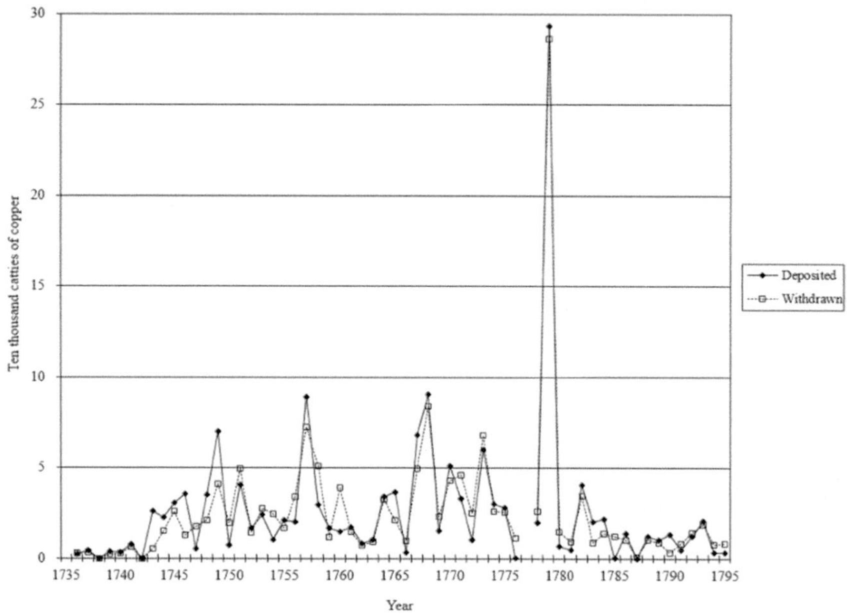

Sources: 'Monthly Reports of the Office of Privy Purse of the Imperial Household Department' of the Qianlong Reign (*Qianlong chao Neiwufu Guangchusi yuezhe dang* 乾隆朝內務府廣儲司月摺檔). For the Ceramic Vaults no data are available for the 1777; for the Yangxin Palace in sum 24 of 60 years of Qianlong reign lack data.

more than 100,000 catties of high-quality copper might have been needed for this project. These two projects increased the consumption of copper in the year 1779 to over 250,000 catties, in contrast to an average 50,000 catties of copper which the Imperial Household Department handled every year.

The Management of Gilding Artisans

The Imperial Workshops, established during the Kangxi reign, were responsible for manufacturing the artefacts commissioned by the Inner Court. According to *Precedent Cases of Statutes of the Great Qing* compiled in the Guangxu reign (1874-1908), Qianlong gave permission to add another two assistant directors (*yuanwailang* 員外郎) to the Workshops in 1758, one specifically to run the Casting Workshop (*Zhuluchu* 鑄爐處) and the other in charge of manufacturing.[38] In 1778, the staff list of the Imperial Workshops were fixed with three directors (*langzhong* 郎中), two assistant directors (*yuanwailang* 員外郎), one secretary (*zhushi* 主事), one deputy secretary (*weishu zhushi* 委署主事), six vault managers (*kuzhang* 庫掌) and fifteen clerks or scribes (*bitieshi* 筆帖式). The workshops and their personnel were responsible for the manufacture of objects for imperial use, as well as supervising the work involved and managing the associated storehouses.[39]

Five of the workshop's units were related to fire gilding in one way or another: the Casting Works, the Goldwork and Stonework Works (*Jinyu zuo* 金玉作), the Gilding Works (*Dujin zuo* 鍍金作), the Engraving Works (*Zanhua zuo* 鏨花作), and the Intarsia Works (*Xiangqian zuo* 鑲嵌作). In addition to the artisans in the Imperial Workshops, more than 1,000 artisans worked in the Six Vaults of the Grand Storage Office, at least some of whom would have handled gold and silver work. Moreover, fourteen silver refinery workers (*huayin jiang* 化銀匠), seven gold refinery workers (*lanjin jiang* 煉金匠), 24 filigree artisans (*leisi jiang* 纍絲匠) and 22 engraving artisans (*zanhua jiang* 鏨花匠) worked in the Bullion Vaults, either extracting or purifying gold and silver or producing goldware and silverware.[40] Like all the court artisans (*jianei jiangyi* 家内匠役) or salaried artisans (*shiliang jiang* 食糧匠) of the Imperial Workshops, they were recruited from the left and right wings of the upper three banners and their salaries were paid for by the banners to which they belonged.[41] Recruitment and terms of apprenticeship followed those of other craftsmen in the Imperial Workshops. Artisan posts were handed down by hereditary or from master to apprentice, making it almost impossible for outsiders to enter the workshops on a permanent basis. However, just like quality gold and copper, skilled and reliable artisans were also in short supply during the years of peak demand, so the Imperial Household Department also hired contract artisans (*waigu jiangyi* 外僱匠役) from the commercial market. Those with extraordinary skills might have their status converted into a permanent court artisan, whilst others made their daily entrance through the Jingyun Gate (*Jingyun men* 景運門) of the Forbidden City instead of lodging within the palace precinct. Overall, contract artisans were paid more and enjoyed better living conditions than the court artisans. It is interesting to note in passing that the salary of

38 *Da Qing huidian shili* (Guangxu ban), vol. 12, *juan* 1173, 674-1; *juan* 1176, 706-1.

39 Ji Huang 嵇璜 et al., *Qing chao Tongdian* 清朝通典 (Qing dynasty Comprehensive canon of laws and institutions) (Taipei: Taiwan shangwu yinshuguan, 1987), *juan* 29, 2189.3.

40 *Zongguan Neiwufu xianxing tiaoli: Guangchusi* 總管內務府現行條例: 廣儲司, in *Jindai Zhongguo shiliao congkan* 近代中國史料叢刊 (Taipei: Wenhai chubanshe, 1972) *juan* 1, 22.

41 *Da Qing huidian shili* (Guangxu ban), vol. 12, *juan* 1214, Jan. – Feb. 1079.

contract painters (*caizi jiang* 彩子匠) was higher than that of other artisans, amounting to 1.8 *qian* daily.[42] (See Chapters One and Two for more on the humans that powered the palace machine.)

In most cases regulations and archival documents only talk about 'artisans' (*jiang* 匠) without specifying their profession.[43] Only rare instances afford a glimpse of their positions within the work and craft flows that made the Gilding Works and its gilding projects efficient.[44] One unusual incident, triggered by a disruption to the flow of workers and materials as laid out in the 'Regulations and Precedents', inadvertently revealed how the chain of command in management and staffing of the workshop worked on a daily basis. On the night of the 14th day of the third month, 1773, the Gilding Workshop was burgled. Twelve small miscellaneous copper items for decorating arrow-carrying bags (*sadai* 撒袋), 79 brass screws and one golden ornament on a Buddhist scripture board weighing two *qian* were stolen. The Magistracy of the Household Department (Yamen) interrogated five artisans, including Liangsan Dazi 梁三達子 and Lü Mingde 呂明德, who testified that, before leaving at night, the Vault Manager Wude 五德, along with the artisans had locked the gates of each workshop and the courtyard before leaving. In the morning of the 15th, the artisans arrived and found that while the lock of the Gilding Works remained intact, its windows had been broken. The Magistracy concluded that the artisans should have stayed in the workshop overnight while there were unfinished artefacts inside. The artisans Liangsan Dazi, Lü Mingde and others received 60 lashes of a whip each as punishment for neglecting their storehouse duties. The deputy foremen (*fu cuizhang* 副催長) Xiande 憲德 and Jinjiang 金江 were found guilty for not arranging any security for the unit and for failing to secure the artefacts properly. They therefore also received 60 lashes each for dereliction of duty. One level higher in the administrative responsibilities, the vault manager Wude was accused of having overlooked implementing precautionary measures,

42 Jiang Yasha et al., eds., *Qingdai gongyuan zeli huibian*, vol. 5, 215.

43 The only '*jiang*' that receive special attention and are named in the documents are jade and ivory artisans (*yu jiang* 玉匠, *ya jiang* 牙匠) from the Chinese south and painters (*huahua ren* 畫畫人). Interestingly, the sometimes very high value attached to these painters' work was not reflected in a higher salary. Those paid most generously were the southern jade artisans, who received 12 or even 13 taels of silver per month. See Chi Jo-hsin 嵇若昕, 'Qing zhong houqi (1821-1911) Neiwufu Zaobanchu nanjiang jiqi xiangguan wenti 清中後期 (1821-1911) 內務府造辦處南匠及其相關問題', *Gugong xueshu jikan* 故宮學術季刊, vol. 32, no. 3 (2015): 63-89.

44 Two archival documents dated 1745 and 1746 indicate that fire-gilding workers were differentiated into salary levels (see *Qing gong Neiwufu Zaobanchu dang'an zonghui*, vol. 13, 575 and vol. 14, 365).

while the foreman (*cuizhang* 催長) Shaode 邵德 and the commandant of Fleet-as-cloud Cavalry (*Yunqi wei* 雲騎尉) Fushan 福山, who were both on shift that night, were found guilty of failing to patrol properly to protect imperial objects. All three were fined one year's worth of salary for neglecting their duties.[45]

The hierarchy of punishment meted out suggests that, at least in security-related matters, the chain of command within an individual workshop consisted of three layers: first, the vault manager (responsible for implementing safety measures) and foreman (on night shift); second, the deputy foreman (in charge of storing artefacts); and third, the individual artisans at work (liable for guarding the artefacts). The responsibility for physically securing the precious materials was shared across the ranks. Outside the workshop structure, the commander of the palace guard was also implicated.

Fire Gilding Techniques and Examples in the Qing Court

As outlined in the Introduction, the palace machine accommodated versatile technological processes and was vitally concerned with the minutiae of material production. The archival documents of the Imperial Household Department record the complicated process of manufacturing gilded copper alloy objects in meticulous details, which involved many preparatory and finishing steps as well as an array of materials and techniques. The first step was producing an object in the desired shape with an utterly smooth surface. In the Department's 'Regulations and Precedents', an often-mentioned technology in relation to gilded objects is a special Tibetan forging technique called *sisa* 鉛鈒. *Sisa* emerged in Tibet as an improvement on the traditional lost-wax technique used to produce Buddhist statues. According to historical metallurgist Yuan Kaizheng, the lost-wax casting technique used in Tibet and Nepal limited the size of Buddhist statues. Most lost-wax statues cast in a single mould were modest in height, about 0.34 metres including the base. Statues whose bases were manufactured separately, those which did not have a base, and those cast in separate parts, were still only about 0.65 metres tall. However, statues of any size could be made using the *sisa* technique, from as small as 0.3 metres high to as tall as 30 metres. However, while the majority of Buddhist brass statues in Tibet were, in fact, made with this technique, small statues were primarily still made from lost-wax

45 *Qianlong chao Neiwufu zouxiao dang*, vol. 324, 152-157 (QL38/12/20 = 31 Jan. 1774).

casting.[46] Two materials central to the *sisa* technique were glue paste and rosin. An investigation into a traditional metal workshop in modern Tibet by Yuan Kaizheng help us to reconstruct how these materials would have been used in the imperial court of the Qing. The artisan demonstrated how, before carving the details of a copper or brass object, a layer of rosin and glue paste would have been applied from the back to fill the hollow parts, in order to maintain the object's shape during the process. The filling would then be covered by a wooden plate with glue paste applied to the edges of the brass plate. After carving, the rosin glue could easily be peeled off by tapping the plate, then melted and reused repetitively.[47] According to the archival documents of the Imperial Household Department, 100 catties of glue paste were used for every square *chi* 尺 (1,024 square centimetres) of brass plate. With each melting 2 taels' worth would wear off. For every catty of glue paste, 11 taels of rosin and 3 taels of sesame oil were used.[48]

When the desired shape and ornamentation of the object was achieved, an object's surface would be burnished and polished, especially on the carved parts. It would first be filed, then be treated with sand-cloth of various grits, followed by fine sandpaper, and subsequently by charcoal made from tilla wood mixed with clean water. For especially high-quality objects rosin paste would be used as an abrasive, which would be pressed and spread using a steel grater to achieve maximum smoothness and shine.[49]

In the next step welders would join the burnished parts together. Borax flux was pressed into the weld to melt the fusible metal so that it would stick to the refractory metal. For every *chi* of welding seam, one catty of white charcoal, 3 *qian* of borax and 2 *qian* of oxidized silver were mixed to make borax flux.[50] For the actual welding itself, every *chi* of welding seam would require another 8 *qian* of welding agent, 3 *qian* of borax, and 80% of a smith's working day.[51] Another variation of a welding agent recipe can be

46 Yuan Kaizheng 袁凱錚, 'Shilun cangchuan fojiao tongfoxiang waibu tezheng yu qi zhizuo gongyi 試論藏傳佛教銅佛像外部特徵與其製作工藝', *Xibei minzu daxue xuebao* (Zhexue shehui kexue ban) 西北民族大學學報 (哲學社會科學版), vol. 5 (2009): 82-89.

47 Yuan Kaizheng 袁凱錚, 'Shixi cangzu liangzhong chuantong zhuzao gongyi de cunzai---you chuantong tong foxiang zhizuo yinfa de sikao 試析藏族兩種傳統鑄造工藝的存在—由傳統銅佛像製作引發的思考', *Zhongguo cangxue* 中國藏學 (March 2012): 175-185. See especially the photographs on page 181 showing the application of the rosin paste to the back of the copper object and the shaping and carving of ornaments into the surface of the object.

48 Wu Shan, ed., *Zhongguo gongyi meishu dacidian*, 1021.

49 Yanagisawa Koji 柳澤光治 and Jo Ryoku Kinzo 定力金藏, trans. Xisheng 西生, 'Dianqi dujin gongye 電器鍍金工業', *Hebei gongshang yuebao* 河北工商月報, vol. 1, no. 9 (1929): 181-190.

50 *Qingdai gongyuan zeli huibian*, vol. 5, 447.

51 Ibid., vol. 18, 374-375.

found in the 'Precedents and Regulations of the Brass Unit of the Ceramic Vault of the Inner Grand Storage Office of the Old Summer Palace' (*Yuanming yuan neigong guangchusi ciqiku tongzuo zeli* 園明園內工廣儲司磁器庫銅作則例), which asserts that, for every *chi* of welding seam, one catty of white ash, 2 *qian* of silver welding agent and 2 *qian* of borax were used.[52]

After polishing the object's surface, a gilding stick was prepared. This stick played an important role in the key step in fire gilding, as it was used to ladle the gold paste onto an object and spread it evenly over it. The process started with choosing a copper stick of the right size for the artefact. The front end of the stick would be flattened and raised slightly into the shape of a small shovel. After cleaning and polishing the stick further, a broth made from smoked *Prunus mume* fruits would be applied to its front end and the shovel subsequently submerged in mercury. This process of applying broth and submersion in mercury was repeated a few times until the front end of the stick had accumulated sufficient mercury.[53]

Subsequently, gold threads were placed in a stainless-steel crucible and heated until they turned red, that is, to about 700-800° Celsius. Mercury was then quickly poured into the crucible and mixed with a charcoal stick. The archival documents of the Imperial Gilding Works show that melting 3 taels of gold required one mercury-pouring jar and a whole day's work by a gold-melting worker.[54] Stirring and slowly cooling the substances in the crucible formed a gold amalgam (Au_2Hg), an intermetallic compound. At 122° Celsius the gold paste would consist of 18.66% of Au_2Hg and 81.34% of mercury.[55] A metal technologist working today, Liu Wanhang, explains that the ratio of gold to mercury in weight should be three to eight. Nevertheless, the ratio used by the Imperial Household Department was only one to seven.

The evenly-combined gold paste was then applied on the prepared metal objects using the gilding stick. The objects were subsequently heated to evaporate the mercury, so that only the gold or gold amalgam remained. Heating was another essential step in the process: at room temperature mercury would evaporate too slowly and no diffusion bond would be formed between the substrate and the gold.[56] A higher temperature would mean that a lower percentage of mercury would remain in the gilding – and therefore a higher percentage of gold – but if the temperature went above

52 *Qingdai gongyuan zeli huibian*, vol. 18, 327.

53 Liu Wanhang, *Jinyin zhuangshi yishu*, 26-28.

54 *Qingdai gongyuan zeli huibian*, vol. 5, 417.

55 Wu Yuankang and Chu Rongbang, 'Liudu', 7.

56 Kilian Anheuser, 'Cold and hot mercury gilding of metalwork in Antiquity', *The Bulletin of the Metals Museum*, vol. 26 (1996): 48-52.

350° Celsius a black copper-oxide layer would form beneath the gilding layer, the thickness of which would increase with heating temperature and duration. That layer would eventually cause the gilding to flake off.[57] Thus, during the evaporation process, the heating was stopped for a short time once white smoke appeared. Subsequently, a brush made from coarse animal hair was used to pat the surface and flatten the gilding. During heating, goldsmiths also had to wipe off the condensation from vaporized mercury from the surface with a cotton cloth. This process encouraged better bonding between the substrate and gold. The gilding would appear after much of the mercury had evaporated.[58]

When the gold amalgam had changed from a whitish-grey to a dull yellow, the object had been sufficiently heated and was ready for a final burnishing and polishing. Having lost more than two-thirds of its weight through the mercury evaporation, the gilding would have a matte appearance and be porous and uneven. An agate or steel burnishing tool was used to create a smooth, shiny surface. The burnished surface was then gently etched with pastes containing chlorides, nitrates, or sulphates to improve its appearance.[59] The 16th-century Italian goldsmith and sculptor Benvenuto Cellini also suggested the use of corrosive agents such as dilute nitric acid.[60] In China, a broth made from smoked *Prunus mume* fruits, soap water, and plain water were used instead to clean the gilded artefacts.[61] A manual used by artisans in the Imperial Workshops, *The Method of Burnishing Copper-alloy Buddhist Statues with Plum Wash* (Foxiang tongtai shenjin meixi jianxin fa 佛像銅胎滲金梅洗見新法),[62] stipulated that 2 *qian* of alkali and 2.5 *qian* of smoked *Prunus mume* fruit were required for every *chi* of gilded

57 Anheuser, 'Archaeotechnology', 58-62.

58 The Imperial Workshops used black charcoal, white charcoal and coking coal as fuel for this process, see *Qing gong Neiwufu Zaobanchu dang'an zonghui*, vol. 33, 687-688 (Guangmu zuo 廣木作, 'Canton wood Workshop'; QL35/1) and ibid., vol. 47, 468-469 (Zhuluchu; QL49/5). Liu Wanhang proposed the use of charcoal, while Wu Yuankang suggested high-quality charcoal or coke, see Liu Wanhang, *Jinyin zhuangshi yishu*, 26-28; Wu Yuankang and Chu Rongbang, 'Liudu', 7.

59 Martin Chapman, 'Techniques of Mercury Gilding in the Eighteenth Century', in *Ancient and Historic Metals: Conservation and Scientific Research,* eds. D. A. Scott, J. Podany, and B. B. Considine (Marina del Rey, Ca.: Getty Conservation Institute – Virtual Library, 1994), 229-238.

60 Benvenuto Cellini, *Due trattati, uno intorno alle otto principali arti dell'oreficeria, l'altro in materia dell'arte della scultura* (Fiorenza: Per Valente Panizzii & Marco Peri, 1568), chapter 26.

61 Liu Wanhang, *Jinyin zhuangshi yishu*, 27.

62 The document is dated QL33/5/4 (18 June 1768), see the facsimile in *Qing gong Neiwufu Zaobanchu dang'an zonghui*, vol. 31, 439-440 (Jinyu zuo 金玉作, 'Goldwork and Stonework Workshop'). See also Li Dou's 李斗 (1749-1817) *Yangzhou huafang lu* 揚州畫舫錄, ann. Wang Beiping 汪北平 and Tu Yugong 涂雨公 (Beijing: Zhonghua shuju, 1961), *juan* 4, 98.

surface produced, and for 'plum-washing' a third of one artisan's (*meixi jiang* 梅洗匠) working day needed to be calculated into the production costs for every three *chi*.[63]

The two case studies of fire gilding projects and their burnishing and alloying technologies introduced below will reveal how studies of the histories of China's economy and technology can benefit by scrutinizing the copper-related archival evidence from the Imperial Workshops of the Qing court. By accounting for deposits and withdrawals of gold and copper and the additional costs for their use in prestigious projects, these resplendent materials reflect the workings of the palace machine.

(A) The gilded roofs of the Dugang Hall of the Putuo Zongcheng Temple in Chengde

In 1776, a total of 4,838.76 taels of gold were used to gild the roof of the new Dugang Hall of the Putuo Zongcheng Temple 普陀宗乘之廟都罡殿 in Chengde. All the visible parts of the 4,400 pieces of scale-shaped roof tiles for the upper and lower eaves, 532 roof ridge tiles, 616 pieces of decorative end tiles for the rain pipes (*ruyi dishui* 如意滴水), and the decorative monster and animal figures were gilded (see Figure 5.2). In addition, remaking the rain pipes on the roof required 3,142.11 catties of copper as well as 8,678.63 taels of silver for materials and labour costs.[64] Qianlong spent more money on constructing the Putuo Zongcheng Temple at the Qing imperial summer residence than on any other Tibetan Buddhist temple near Beijing, despite its location in the north, beyond the Great Wall. The Qianlong emperor's annual travels to Chengde had established a smooth working connection between centre and periphery of the Qing empire that also allowed for easy transfer of gilding artisans back and forth between workshops in the Forbidden City and the Chengde construction sites. Upon completion, this magnificent temple was comparable to the Potala Palace in Tibet and attracted Mongol princes to make pilgrimages to Chengde.

Also in 1776, a pagoda made from imported dark *zitan* 紫檀 wood was newly built at the South Pavilion of the Putuo Zongcheng Temple complex. Inside the pagoda were 2,160 gilded brass *Amitabha* statues that were 3 *cun* high, 2.3 *cun* wide, and 1.3 *cun* deep. Gilding these statues used up 133 taels of pure gold, 1,869 catties and 12 taels of copper bars, as well as 4,844.11

63 *Qingdai gongyuan zeli huibian*, vol. 18, 239-240.

64 *Qianlong chao Neiwufu zouxiao dang*, vol. 344 (QL41/11/28 = 28 Dec. 1776).

Figure 5.2 The gilded roof of the Dugang Hall of the Putuo Zongcheng Temple, Chengde

Photos taken by the authors, 2008

taels of silver that were needed to pay the workers. In the following year these objects were again gilded twice more, using another 133.6 taels of gold leaf.[65] The many mandalas and pagodas, together with the numerous offerings, European objects and treasures inside the temple complex were extraordinary sights. According to an inventory list compiled in 1789, there were 1,214 decorative objects altogether in the Putuo Zongcheng Temple complex.[66]

In order to enhance their lustre, the gilded statues had to be burnished with a plum wash solution. The standard procedures for this were outlined in the Imperial Workshops' manual, 'The Method of Burnishing Copper-alloy Buddhist Statues with Plum Wash', mentioned above. The method was also used outside the court. The 18th-century writer Li Dou described a similar practice for private Chinese workshops: 'Plum wash artisans burnish gilded brass Buddhist statues with plum wash, using alkali, smoked plum fruits, firewood, and unrefined white cloth'.[67] Safflower, a common Tibetan dye, and gardenias were also used for burnishing, again both being used inside and outside of the court. Another common practice was to add fragrances such as rues, cloves, nutmegs, and alpinia fruits into the solution in what might have been an effort to symbolize the 'five scents' used in the Tibetan

65 *Qianlong chao Neiwufu yinku yongxiang yuezhe dang* 乾隆朝內務府銀庫用項月摺檔 (Monthly expenditure reports of the Imperial Household Department's Bullion Vault from the Qianlong reign), QL41/9, 12 (The First Historical Archives of China, Beijing); *Qianlong chao Neiwufu zouxiao dang*, vol. 346 (QL42/6/15).

66 Xing Yongfu 邢永福 et al., eds., *Qing gong Rehe dangan* 清宮熱河檔案 (Beijing: Zhongguo dang'an chubanshe, 2003), vol. 6, 337-372.

67 Li Dou, *Yangzhou huafang lu*, *juan* 4, 98.

ceremony of installing Tibetan or Sanskrit mantras or scriptures into a statue (*zhuangzang* 裝臓).[68] The use of Tibetan ingredients and burnishing techniques suggests that some aspects of Tibetan practices had already been adopted for gilding the *Amitabha* statues in the Putuo Zongcheng Temple.[69]

According to modern scholar Luo Wenhua, an expert in Tibetan Buddhist art at the Palace Museum in Beijing, the Qing court did not formally adopt the Tibetan fire gilding method until 1781 when the Imperial Workshops sent an inquiry to Zhongba Hotogtu (Chinese: *Zhongba Hutuketu* 仲巴胡土克圖), a lama residing at the Lhakang Serbo Temple (also called West Gelug or *Xihuang* 西黃 Temple), north of the outer city walls of Beijing, regarding gilding techniques. Hotogtu, who had often served as an art consultant to the Qianlong emperor, replied that gilding was performed by Nepalese (*Baerbu* 巴爾布) artisans using the highest-quality pure gold for plating. If there was any flaw in the plating, a second gilding could be performed and the gilded surface could be burnished with a broth made from Indian madder to achieve a darker golden colour. As a rule, four *li* 釐 of gold were used for one *cun* of surface.[70] Workshop artisans were ordered to learn this technique. A workshop record from 1781 shows that two gilded silver pots were produced in Beijing and shown to the emperor, who ordered them to be gilded for a second time and be burnished further with Indian madder broth. The same treatment was also ordered for a statue of Zongkaba 宗喀巴, a great Tibetan Buddhist teacher.[71] By the late Qianlong reign, the Tibetan techniques of multi-gilding and burnishing with Indian madder water had become a familiar and standard procedure at the Qing court.

(B) Buddhist ritual objects in the Temple of Great Benevolence (Hongren si 弘仁寺; *Sandalwood Temple*) in Beijing

The Temple of Great Benevolence stood at the western bank of Lake Taiye in the imperial park, to the west of the Forbidden Palace (see Chapter Seven

68 Gongbuchabu 工布查布 (Qing 清) trans., *Zaoxiang liangdu jing* 造像量度經 (rep. Taipei: Taiwan yinjingchu, 1956), 49. The five scents named in this text are: Indian sandalwood, agarwood, nutmeg, Bahasa keruing, and tulips. This Chinese translation of this originally Tibetan text was commissioned in the Qianlong reign.

69 S. Wells Williams, *The Chinese Commercial Guide, Containing Treaties, Tariffs, Regulations, Tables, etc., Useful in the Trade to China & Eastern Asia: With an Appendix of Sailing Directions for Those Seas and Coasts* (reprint, Taipei: Ch'eng-Wen Publishing, 1966), 84-85.

70 Luo Wenhua 羅文華, *Longpao yu jiasha — Qing gong zangchuan fojiao wenhua kaocha* 龍袍與袈裟—清宮藏傳佛教文化考察 (Beijing: Zijincheng chubanshe, 2005), vol. 2, 451.

71 *Qing gong Neiwufu Zaobanchu dang'an zonghui*, vol. 45, 2-3, 109-113 (QL46/1 = Jan. or Feb. 1781).

for this park and the economic use of its lake) at the site of the Hall of Clear Scent (*Qingfu dian* 清馥殿) of Ming times. When in 1665 the building was transformed into a Buddhist temple, a historic sandalwood Buddha statue, whose history is claimed to go back to the year 401 when it was brought to China from Kucha (in present-day Aksu prefecture in Xinjiang), was moved to its main hall. Its previous location had been the Temple of Eagle Peak (*Jiufeng si* 鷲峰寺) in the Xidan 西單 neighbourhood of Beijing. The temple thus acquired its moniker, Sandalwood Temple (*Zhantan si* 栴檀寺).

The sandalwood Buddha was an important artefact, said to have been brought to China by a monk called Kumarajiva, the great translator of Buddhist sutras into Chinese. As time went by, the statue acquired a legendary status that was both political and religious in nature. In 1289, for example, the Mongol ruler Kublai Khan installed the sandalwood statue inside the Temple of Forever Peace (*Dashengshou Wan'ansi* 大聖壽萬安寺) in Beijing and ordered the imperial preceptor (*dishi* 帝師) and monks from the Western Regions to perform twenty sets of Buddhist rituals in its honour.[72] When the Qing established its lama administration at the Sandalwood Temple, it was following this Mongol precedent. It became customary for Mongol princes and Tibetan lama envoys to worship at the temple during their annual visits to Beijing.

As it was clearly not an option to gild the fabled sandalwood Buddha statue, the Qianlong court sought other means to enhance the lustre of the Sandalwood Temple. In 1767, Changzhu 常柱 and others serving at that time as directors (*langzhong* 郎中) in charge of the Six Vaults of the Grand Storage Office reported that 48 small gilded copper-alloy musical bells and 25 gilded bells were to be made to be used in ceremonies in the Sandalwood Temple using first-grade pure gold.[73] Later, in 1774, the Imperial Workshops made several additional items for use at the temple, which included not only costumes for a Tibetan ritual dance (*tiao buzha* 跳布扎), but also 98 small gilded copper-alloy music bells and 120 jingle bells, which consumed 8.57 *qian* of first-grade gold.[74]

72 Cheng Jufu 程鉅夫, 'Zhantan ruixiang ji 栴檀瑞像記 (Note on the Sandalwood Buddha)', in *Xuelou ji* 雪樓集 (rep. Shanghai: Shanghai shuju, 1994), *juan* 9, 16-20. See also the rubbing of a stone inscription with an illustration of the Buddha made by Qin Yingrui 秦應瑞 et al. in 1586 that is stored at the Fu Sinian Library of the Taiwan Academica Sinica, the 'Zhantan ruixiang bingti ji' 栴檀瑞像并題記, and Xue Zhiheng's 薛之珩 note from 1922: 'Jingshi xiao zhantansi bei ji 京師小栴檀寺碑記'. The sandalwood Buddha was first mentioned in 'The Event of Udayana's Making an Auspicious Sandalwood Statue' (*Yuotianwang suozao zhantan ruixiang liji* 優填王所造栴檀釋迦瑞像歷記), written by monk Shiming 十明 from the Kaiyuan Temple in Jiangdu in 932. The text is reproduced in *Dai Nihon Bukkyō zensho* 大日本佛教全書 (Tōkyō: Daihōrinkaku 大法輪閣, 2007), vol. 114, 309-320.

73 *Qianlong chao Neiwufu yinku yongxiang yuezhe dang*, QL 32 (1767).

74 *Qianlong chao Neiwufu zouxiao dang*, vol. 337 (QL40/10/20 = 12 Nov. 1775).

But it was not only the outside lustre of the gilded bells that show how materials and knowledge from different cultural spheres were aggregated to serve as objects of Qing splendour. The recipes the Imperial Workshops used to cast the resonating bodies of these bells also deserve scrutiny. According to Luo Wenhua, in 1771 both the Changkya 章嘉 Rölpé Dorjé 業西丹畢蓉梅 (1717-1786), who was the Qianlong emperor's preceptor and head of the Gelug lineage of Tibetan Buddhism, and the Jesuit priest Michel Benoist (1715-1774; Chinese name Jiang Youren 蔣友仁) were asked to write down the formulas they had been using to cast bells. The formula eventually used by the Qing court, in Luo's estimation, was a combination of Chinese, Tibetan, and Western formulas.[75]

Table 5.1 Changkya Rölpé Dorjé's formula

Material	Quantity	Percentage
Japanese copper bars	13.87 taels	69.35%
Xian tin (杴錫)	5.95 taels	29.75%
Gold	6 *qian*	0.3%
Silver	6 *qian*	0.3%
Precious stones of three colours (三色寶石)	6 *qian*	0.3%

Table 5.2 Benoist's formula

	Material	Quantity	Percentage
Benoist's formula I	Copper	1 catty (16 taels)	81.97%
	Tin	3.2 taels	16.39%
	Antimony	3.2 *qian*	1.34%
Benoist's formula II	Copper	1 catty (16 taels)	78.74%
	Tin	4 taels	19.68%
	Antimony	3.2 *qian*	1.57%

Benoist especially recommended the addition of antimony, claiming that this would produce bells with a crisper sound. But actually the percentage of antimony he suggested was comparatively low, and his formulas also contained much less tin than that of the Changkya master. After a set of casting tests using different amounts of tin, Qianlong decided that bells with a higher

75 Luo Wenhua, *Longpao yu jiasha*, vol. 2, 401. For the data in Table 5.1 and 5.2 see *Qing gong Neiwufu Zaobanchu dang'an zonghui*, vol. 34, 527-536 (QL36/9 = Oct.or Nov. 1771), and ibid., vol. 48, 143-144 (QL50/2 =March or April 1785).

percentage of tin produced a better sound. He thus ordered future casting to follow the Tibetan formulas which had a high tin content.[76] Luo argued that the emperor adopted Benoist's formulas to produce the new set of bells.[77]

But the Qianlong emperor was ever inventive. In 1785, the eunuch Changning 常寧 conveyed the following imperial order to the Imperial Workshops: 'Take out the damaged self-chiming clocks and remove their chimes to reclaim the copper. Ask Shuwen 舒文 to use the reclaimed copper to make one large- and one medium-sized Tibetan bell according to the old formula'. Although the order did not specify what the 'old formula' was, from the order's context we can construe it to be the antimony-rich one advocated by Benoist. After evaluating the result, the emperor ordered the following: 'The bells produce only a mediocre sound, and thus shall be given to the lamas at Zhongzheng Hall (*Zhongzheng dian*中正殿) to be recast into ten medium-sized Tibetan bells according to the old [Tibetan] formula, with the addition of gold, silver, and precious stones of three colours'.[78] This incident suggests that in Qianlong's estimation, the Tibetan formula produced bells that sounded better. Table 5.3 provides a summary of the alloy elements in the surface of standard copper bells and the so-called Qianlong *jin'gang* copper bells (*jin'gang tongling* 金剛銅鈴, see Figure 5.3) using modern composition analysis method. A comparison with the formulas listed in Tables 5.1 and 5.2 above shows that, whereas the copper bells seemed to have used the Tibetan recipe, the *jin'gang* copper bells used a recipe closer to the Benoist's Western recipe but no traces of antimony were detected.

Table 5.3 Results of a modern alloy composition analysis on the surface of standard copper bells and *jin'gang* copper bells from the Qianlong period (shown in wt.%, Bal. = balance)

Item	Part	Sn	Zn	Au	Pb	Fe	Cr	Ag	Cu
Copper bell	body	24.01	0.19	–	0.38	0.11	0.03	–	Bal.
	bell handle	–	25.33	–	0.85	0.19	0.06	–	Bal.
Qianlong *jin'gang* copper bell	body	15.42	–	–	1.62	0.12	0.04	–	Bal.
	bell handle (gilded)	–	0.87	97.68	–	–	–	0.16	Bal.
	clapper	–	–	–	–	100.00	–	–	–

Data provided by Dr Donghe Chen 陳東和 from the Science Lab at the National Palace Museum, Taipei (台北故宮科技室) (personal communications, 2018)

76 *Qing gong Neiwufu Zaobanchu dang'an zonghui*, vol. 34, 527-536 (QL36/12 = Jan. or Feb. 1772).

77 Luo Wenhua, *Longpao yu jiasha*, vol. 2, 400-401.

78 *Qing gong Neiwufu Zaobanchu dang'an zonghui*, vol. 48, 143-144 (QL50/2 = March or April 1785).

Figure 5.3 Qianlong *jin'gang* copper bell

Gilt bronze bell with *vajra* handle made in the 18th century, Qing Dynasty, China. The lotus pattern at the centre of the bell was formed by casting. The head of the handle is connected to a gilt *vajra* with the depiction of a coronated Buddha with a long face, a wide forehead and a straight nose. The details of the crown were finely carved. During the practice, the bell was shaken for spiritual vigilance and joy. Reproduced with permission from National Palace Museum, Taipei.

Conclusion

During the Qing dynasty, the workflow, materials, and artisans of the Imperial Household Department were coordinated by a systematic, codified form of management that was very similar to tasks of the civil government. Construction offices were set up for specific projects and each would issue 'Regulations and Precedents' detailing the amount and price of raw materials, as well as the number and salaries of artisans required to fulfil the commission. These regulations formed the basis of regular reports and facilitated the monitoring of project expenses and progress.

Court artisans' positions in the Imperial Household Department were hereditary, receiving a salary in addition to daily rations of food items. The department also hired a large number of contract artisans who were appointed by the chief contract artisan, given written contracts and a pass to enter the palace precinct. Contract artisans formed guilds and unions, as evidenced by stele inscriptions. For example, artisans from Shanxi formed a Furnace Commerce Guild (*Lufang shanghui* 爐房商會) in Beijing, which was involved in refining gold and silver. Further investigation into how the government managed such artisan organizations outside the Imperial Workshops system is needed.

The technology of fire gilding and the forging and casting of copper alloy bodies were fields of expertise in which Qing artisans were particularly innovative. Before the Qing, fire gilding was primarily used on silverware. With the shift to gild copper alloys in the Qing, artisans learned to master gilding irregularly-shaped objects and also solved the issue of flaky gilding layers. The gilding technique was further standardized under the influence of methods introduced from Tibet. To learn better techniques, the emperors invited lamas from Tibet to come to Beijing to teach them how to manufacture numerous Buddhist statues and ritual objects using Tibetan techniques. Two techniques stand out. Firstly, by introducing the *sisa* technique from Tibet, the Qing court could manufacture Buddhist statues in larger sizes. Secondly, the smoked plum broth which had originally been used for burnishing, was replaced by broth made from safflowers or Indian madders that also gave the golden surface a redder appearance. The Qing court also adopted a multi-layered gilding technique from Tibet to achieve smoother surfaces but resulted in a higher cost of investment in the gold used for each object. To conclude, the Qing court's craft in making Buddhist statues and ritual objects was the result of considerable experimentation that involved methods and formulas from divergent knowledge cultures, paired with an enormous need for gold and copper, which the court had to secure sufficient sources of. Ultimately, it was the palace machine that assembled skilled labourers and they, in turn, combined techniques, formulas, and precious metals to realize the Qianlong emperor's numerous resplendent projects.

Translated from Chinese by Ju-Yi Chou; edited by Dorothy Ko

About the Authors

Te-Cheng Su 蘇德徵, Ph.D. (Imperial College London, 2019) is an Assistant Professor at the Department of Materials Science and Engineering, National Taiwan University. He specializes in manufacturing technologies and is the co-author of 'Brass consumption in the Qing Empire' in *Living the Good Life: Consumption in the Qing and Ottoman Empires of the Eighteenth Century*.

HUI-MIN LAI 賴惠敏, Ph.D. (National Taiwan University, 1988) is a Research Fellow at the Institute of Modern History, Academia Sinica, in Taiwan. For many years, she has been studying the economic and cultural history of the Qing Dynasty. She is the author of a monograph on *Emperor Qianlong's Purse: Imperial Household Finance in Eighteenth Century China* (Qianlong huangdi de hebao 乾隆皇帝的荷包).

6 Transporting Jade

Objects, Ecology, and Local Bureaucracy in Qing Xinjiang

Yulian Wu

Abstract

This chapter examines jade shipments relating to the Gao-pu case to reconstruct the complex process of transporting from Xinjiang to Beijing in the late eighteenth century. The relay stations in Xinjiang and post stations in northern China served the palace as crucial links to this new territory. Focusing on the route within Xinjiang, this chapter argues that Xinjiang officials' management of jade delivery was developed closely around local ecology: they worked with animal and human resources of the local communities which were familiar with the landscape and knowledgeable about various ecological and environmental elements. Constructing transport infrastructure, therefore, provided a mechanism through which court officials localized and inculcated Qing ruling power into the material and natural world of Xinjiang.

Keywords: jade, Xinjiang, transport infrastructure, ecology, bureaucracy

On 25 December 1780 (QL 45/11/30), a transport troop delivering thirteen pieces of large jade stone that had been newly mined from the Mi'erdai Mountains 密爾岱山 in Xinjiang arrived at Neiqiu 内邱 county in Zhili. This troop had travelled for almost one year from Yarkand before finally reaching the Zhili area, close to their final destination, the Qing capital of Beijing. However, an accident occurred in Neiqiu. The draught animals pulling a four-wheeled wagon slipped down a slope, causing the driver to lose control, breaking an axle and flipping the wagon. This wagon was carrying two jade boulders, weighing 2,300 catties (*jin* 斤) and 4,000 catties (1,149.4 kilograms and 1,999.9 kilograms respectively). The 2,300-catty piece broke into two big chunks and a pile of smaller fragments weighing

Siebert, Martina, Kai Jun Chen, and Dorothy Ko (eds), *Making the Palace Machine Work: Mobilizing People, Objects, and Nature in the Qing Empire.* Taylor & Francis Group, 2021
DOI: 10.5117/9789463720359_CH06

20 catties. The governor-general of Zhili (*Zhili zongdu* 直隸總督) Yuan Shoutong 袁守侗 (1723-1783) immediately ordered an investigation into this accident and proposed punishing local officials for their failure to pave the road in advance.[1]

This episode, which was reported to the Qianlong emperor (r. 1736-1796, unofficially until 1799) in 1781, describes one of several failed efforts to transport jade from the empire's new frontier of Xinjiang to the capital of Beijing.[2] As these accidents indicate, transporting jade over this long distance through difficult territory was a complicated task which involved a number of logistical challenges. The jade had to be carefully wrapped to prevent cracking, and specially manufactured wagons were needed in order to carry the heavy but fragile stones over thousands of miles. In addition, as road conditions varied, with uneven patches and unexpected slips, skilled drivers were essential to steer the draught animals through the many difficult situations. Furthermore, local officials along the transportation route needed to adapt their management, as they were held responsible for the varied logistical and infrastructural conditions on the road. During the 18th century an unprecedented amount of Xinjiang nephrite jade (Chinese: *yu* 玉; Manchu: *gu*), a type of hard semiprecious stone,[3] was delivered from the raw jade's production sites to the Imperial Household Department in Beijing. Local administrative management, along with the transportation facilities, were the backbone that enabled the Qing court to extend its reach all the way to Xinjiang, the so-called 'new frontier', both to acquire

1 Yuan Shoutong's memorial on 12 Jan. 1781 (QL 45/12/18), *Zhupi zouzhe* 朱批奏折 (Imperial Memorials with Red Rescripts), no. 04-01-12-0195-002 (The First Historical Archives of China, Beijing). To facilitate navigation in the archival sources, dates are given according to the Chinese lunar calendar in the format reign year/lunar month/day, e.g. QL 26/7/4. A small letter 'r' before the month indicates an intercalary month (*runyue* 閏月). Western equivalents are given in the order day, month, year, e.g. memorial on 3 Aug. 1761.

2 This was not the only incident. One month later, the staff supervising the shipment discovered that the leather wrapping of a package was damaged and the jade inside had broken in half. For this accident, the transport staff suspected that the rough roads from Tongguan in Shaanxi province might have cracked the jade. Yuan Shoutong's *ziwen* 咨文 (official despatch) on 8 Feb. 1781 (QL 46/1/16), *Junjichudang zhejian* 軍機處檔摺件 (Archives of the Grand Council), no. 029526 (National Palace Museum, Taipei).

3 There are two mineralogical different forms of jade: nephrite and jadeite, identified in Chinese as *yu* 玉 and *feicui* 翡翠. For scientific information on jade, see Jill Walker, 'Jade: A Special Gemstone', in *Jade*, ed. Roger Keverne (Boston: Springer, 1991), 19-41. In Xinjiang, most of the nephrite jade pieces were picked out of shallow river beds by miners in Khotan and Yarkand, while larger boulders, such as the ones from the Mi'erdai Mountains in Yarkand, were mined from quarries.

its material resources and to display Qing ruling power, both of which comprised essential aspects of the functioning and materiality of the palace machine.

The Qianlong era marked the apex of jade production and consumption in Chinese history. The Qing court's extensive access to jade was a direct consequence of the Manchus' occupation of the Inner Asian territory of Xinjiang. The areas Yarkand and Khotan had especially rich deposits of nephrite jade, both in rivers and in mountain quarries. From 1759, when the Qing troops first occupied this northwestern frontier, the court immediately began mining. From 1760 to 1799, vast amounts of jade were quarried and transported overland from Xinjiang to Beijing. After its arrival at the palace, the Imperial Household Department arranged to have these stones crafted into beautiful objects. The number, size and quality of existing jade objects from the Qianlong era demonstrate the emperor's unfailing zeal for jades and his passion for large jade sculpture, which were the largest ever produced in China.[4] While the material, aesthetic and representational roles these jades played in the palace machine still appear discernible for us today in the form of the remaining objects, we know very little about how the jade stones actually travelled from the empire's new frontier to the capital of Beijing – an incredibly long route across a rugged landscape.[5] This chapter reconstructs the practical and logistical side of the long-distance transport procedures during the High Qing period, as well as the political function and display of power and control associated with the successful conveyance of this rare, state-monopolized material.

The famous Gao-pu 高樸 (?-1778) incident of illegal private jade mining and selling in 1778 will serve here as a particularly revealing case, as it marked a turning point in the court's dealing with Xinjiang jades and their transportation. Following the detection of Gao-pu's illegal

4 The most representative example of this is the jade sculpture named 'Great Yu Regulating the Waters' (*Da Yu zhishui tu* 大禹治水圖), which is currently displayed in the Forbidden City. This sculpture is 224cm tall, 96cm wide, and weighs about 11,684 lbs. See introduction on the Palace Museum website: https://www.dpm.org.cn/collection/jade/228206.html (accessed 15 Feb. 2021)

5 Some Chinese scholars have provided a general picture of jade mining procedures in Xinjiang at this time. See Cai Jiayi 蔡家藝, 'Qingdai Xinjiang yushi de kaicai yu shuchu 清代新疆玉石的開採與輸出', *Zhongguo bianjiang shi di yanjiu*, vol. 20, no. 3 (2010): 121-130. Their research mainly uses sources from Qing official histories, gazetteers, and elites' writings. The sources from memorials and edicts, particularly those written in Manchu, have not been incorporated. Peter Perdue also briefly discusses the Chinese state's capacity for moving bulk goods over long distances. See Peter Perdue, *China Marches West: The Qing Conquest of Central Eurasia* (Cambridge, MA: Harvard University Press, 2005), 539-540.

activities, an unprecedented large number of already-mined jade stones had to be carried from Yarkand to the palace in Beijing between 1779 and 1781, emphasizing the Qianlong emperor's claim that jade was a 'court-monopolized object' (*guanwu* 官物 or *guanyu* 官玉).[6] Moreover, the large scale of this shipment provides a perfect case for historians to investigate how the shipping loads and the material qualities of jade itself, especially its weight, size and fragility, posed a series of challenges to the transport officials.

While I provide a general picture of the route and methods for the entire journey, my research has led me to focus on the officials in southern Xinjiang who managed the transportation on the first segment of the jades' long journey, namely from the original production site in the Mi'erdai Mountains to the Jiayu Gate (*Jiayu guan* 嘉裕關) in Gansu province. By analysing the two types of Gao-pu jade shipments Xinjiang officials needed to carry out – one for a large quantity of jade stones and the other for jade stones of enormous size – I demonstrate that many of these challenges were rooted in the natural environment of this new frontier and the Qing empire's unfamiliarity with it. Previous scholarship has convincingly demonstrated how ecological features impacted on Qing frontier policies in general.[7] This research further shows how Xinjiang officials' management of jade delivery was developed closely around Xinjiang's local ecology: Qing officials worked with animal and human resources from the local communities which were used to the landscape and knowledgeable about various ecological and environmental elements, such as climate, flora and livestock. Constructing a transport infrastructure in Xinjiang, in this context, provided a mechanism through which court officials localized and inculcated Qing ruling power into the material and natural world of southern Xinjiang. By examining how one particular native natural resource was conveyed within Xinjiang territory on its way to the Beijing palace, this chapter reveals the dynamic interconnections between Qing ruling power, jade as a court-monopolized material and Xinjiang as a new environmental challenge.

6 It is important to note that *guan* could also be translated as 'government' or 'official'. I translate it here as 'imperial court' because most of the jade stones from Xinjiang were delivered to the Imperial Household Department, not to the Board of Revenue (*Hubu* 户部), which indicates that jade was considered the emperor's personal property.

7 David Bello, *Across Forest, Steppe, and Mountain: Environment, Identity, and Empire in Qing China's Borderlands* (Cambridge: Cambridge University Press, 2015); Jonathan Schlesinger, *A World Trimmed with Fur: Wild Things, Pristine Places, and the Natural Fringes of Qing Rule* (Stanford, CA.: Stanford University Press, 2017).

'Transport Them All': Jade as Court-Monopolized Object (*guanyu*)

Immediately after its occupation of Xinjiang in 1759, the Qing court began quarrying jade. Until the Gao-pu case in 1778 the court had gradually increased its access to these precious stones and the amount mined. After 1778 the Qing court's policy on how to handle jade changed. While before 1778 only part of the mined jade had been delivered to the capital, after the Gao-pu case the Qianlong emperor ordered all jade to be first transported to the court. The emperor's new policy confirmed the status of jade as a 'court-monopolized object' (*guanyu*), reclaiming the emperor's exclusive ownership of this natural resource. At the same time, the Gao-pu case led to one of the largest jade shipments in Qing history, testifying (and testing) the ability of the transport infrastructure from the 'new frontier' to the capital.[8]

The Gao-pu case was one of the most famous and far-reaching corruption cases in Qianlong reign. Gao-pu, a trusted bondservant from the Bordered Yellow banner, was sent to Yarkand by the court to act as the grand minister superintendent of Yarkand (*Ye'erqiang banshi dachen* 葉爾羌辦事大臣) in 1776. Two years later, in 1778 (QL 43), a local beg[9] reported to Yunggui 永貴 (?-1783; Chinese in *pinyin*: Yonggui), who was at that time the grand minister consultant of Ush (*Wushi canzan dachen* 烏什參贊大臣), that Gao-pu had secretly sent 3,620 local Muslim labourers to the jade reserves at Mi'erdai Mountains to mine the precious stone,[10] which he had hidden and then sold to three Jiangnan merchants for personal profit. The Qianlong emperor was furious, as Gao-pu's behaviour had not only caused the loss of state revenue but could have also incited local revolts due to his forced conscription of local Muslim labourers. For these charges, in addition to others such as extorting gold and money from local begs, the Qianlong emperor immediately ordered Gao-pu to be publicly beheaded in Xinjiang.

8 This function appears to have been similar to herding South-Asian elephants to Beijing to serve as officials in palace rituals. See Chapter Nine in this volume.

9 Begs were local chief administrators of Xinjiang during the Qing. Hakim begs were in charge of civil administration of the Muslim communities in a given region. See L.J. Newby, 'The Begs of Xinjiang: between Two Worlds', *Bulletin of the School of Oriental and African Studies* no. 61.2 (1998): 278-297; and Wang Ke 'Between the "Ummah" and "China": The Qing Dynasty's Rule over Xinjiang Uyghur Society', *Journal of Intercultural Studies* (Kobe University), no. 48 (2017): 183-219.

10 The Mi'erdai Mountains contained rich jade reserves. Cen Zhongmian locates the Mi'erdai Mountains southwest of Yarkand. See Cen Zhongmian 岑仲勉, *Hanshu Xiyu zhuan dili jiaoshi* (shang) 漢書西域傳地理校釋（上）(Beijing: Zhonghua shuju, 1981), 86.

Gao-pu was executed on 25 November 1778, only one and half months after Yunggui reported the case to the emperor.[11]

While the court continued to investigate the details of how had Gao-pu mined and traded jade, and where all his jade had gone, the local officials in Xinjiang were facing another problem: how should they deal with the jade that Gao-pu had already mined but had not yet sold? Those jade stones were all confiscated by the court. At the time of his execution, it was recorded that 107,763.9 catties (approximately 53,881.3 kilograms) of jade had already been taken from the mines to Yarkand, and an additional nine large jade stones were still sitting in the mountains, each weighing 1,000 to 10,000 catties.[12]

In his report in December 1778, the court official Yunggui first proposed dividing these confiscated jade stones into two parts as usual: the first part, amounting to 31,212 catties (15,606 kilograms), was considered high-quality 'tribute jade' (*gongyu* 貢玉) that should be transported to the court, while the rest of the much larger portion of 76,551.9 catties (38,275.9 kilograms) minor jade stones should be handled in Xinjiang. He suggested that sixty percent of those should be 'commissioned the sale to merchants' (*zhaoshang chengmai* 招商承買), and the other forty percent should be distributed among the local Muslim labourers as a reward for their hard work in mining and transporting the jade.[13] The Qianlong emperor, however, seems to have struggled with how best to deal with this – and future – sales of Xinjiang jades. Even before Yunggui submitted this memorial, the emperor had expressed his concern that officials might deliberately submit lower-quality jade stones to the court so that they could sell higher-quality ones to merchants. In order to prevent this, he ordered to 'forever ban the selling of jade to merchants for cash' and demanded that all low-quality jade should be thrown into the rivers.[14] This practice, similar to the treatment of inferior porcelain at the Porcelain Manufactory, reflected a desire to control the quality of imperial production as much as stamping out corruption.

11 For more details on this case, see Andrea McElderry, 'Frontier Commerce: An Incident of Smuggling', *The American Asian Review*, no. 5 (1987): 47-82; Lai Hui-min 賴惠敏, 'Cong Gao-pu an kan Qianlong chao de Neiwufu yu shangren 從高樸案看乾隆朝的內務府與商人', *Xin shixue* 新史學, vol. 13, no. 1 (2002): 71-133.

12 Yunggui and Maxing'a's 瑪興阿 memorials on 1 Dec. 1778 (QL 43/10/13), *Zhupi zouzhe*, no. 04-01-36-0092-018 and on 17 Dec. 1778 (QL 43/10/29), *Zhupi zouzhe*, no. 04-01-36-0092-019. The number and weight of these stones were also recorded in the inventory lists submitted to the emperor by the Grand Council. See documents 441 and 442, *Qianlong chao zhengban tanwu dang'an xuanbian* 乾隆朝懲辦貪污檔案選編 (Beijing: Zhonghua shuju, 1994), vol. 1, 932-933.

13 Yunggui and Maxing'a's memorial 1 Dec. 1778 (QL 43/10/13), *Zhupi zouzhe*, no. 04-01-36-0092-018.

14 Qianlong's edict on 17 Nov. 1778 (QL 43/9/29), *Junjichu dang'an* 軍機處檔案 (Documents of the Grand Council), no. 03-135-2-048 (The First Historical Archives of China, Beijing).

One month later, and in direct reply to Yunggui's suggestion, the emperor changed his instructions. On 11 December 1778 (QL 43/10/23), the emperor sent another edict to Yunggui, in which he ordered the officials to deliver all of the jade stones to the court, including those Yunggui had planned to throw into the river, because some dishonest people might still obtain the pebbles in the future.[15] Ten days later, the Qianlong emperor issued another edict, in which he stressed his disapproval of Yunggui's proposal to sell jade to merchants and further explained why all jade needed to be transported to Beijing instead. In the emperor's opinion, selling court-monopolized jade (*guan yu*) to merchants had been the fundamental reason for jade smuggling, as the significant price difference between buying jade from the government in Xinjiang and buying jade on the market had tempted both officials and merchants to indulge in that illegal trade.[16]

This shows how the Gao-pu case spurred the Qianlong emperor to reconsider, and finally change, his policy for handling jade in Xinjiang. Before 1778, newly-mined lower-quality jade could be sold to merchants and soldiers to make a profit for the court;[17] at that time the emperor had even had a more lenient attitude towards jade trading and generously 'allowed' small amounts of jade to slip through state control to profit petty merchants.[18] After the Gao-pu case, however, the emperor strictly dismissed all these secondary or illegal businesses. As Yunggui confirmed in his memorial, the 'new permanent policy' was for officials to submit all of the Yarkand and Khotan jades directly to the court.[19]

15 In the same edict, the emperor also pointed out that it would not be difficult to deliver these jade pebbles to Beijing because there would not be any large pieces among them. Qianlong's edict on 11 Dec. 1778 (QL 43/10/23), *Junjichu dang'an*, no. 03-135-2-063.

16 Qianlong's edict on 21 Dec. 1778 (QL 43/11/3), *Qing shilu* 清實錄 (Beijing: Zhonghua shuju, 1986), Gaozong 高宗, vol. 22, *juan* 1070, 17.

17 According to Yunggui's memorial on 10 Feb. 1779 (QL 43/12/24), in 1770 (QL 35), official Qicheng'e 期成額 had decreed that all of the newly mined jade except for the tributary jade could be sold to soldiers at a price of 1 *qian* 錢 per *jin*. *Junjichudang zhejian*, no. 022492.

18 As the emperor stated in another edict, he had long known that some merchants smuggled jade from Xinjiang and that was why beautiful large white jade objects were available in the Suzhou markets. However, because his empire contained abundant beautiful jade, he was content to 'take the advantage of this rich resource from nature to benefit the petty commoners' trading'. Qianlong's edict on 19 Dec. 1778 (QL 43/11/1), *Qing shilu*, Gaozong, vol. 22, *juan* 1070, 3-4.

19 See Yunggui's memorial with the emperor's comments on, dated 4 Feb. 1779 (QL 43/12/18). This document was originally written in Manchu but I have only found a translated copy. The Chinese translation of the quotation is '*yongjiu weili* 永久為例'. *Qianlong chao zhengban tanwu dang'an xuanbian*, vol. 1, 841-843. It is important to point out that the practice of selling jade to merchants with government licenses (*zhaopiao* 照票) continued throughout the rest of Qianlong

Through this new policy, the emperor declared and confirmed the status of jade as a 'natural resource' to be commandeered exclusively by the court. Despite the use of the term *guan* (government or official) or *gong* (tribute) when mentioning Xinjiang jade in edicts, the final destination of all transported jade was the Imperial Household Department. The involvement of the Imperial Household Department further confirmed the status of jade as a court-monopolized object.[20]

The new policy also directly led to the difficult task of transporting all of the jade across the country. This not only dramatically increased the volume of jade that needed to be transported from Xinjiang to Beijing, but meant that the shipments also often included several enormous pieces of jade,[21] both of which posed new challenges to local officials.

Court-monopolized Jade and State-Sponsored Transport Infrastructure

Although all jade stones were ultimately delivered to the Imperial Household Department, that final recipient was not directly involved in managing its carriage. It was the task of local Qing officials in Xinjiang and the inner provinces to take full responsibility over every aspect of the transportation process, including making schedules, arranging labour and materials, maintaining roads and implementing security systems. In other words, the bureaucratic structure and the physical infrastructure of the Qing state provided the managerial and material foundations upon which the transport mission was operated and, in so doing, enabled the palace machine to extend its reach to this distant new frontier.

reign. But, at least straight after the Gao-pu case, the emperor was determined to ban this practice. Very few scholars have studied the history and operation of this license system. See, for instance, Zhang Jianhong 張劍虹, 'Cong Gao-pu an kan Qingdai yushi kuangchan zhengce 從高樸案看清代玉石礦產政策', *Zhongguo kuangye* 中國礦業, vol. 25, no. 10 (2016): 169. More research needs to be done on this subject.

20 The question of how to define the nature of jade – whether as '*guanwu*' or '*gongwu*' – and what the meanings associated with these categories were, deserves separate research.

21 It is important to note that, while progressing with their investigation of the Gao-pu case, officials in Xinjiang and the northern provinces were ordered by the court to look for, and submit, any smuggled jade that they could find. As court documents show, a large number of jade shipments from Xinjiang and other provinces made their way to Beijing. For memorials about searching for smuggled jade, see the Gao-pu case documents in *Qianlong chao zhengban tanwu dang'an xuanbian*, vol. 1, 373-937. Among these documents, for the inventory lists, see documents no. 440, no. 441, no. 442, no. 931-933.

The mission of moving the enormous load of jade stones confiscated from Gao-pu was divided into two shipments based on the weight and size of the jade pieces and the logistic requirements involved in their transport. In the first shipment, which set out on 13 June 1779 (QL 44/4/29), the officials conveyed a total of 78,528.65 catties (approximately 39,263.9 kilograms) of jade in thirty-three allotments from Yarkand to Beijing.[22] This shipment contained only one jade stone weighing over 1,000 catties, the rest were smaller stones. The second shipment delivered thirteen large pieces of jade quarried out of the Mi'erdai Mountains, and set off as soon as the last portion of the first shipment had left Yarkand.

The shipments relied on the transport system constructed and maintained by the Qing state. As the officials proposed and the emperor approved, 'Outside Jiayu Gate, [all jade stones] were transported through military stations; inside Jiayu Gate, [they] were delivered through post stations (*kouwai juntai, kounei yizhan* 口外軍台，口内驛站)'.[23] In other words, the route from Yarkand to Beijing consisted of a western and eastern part divided by Jiayu Gate: the outer lands (*waiyu* 外域) route from Yarkand to Suzhou 肅州 in Gansu province, and the inner lands (*neidi* 内地) route from Suzhou to Beijing. The first part of this route, which relied on the system of military stations (*juntai* 軍台), is of especial interest in this chapter. These stations were part of a system of relay stations along the main roads that had originally been founded in Xinjiang for military purposes such as the speedy delivery of warfare-related messages. But, from early on, they also functioned as post stations for delivering letters and objects between the court and local administration as well as between the inner lands and the new frontier.[24] When passing the Jiayu Gate to the west, the road that connected China's inner provinces to Xinjiang extended to Hami 哈密 and

22 As mentioned in the previous section, the total amount of jade already stored in Yarkand city when the Gao-pu case was discovered was 107,763.9 catties. Yunggui first selected 31,212 catties of high-quality stones to be transported to the court. These stones arrived at the palace in around October and November of 1779 (QL 44/9). See document no. 441, *Qianlong chao zhengban tanwu dang'an xuanbian,* vol. 1, 932.

23 Fuxing 復興 and Shutai's 舒泰 *ziwen* to the Grand Council on 2 June 1779 (QL 44/4/18), *Junjichudang zhejian*, no. 023754.

24 Whereas in the original design, the military and postal stations were differentiated, in practice, these categories were not strictly separated, since many stations served multiple functions, as mentioned above. See Jin Feng 金峯, 'Qingdai Xinjiang xilu taizhan (yi) 清代新疆西路台站 (一)', *Xinjiang daxue xuebao* 新疆大學學報, no. 1 (1980), 60. For an introduction to the system of relay stations in Xinjiang, also see Liu Wenpeng 劉文鵬, *Qing dai yichuan ji qi yu jiangyu xingcheng guanxi zhi yanjiu* 清代驛傳及其與疆域形成關係之研究 (Beijing: Zhongguo renmin daxue chubanshe, 2004), 104-114.

Map 6.1 Main post routes in southern Xinjiang during the late Qing

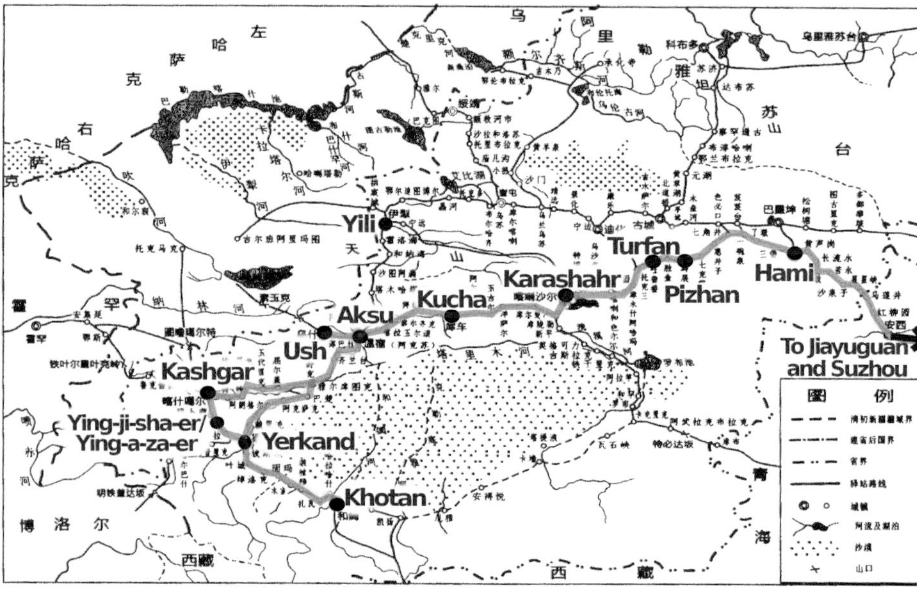

The original map was published in 1890 (GX16). See Wang Qiming 王啟明, *Tianshan langdao: Qingdai Tianshan daolu jiaotong yu yichuan yanjiu* 天山廊道：清代天山道路交通與驛傳研究 (Xi'an: Shaanxi shifan daxue chubanshe, 2016), 182. I have highlighted the roads and cities used to transport jade that are discussed in this chapter.

then split into a northern and southern route; the southern route then further divided into two roads after Turfan 吐魯番.[25] It was this southern route which eventually reached the jade mining areas of Yarkand and Khotan. The movement of jade discussed in this chapter took the southern route from Yarkand, passing through seven major cities, namely Kashgar 喀什噶爾, Ush 烏什, Aksu 阿克蘇, Kucha 庫車, Karashar 喀喇沙爾, Pizhan 闢展 and Hami, to eventually arrive at the Jiayu Gate (see Map 6.1). A varying number of military relay stations were established between these main cities, depending on the distance between them. Beyond the Jiayu Gate to the east, the shipments relied on the post station system to travel through Gansu, Shaanxi, Shanxi, Henan and Zhili provinces on their way to the palace in Beijing. The roads and relay stations used in the jade shipments

25 Using Turfan as a break point, the route was divided into two parts of unequal length: the shorter east road (*donglu* 東路) which extended from Turfan to Hami, and the west road (*xilu* 西路) which included the stations between Turfan and Kashgar and measured 7,240 *li* in total. See Liu Wenpeng 劉文鵬, 'Lun Qingdai Xinjiang taizhan tixi de xingshuai 論清代新疆台站體系的興衰', *Xiyu yanjiu* 西域研究, no. 4 (2001): 29-38.

had been established and in use before 1778, but the high-profile shipment of the Gao-pu jades attracted the court's special attention to their functionality and efficiency.[26]

In addition to the physical transport infrastructure consisting of roads and relay stations, successfully transporting the jade stones in a timely manner also relied heavily on local officials' management. In other words, the local administrations provided the intangible part of the infrastructure by managing and supervising the shipments. It was the high-ranking official Yunggui who planned and oversaw the entire procedure of conveying the Gao-pu jades on their way through Xinjiang territory, in consultation with the emperor. Yunggui in turn depended on the local officials to carry out the proposed plan and to appoint 'trusted and capable people' (*tuoyuan* 妥員) to guard the jade in transit. After passing the Jiayu Gate the administrative bureaucrats of the inner provinces took over responsibility, involving local officials of different ranks, from county magistrates (*zhixian* 知縣) to provincial governors (*xunfu* 巡撫) and governor-generals (*zongdu* 總督) in the task. In each province, every post station assigned 'civil and military officials' (*wen wu yuan bian* 文武員弁) along with soldiers (*bingyi* 兵役), to make sure that the shipment travelled through an area safely.[27]

While the general plan for transporting the Gao-pu jade shipment was discussed and negotiated with central authorities, the concrete procedures lay in the hands of local officials and their actual management of the specific logistical and local issues. Particularly in Xinjiang, many challenges were rooted in the ecological environment, which determined the availability of transport resources as well as the necessary precautions against failure. The officials needed to consider the specific conditions and material characteristics of the shipment in order to choose the most economic and practical methods to deliver the stones in the context of their own localities. The actual operation of the palace machine was thus, by necessity, often ad hoc and open-ended.

26 The roads connecting Xinjiang with the inner provinces were certainly not a completely new creation of the Qing. Indeed, part of the road followed the heavily-trafficked Silk Road which had been formed by Chinese and Inner Asian tradesmen long before the Qing occupation of this area. During the Qing conquest of Xinjiang, the court developed the roads and stations into an effective system for military use. For instance, from 1758 to 1759, in the process of pacifying the resistance from southern Xinjiang, the court established additional relay stations between Karashahr to Kashgar, in order to effectively circulate military information in southern Xinjiang. See Jin Feng 金峯, 'Qingdai Xinjiang xilu taizhan (er) 清代新疆西路台站 (二)', *Xinjiang daxue xuebao* 新疆大學學報, no. 2 (1980): 93.

27 Fuxing and Shutai's *ziwen* to the Grand Council on 2 June 1779 (QL 44/4/18), *Junjichudang zhejian*, no. 023754.

A Shipment of 33 Allotments: Transporting Large Quantities of Jade

The first part of the shipment, as outlined above, contained total 78,528.65 catties of jade. Xinjiang officials were assigned the responsibility to manage the transportation of this huge amount of jade stones in one shipment all the way from Yarkand to the Jiayu Gate. This was both a logistical and a technical challenge. The limited availability of draught animals at the various relay stations emerged as one of the key problems.

Because of the large number of stones included in the shipment, the officials divided the load into 33 portions. Whereas the first thirty-one, each weighing 2,500 catties, contained the jade confiscated during the Gao-pu case, the jade in the last two allotments was partly derived from other sources. The thirty-second allotment contained 1,280 catties of Gao-pu jades plus 616.9 catties of stones recently submitted by local begs, while the thirty-third portion was an added-on shipment of 2,980 catties of jade stones obtained from breaking up a large jade stone that had recently been discovered in the Mi'erdai Mountains.[28] After a series of careful calculations and preparations they decided to schedule one allotment leaving Yarkand every four days (*siri yiyun* 四日一運). The first allotment left Yarkand on 13 June 1779 (QL 44/4/29) and reached Chang'an county in Shaanxi province on 1 October 1779 (QL 44/8/22); the thirty-third allotment was transferred from Shaanxi to Henan on 21 March 1780 (QL 45/2/16).[29]

In order to guarantee that these jade stones were transported from Xinjiang to Beijing with no unexpected damage or losses, transporting certificates (*chuanpai* 傳牌) were issued for each allotment. According to this system, after all the jade had been packed into boxes, the officials recorded the number of boxes contained in the allotment and the weight of each box on an inventory list (*niandan* 粘單) and attached this list to the certificate (Figure 6.1). When the officials who were assigned to supervise the shipments received an allotment from the previous station, they were

28 To make sure that the jade stones submitted by the local begs did not get mixed up with the Gao-pu stones, Fuxing and Shutai filed two separate transport certificates (*chuanpai* 傳牌) for the thirty-second allotment. Following court instructions, the large jade stone discovered in the Mi'erdai Mountains that weighed over 20,000 catties was broken into smaller pieces. Whereas the largest of its fragments, still weighing 3,000 catties, was to be transported together with other large jade stones in the future, the 146 smaller pieces of jade weighing 2,980.6 catties in total were added as the thirty-third allotment of the shipment. See Fuxing and Shutai's *ziwen* to the Grand Council on QL 44/10/8, *Junjichudang zhejian*, no. 025263.

29 Liu Bingtian's 劉秉恬 memorial on 2 April 1780 (QL 45/2/28). See *Zhupi zouzhe*, no. 04-01-14-0043-064, and *Junjichudang zhejian*, no. 026590.

Figure 6.1 Transporting certificate (left part of the document) and its attached inventory list (right part) of the third allotment that left Yarkand on 23 June 1779 (QL 44/5/10) with 22 packages in total

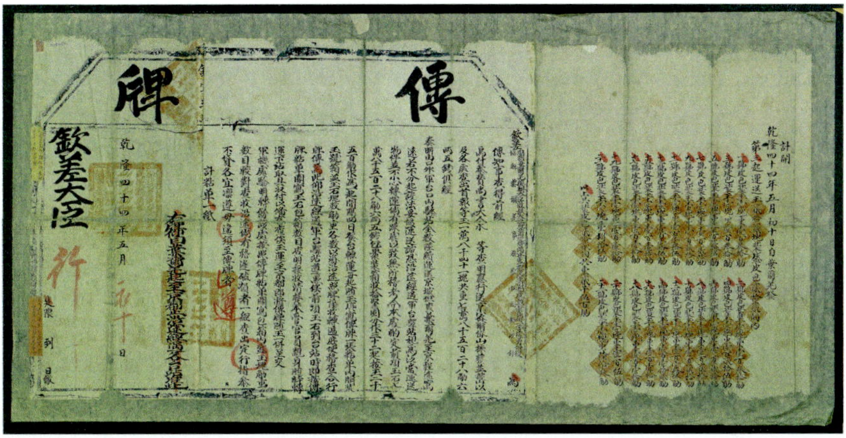

Source: *Junjichudang zhejian*, no. 025232 (National Palace Museum, Taipei)

responsible for checking all the boxes, including the number, weight and package condition before they sent it on its way to the next station. At the end, all thirty-three allotments of the first shipment, along with their certificates and inventory lists, were delivered to the Grand Council (*Junjichu* 軍機處) for a final check and then sent on to their final destination, the Imperial Household Department.[30]

Whilst in transit, the stones were packed in fur and placed on wagons pulled by draught animals. Wagons and draught animals were the only transporting facility able to move weights as heavy as those of the 33 allotments. Human labour would have been an inefficient method, and jade pieces of over 165 catties (approximately 82.6 kilograms) were considered too heavy for pack animals to carry.[31] Therefore, the majority of jade stones

30 I found one *ziwen* containing a side note stating that '[the Grand Council] had received [this *ziwen*] on QL 46/1/19 and had sent it on to the Imperial Household Department'. See Yuan Shoutong's *ziwen* dated 8 Feb. 1781 (QL 46/1/16), *Junjichudang zhejian*, no. 029526. The fact that the Grand Council sent this *ziwen* to the Imperial Household Department indicates that it was the Imperial Household Department itself which ultimately received these jade shipments. For an overview of the amount of jade that was delivered to the Imperial Household Department during Qianlong and Jiaqing reigns, see Lai Hui-min, 'Cong Gao-pu an kan Qianlong chao de Neiwufu yu shangren', 129-131.

31 The decision about whether to use wagons or pack animals must have depended on the size and quantities of a shipment. I have not found an exact rule for this, but the documents on transporting jade stones for making individual chimes in 1762 provide some clues. According to

were carried by wagons. Moreover, since the roads were already established, using wagons and draught animals was much faster than human or animal carriers. As a result, draught animals – including horses, bullocks and camels – and wagons played a central role in all jade transports and were also in the focus of the shipment of Gao-pu's jades.

Indeed, for this first part of the shipment, the main obstacle was the limited resources of draught animals available to deliver such a large number of stones. To resolve this problem, the officials developed a variety of strategies to secure enough draught animals at each relay station from Yarkand to Hami. But animal wealth was not equally distributed between stations, thus the officials needed to adjust their operations for each individual station. Along the road from Yarkand to Aksu, for example, there were more than 20 stations, all with limited transport supplies: some had two or three wagons, while others did not have a single camel. In addition, these stations were also responsible for another task, shipping cloth (*bupi* 布匹) to Yili 伊犁, which made resources on this leg of the route even scarcer. In this context, Yunggui submitted two suggestions for approval almost six months before the first allotment left Yarkand. He first suggested commandeering more draught animals from, for example the cities of Yarkand and Aksu to increase the stations' general transport capacities. His investigations had revealed that there were more than 100 mules, camels, bullocks and donkeys that were not in use there, 'instead leisurely put out to pasture' (*xiankuang mufang* 閒曠牧放). Yunggui suggested, 'it is more useful for the state to seize them and redistribute them [among the stations] that deliver cloth'.[32]

About three weeks later Yunggui filed another suggestion, this time directly connected to transporting the Gao-pu jades. After Gao-pu had been executed, Yunggui went to Kashgar to make an inventory of the possessions of beg Edui 鄂對, who was accused of assisting Gao-pu with his jade smuggling, so his possessions were being confiscated. According to his

his memorial, the court official Dorge, who was in charge of shipping the newly-mined Yarkand jade to Beijing, reported that the 'jade pieces are so large' that 'camels and horses were not able to carry them' (Manchu: *temen morin de acifi inu eterakū*). Therefore, Dorge proposed using one big and five small wagons (Manchu: *sejen*) pulled by six mules to transport these stones. See Qianlong's edict on QL 27/10/13 (28 Nov. 1762), *Junjichu dang'an*, no. 03-129-5-005. According to the memorial submitted by Sinju five months later, the uncut jade pieces for making the chimes (*yuqing caopi* 玉磬糙坯) were of various sizes and weights, ranging from 5 catties to 165 catties. Sinju's memorial on 19 July 1762 (QL 27/r5/28), *Manwen yuezhe bao* 滿文月摺包 (Manchu Monthly Archive), 1951-010, 063-2206 (The First Historical Archives of China, Beijing).

32 Yunggui's memorial on 11 Jan. 1779 (QL 43/11/24), *Zhupi* zouzhe, no. 04-01-01-0367-022.

report, Edui owned approximately 6,000 sheep.[33] While sheep were often used as food for soldiers in Xinjiang, Yunggui proposed managing them differently in this instance:

> There are 6,000 sheep which normally would be placed in the state-owned pasture (*guan chang* 官場) and be used as soldiers' food. I consulted with Dunfu 敦福 about this situation.[34] He told me that the state-owned pastures in Kashgar and Ying'azaer 英阿雜爾 had already raised more animals than those needed as food. There are currently over 7,000 sheep which provide more than enough food [for the soldiers stationed in the area]. If the state pasture accepted these newly-confiscated 6,000 sheep, there would be too many. These additional animals would not only be useless, but would also exhaust the pasture which only receive a normal amount of rainfall and grass production (*shui cao xiang shu pingchang* 水草向屬平常). Therefore, it seems that [I] should find another way to manage these sheep so that they can be of practical benefit (*she fa banli, yi gui shi ji* 設法辦理，以歸實濟) ... [I, your servant,] and Dunfu inquired with the begs who were in the city about the market price for sheep and horses, considering that we would exchange our sheep for their horses. Categorizing these 6,000 sheep (*dayang* 大羊) and lambs (*yanggao* 羊羔) into three ranks, they could be exchanged for a total of 549 small-toothed strong and young gelded horses (*chi xiao jian zhuang er shanma* 齒小健壯兒騸馬).[35]

The list attached to this memorial detailed the number of horses that the official could exchange with the begs according to each rank of sheep (see Table 6.1). Ultimately, by following this plan, Yunggui calculated that at least twenty camels, horses, mules and bullocks could be distributed to each station along the route from Yarkand to Aksu. By adding two extra wagons to each station, the officials in charge would be able to speedily ship cloth to Yili and also transport the jade shipments to Aksu. It might also be possible to transport some, or even the entire lot of large pieces of jade that were still sitting in the Mi'erdai Mountains.[36]

33 Edui had already passed away when the case was discovered but his property was still confiscated by the court as a punishment. See Qianlong's edict on 4 Dec. 1778 (QL 43/10/16), *Junjichu dang'an*, no. 03-135-2-060.

34 Dunfu served at that time as the grand minister consultant of Kashgar (*Kashika'er canzan dachen* 喀什卡爾參贊大臣).

35 See the Grand Council's copy of Yunggui's memorial on 31 Jan. 1779 (QL 43/12/14), *Junjichudang zhejian*, no. 022389 (National Palace Museum, Taipei).

36 Yunggui's memorial on 31 Jan. 1779 (QL 43/12/14), *Zhupi zouzhe*, no. 04-01-01-0367-003.

Table 6.1 Exchange rate between sheep and horses as reported by Yunggui in 1779

Sheep rank	Number	Cost per sheep	Number of sheep per horse	Number of horses exchanged for sheep
First rank	3,000	0.65 tael	9	333
Second rank	2,000	0.5 tael	12	166
Third rank	1,000	0.3 tael	20	50

According to this list, one horse was worth 6 taels of silver. Xinjiang officials exchanged 5,889 sheep for horses and the remaining 11 sheep (3 first rank and 8 second rank sheep) were placed in state pastures. This list was attached to the Grand Council's copy of Yunggui's memorial on 31 Jan. 1779 (QL 43/12/14). *Junjichudang zhejian*, no. 022389.

This memorial tells us not only how Yunggui and Dunfu planned to obtain horses, but also the logic and deliberations behind their plan. A variety of ecological elements had been carefully evaluated and calculated into their proposal. They had considered the quality of the draught animals needed, judging only 'small-toothed strong and young gelded horses' as suitable for their needs. They also took into account the environmental conditions of the state pastures: climatic conditions and rainfall determined the growth rate of grass and thus the number and survival rate of the animals raised. Yunggui and Dunfu adjusted their strategy to the ecological limits and, in the same time, tried to maintain a healthy ecosystem. Based on these ecology-related considerations, the officials came up with a plan that, as Yunggui proposed, could make the best use of these 6,000 sheep for 'practical benefits' (*shiji* 實濟).

In contrast to the conditions between Yarkand and Aksu, the stations on the next leg of the transport route from Kucha to Hami were not involved in shipping cloth to Yili. Therefore, Yunggui and Jingfu 景福, an imperial agent in Yarkand (*Ye'erqiang lingdui dachen* 葉爾羌領隊大臣), originally planned to distribute just ten horses and two wagons to each of these stations.[37] However, after the detailed shipping schedule from Yarkand had been fixed, they realized that on this part of the road there might also be insufficient draught animals to guarantee smooth passage. They were especially concerned about the natural conditions, including geographical features and climate, of the route between Kucha and Hami. The schedule stated that, starting on 13 June 1779 (QL 44/4/29), thirty-two allotments of

37 According to the list attached to Yunggui's memorial they proposed distributing 300 horses and 60 wagons to the 30 stations between Kucha and Hami. See Yunggui's memorial on 12 May 1779 (QL 44/3/27), *Junjichudang zhejian*, no. 023436. This distribution plan was also mentioned in Yunggui and Jingfu's memorial on 24 June 1779 (QL 44/5/11), *Zhupi zouzhe*, no. 04-01-01-0371-051.

jade would come through every four days. If the stations in the Aksu area could smoothly deliver these allotments, the last allotment would arrive in Kucha by December 1779 (QL 44/11). But the stations belonging to Kucha 'were located in a remote area of the desert' (*gebi yaoyuan zhi chu* 戈壁遙遠之處). The officials worried that the heat and long distances between stations in this area would exhaust the horses, and as a consequence, the animals might become too tired to keep up the four-day round trip schedule.[38] The local official Changxi 常喜, therefore, tried to negotiate with the Yarkand officials to slow down the delivery schedule there (*kuanzhan qixian* 寬展期限). For reasons analysed in detail in the next section, the transport schedule from Yarkand could not be changed. Therefore, Yunggui and Jingfu finally proposed distributing more draught animals to the stations between Aksu to Hami as well.[39] As this case shows, the officials needed to adjust their original transport plan to suit the geographical and climatic conditions of each area.

For the seven stations along the next leg of the route from Hami to Suzhou, Yunggui proposed a different operation from the first two legs. Instead of using soldiers to accompany the allotment, he suggested a method put forward by the grand minister superintendent of Hami (*Hami banshi zhifu zhixian* 哈密辦事知府職銜) Fode 佛德 – contracting merchants to take responsibility for all aspects of the transport.[40] Yunggui explained this more economical approach in a memorial:

> If we hire transporting labourers (*gujiao* 僱脚), it would cost 3,000 taels. [Alternatively], if we add more horses [to the stations and arrange the transport ourselves], we would need to provide at least 200 horses which would consume fodder worth about 1,000 taels in one month alone. It is thus better to follow the previous practice and hire merchant contractors. Compared to the fodder that each station would consume with these additional horses, this way will [help the state to] save money.[41]

38 Yunggui and Jingfu's memorial on 24 June 1779 (QL 44/5/11), *Zhupi zouzhe*, no. 04-01-01-0371-051.

39 The officials first considered getting more horses and bullocks from Kashgar but, because it was too far to transport these animals from Kashgar to Kucha, Yunggui and Jingfu suggested that the station officials should borrow animals from local Muslim people and return them after the tasks were finished. See Yunggui and Jingfu's memorial on 24 June 1779 (QL 44/5/11), *Zhupi zouzhe*, no. 04-01-01-0371-051.

40 Fode was the grand minister superintendent of Hami, who also had a rank four prefect (*zhifu* 知府) title.

41 Yunggui's memorial on 24 June 1779 (QL 44/5/11), *Zhupi zouzhe*, no. 04-01-01-0371-050.

For these last seven stations east of Hami until the shipment was handed over to the postal relay system of the inner province, this memorial shows that it was more economical to contract with trusted merchants than to arrange the shipment themselves, and that feeding these animals was an important cost factor in this. As animals, particularly horses, became a high-demand resource during the time of the jade transportation, it might have pushed up the cost of fodder and thus put an unexpected additional fiscal pressure on the officials in charge.

Animals were a central element of the material world in Xinjiang, and were deeply embedded in the ecological structures. They provided food and clothing and enabled necessary transport across large areas, connecting Xinjiang with the rest of the Qing empire. Acquiring and managing draught animals, in this context, required officials to incorporate issues of livestock availability, as well climatic and geographical features into their planning in general and especially in case of transporting the large number of Gao-pu jade, as well as dealing with fluctuating supply and demand. As shown above, Xinjiang officials were able to not only redistribute animal resources along the transportation route by making use of local communities, but also to adapt their management plans to fit the local ecology. This shows the significant adaptability of the palace machine, not to mention its substantial power.

Transporting Large Jade Pieces

The second part of the shipment contained thirteen pieces of large jade stones that Gao-pu had mined but left in the Mi'erdai Mountains. The weight of these boulders ranged from 1,000 to 10,000 catties. Their enormous size and weight posed a different set of challenges to the officials, meaning that they needed to adjust their management plans for this shipment – largely the arrangement of human resources. While the officials mainly used their own soldiers to deliver the tons of smaller pieces of jade in the first shipment, they heavily relied on Muslim begs and Muslim labourers to complete the second one. But common to both shipments was the need for a close consideration of environmental conditions. By examining how the Xinjiang officials cooperated with the local Muslim population to organize the transport, this section shows that, once again, the central principle behind their strategies was to make best use of, and adapt to, the local ecology.

This second shipment consisted of two undertakings: firstly to carry the stones out of the Mi'erdai Mountains; secondly to transport them to Yarkand

and from there on to Beijing. I have not found the exact starting date of this shipment of large pieces in the archival documents, but by the autumn of 1779 (QL 44) at the latest, straight after the thirty-third allotment had left, they were on the road. By December 1780 (QL 45/11), they had arrived in the Zhili area.[42]

The Qing officials incorporated Muslim begs and labourers into this transportation project from the very beginning. As soon as Gao-pu's cases had been discovered, the Qianlong emperor told Yunggui to dismiss over 3,000 Muslim labourers who had been illegally drafted in by Gao-pu to mine and convey jade in the Mi'erdai Mountains. Yunggui sent back 2,300 Muslim workers at first, followed by another 800, but deliberately retained 500 of them 'for transporting large jade stones'. These labourers successfully delivered a piece of jade weighing around 5,000 to 6,000 catties (2,499.7 to 2,999.9 kg) to Yarkand on 30 November 1778 (QL 43/10/12).[43]

At around the same time, Yunggui and Maxing'a 瑪興阿 (Manchu: Mahingga) were trying to figure out a way to haul an even larger piece of jade, weighing over 10,000 catties, out of the Mi'erdai Mountains. But, as they stated in their memorial, extracting this piece was more difficult:

> The people and begs who had delivered the previously found jade stones to [Yarkand] today reported that this jade piece is located further back in the mountain (*ceng shan zhi hou* 層山之後). Therefore, [the labourers] would need to first fill up the gullies on the mountain (*tianping gouhe* 填平溝壑), so that they could drag this piece to the foot of the mountain (*yunzhuai xiashan* 運拽下山). The colour of this piece is not pure and thus not beautiful. There are six other large jade stones in the mountain. Their weights vary from 1,000 to 8,000 or 9,000 catties. However, because of winter, [these pieces] could not be pulled out.[44]

This memorial first highlights the difficult geographical conditions faced in extracting jade boulders out of the Mi'erdai Mountains, which were much more complex and difficult than those for smaller stones. Based on the begs'

42 Qianlong's edict on 12 Jan. 1781 (QL 45/12/18), *Zhupi zouzhe*, no. 04-01-12-0195-002. The number of these pieces of mountain jade was recorded differently in the memorials, ranging from nine to thirteen. In July 1779 (QL 44), eight pieces were hauled out of the Mi'erdai Mountains. In September that same year, those pieces were delivered to Yarkand city. But eventually there were thirteen pieces that were sent on the long journey to Beijing and delivered to the palace.

43 The court rewarded them with 50 sheep and 1,000 catties of flour. Yunggui's memorial on 1 Dec. 1778 (QL43/10/13), *Junjichudang zhejian*, no. 021386-1. Also see Yunggui and Maxing'a's memorial on 1 Dec. 1778 (QL 43/10/13), *Zhupi zouzhe*, no. 04-01-36-0092-018.

44 Ibid.

evaluation of the location and size of the first jade piece, the officials estimated that a path to drag the jade to the foot of the mountain would need to be constructed by filling in gullies. The six other pieces could not be hauled out at that point due to the wintry conditions. In a report filed the following spring, beg Sedib aldei provided some more explanation about this endeavour.

In March 1779 (QL 44), beg Sedib aldei went back to the mountains to examine the condition of these large pieces. His report confirmed that, during winter, cold weather would reduce the efficacy of the labourers' work (*renli nanshi* 人力難施) and the snow that accumulated on the mountain paths would make it difficult for them to drag these large pieces out of the mountains. In spring, however, labourer availability would be scarce because the Muslim people would be busy with their agricultural affairs then. Having considered these seasonal factors, the beg thus suggested that it would be easier to remove these pieces after the snow had melted and the busy spring planting period (*chunzhong* 春種) had finished. The Xinjiang officials accepted the beg's suggestions and recommended that the emperor should wait until the slack farming period (*nongxi* 農隙) to ask the Akim begs[45] to arrange for the necessary labourers. The Qianlong emperor paid special attention to the payment of these Muslim workers. As he emphasized in two consecutive edicts on this issue, the court officials must distribute 'some money for food' (Manchu: *buda jiha*) to these men.[46]

Following an edict in July 1779 (QL 44/5), in September that same year Muslim labourers who lived near the Mi'erdai Mountains finally pulled these eight pieces of jade out of the mountains.[47] For the next step, namely shipping the pieces to Yarkand, the court official Fuxing reported to the emperor that beg Sedib aldei and other local begs 'volunteered to make wagons by themselves' (Manchu: *cihangge beyei hūsun i sejen weile*[*mbi*]) to transport these pieces.[48] Edicts filed in April 1780 (QL 45/2 and 3) reveal that the local begs from several major cities on the route had successfully delivered a total of thirteen pieces of large mountain jade from Yarkand to

45 Fuxing and Maxing'a's memorial on 30 March 1779 (QL 44/2/13), *Zhupi zouzhe*, no. 04-01-36-0092-024. Akim beg was the highest official title for begs in Qing Xinjiang. Akim begs often oversaw the affairs of the Muslim population in the city that they were in charge of.

46 Edict on 11 July 1779 (QL 44/5/28), *Junjichu dang'an*, no. 03-135-3-039.

47 According to the edict on 11 July 1779 (QL 44/5/28), *Junjichu dang'an*, no. 03-135-3-039, the number of pieces of jade dropped from nine to eight. This edict does not explain why the number changed, but a later edict mentioned that one large piece had fallen in a back valley of the mountain and broke into smaller pieces. This broken piece might be the ninth one mentioned in the previous memorials. See the edict on 6 Aug. 1779 (QL 44/6/25), *Junjichu dang'an*, no. 03-135-3-045.

48 Edict on 17 Sept. 1779 (QL 44/8/8), *Junjichu dang'an*, no. 03-135-3-057.

Aksu by that time. To accomplish this, they had paved the road in advance and arranged for local begs along the travel route to guard the shipments on the way. The success of this mission led the court to decide that, on the next part of the journey, from Aksu to Kucha and from Kucha to Karashahr, local begs should also deliver these stones. The Akim begs from these cities followed the model of the begs responsible for the earlier shipment by paving the road and appointing trusted begs to guard the conveyance.[49]

I have not found detailed information on how the local Muslim begs and labourers transported these stones. The fact that it was the Muslim population rather than the Qing officials who carried out the actual delivery procedure might have made it unnecessary for the officials to report the details to the emperor. One issue, however, was important enough to emerge from the brief reports officials submitted to the emperor: the fact that the begs continued to pave the roads. While none of my sources described how they did this – and whether they received any kind of compensation for this – the importance of road conditions for transporting these large jade pieces is frequently mentioned in the sources. When Changxi, the local official in charge of the first shipment on its way from Aksu to Hami, tried to negotiate with Yarkand official Fuxing to delay the starting date of the second shipment:

> Fuxing immediately responded that the original delivery schedule had been already submitted to the Grand Council. If they postponed the starting transport date in Yarkand, they definitely could not accomplish the transportation of the large pieces of jade that were waiting to be conveyed from Yarkand this year. If they waited for the spring of the coming year when the snow has melted, the *qi* of the ground would be mild and the road would become waterlogged (*diqi yi he, lujing bian chao* 地氣已和，路徑遍潮), and [the transport further delayed, as] it would be impossible for the heavy wagons to drive on it (*dache nanyi zaidao* 大車難以載道). Therefore, the schedule for transporting these jade stones could not be changed.[50]

According to Fuxing's explanation, they had to stick to the original shipping schedule to avoid delaying the second shipment. Given that these jade pieces were much heavier than the weight on the wagons in the first shipments, the

49 Edict on 2 April 1780 (QL 45/2/28), *Junjichu dang'an*, no. 03-135-4-020 and edict on 19 April 1780 (QL 45/3/15), *Junjichu dang'an*, no. 03-135-4-030. I could not find evidence of how these large pieces were then transported on the last leg of the Xinjiang route from Karashahr to Jiayu gate.
50 Yunggui and Jingfu's memorial on 24 June 1779 (QL 44/5/11), *Zhupi zouzhe*, no. 04-01-01-0371-051.

second shipment placed higher demands on weather and road conditions. The officials therefore needed to consider ecological reasons, including snow melting and its consequent effects on roads, to make a practical transit schedule.

In addition, the officials' decision to rely on Muslim begs and labourers, instead of running the project themselves, was influenced by a consideration of the local environmental features. First of all, outlined in the discussion of the first shipment above, Xinjiang officials struggled to locate transport resources. With the first shipment still on the way, the officials probably also had to assign this task to local begs because they owned a separate set of resources. Indeed, the begs took over all of the tasks involved in the second shipment: they manufactured wagons, provided draught animals, paved roads and arranged for the hire of Muslim workers.[51] In this sense, incorporating the Muslim population into the project provided a more effective way for Xinjiang officials to deal with the problem of securing sufficient transport and labour supplies. Second, the local people's practical skills, knowledge of mining and conveying large pieces of mountain jade made them indispensable contributors, especially for the second shipment. Long before the Qing conquest of Xinjiang, the Muslim population had already been mining jade in the Mi'erdai Mountains, thus were knowledgeable about the local geography and environmental features. The Qing state, on the other hand, had occupied the territory for less than twenty years and most of the Qing officials involved in the Gao-pu jade transport had been sent to that place only a few years earlier, so were unfamiliar with the natural conditions of the new territory. As explained above, they relied on the local begs to tell them where jade pieces were located, what the appropriate season was for moving them, and which specific methods were best to carry them out of the mountains.[52] Incorporating Muslim begs and labourers into the process was, on the one

51 I have not found any sources describing exactly what these wagons were. In Chinese memorials, they were called *tiewa che* 鐵瓦車, which can be roughly translated as 'iron-reinforced wagon'. *Tiewa* refers to the iron rim used to wrap the outside surface of the wheels. It is possible that, in the wagons used in the jade transport, not only the rim, but also the spokes of the wheel were wrapped in iron, which would have enabled this type of wagon to carry heavy loads.

52 In his description of Xinjiang, Qing elite Chun Yuan also pointed out that, in order to mine 'pure and flawless jade' from the mountain, 'local Muslims often carried their tools with them when they rode a yak [to climb the mountain] to chisel jade pieces (*Huiren xieju chengniu, panyuan chuizao* 回人攜具乘牛，攀援鎚鑿)'. Chun Yuan 椿園, *Xiyu wen jian lu* 西域聞見錄 (1777), *juan* 3, 15b (rep. Lanzhou: Lanzhou guji shudian, 1990). Xu Song also stated that the jade miners would let jade pieces fall first and then collect them. According to Xu Song, 'Jade miners needed to ride yaks (*liniu* 犁牛) to go to peaks (*yan* 巘) to chisel out (*zao* 鑿) jade. After the pieces had fallen, the miners would pick them up… The jade often weighed thousands and tens of thousands of *jin*'. Xu Song 徐松, *Xiyu shuidao ji* 西域水道記 (1819), *juan* 1, 19a (rep. Shanghai: Zhonghua shuju, 2005).

hand, a pragmatic decision, due to the Qing officials' limited access to material resources and their unfamiliarity with the local environment. On the other hand, by incorporating local human workers and adapting their management to the ecological system of Xinjiang local society, the officials enabled the Qing palace machine to successfully extend its reach far into this remote, difficult to access area and thus efficiently serve the material needs of the emperor.

Conclusion

This chapter has examined the complex process of transporting jade from Xinjiang to Beijing in the late-18th century, using the two large-scale shipments that arose from the Gao-pu jade case as an example. The state-sponsored transport infrastructure of relay stations in Xinjiang and post stations in northern China served the Qing palace as crucial links to this remote, newly-vanquished territory. While the system of stations and routes between the two areas had been established before 1778, local officials needed to update and reinforce both to meet the specific needs of the Gao-pu jade shipments. Especially in relation to the transport route within Xinjiang, Qing officials developed multiple strategies to secure the animal resources necessary to deliver this large amount of jade from this distant frontier to Beijing, a feat never before attempted at such scale. In addition, they cooperated with Muslim begs and labourers to extract large pieces out of the Mi'erdai Mountains and transfer them on the roads. In the process of wrestling with these challenges, Xinjiang officials adjusted their management to local ecologies and, in a sense, localized Qing ruling power in the new territory.

This case study also reveals the surprising absence of the Imperial Household Department in the actual quarrying and transport procedure. While it is clear that all the jade stones were eventually delivered to the Imperial Household Department for future design and carving, it was the local officials of the regular bureaucratic system who carried out the tasks. This was not only the case for the Gao-pu jades, in fact, all of the raw jade stones the imperial court regularly received through a newly-developed tribute system were handled by local officials in Xinjiang and conveyed using the state-run infrastructure. In other words, the Imperial Household Department had joined forces with the ordinary bureaucratic system in Xinjiang and with state institutions to fulfil its material needs. This jade transport case demonstrates that the palace machine, although centred on the Imperial Household Department, could neatly and smoothly use the state infrastructure to deal with its logistic operations concerning natural resources on the new frontier.

This raises the question: what was the Imperial Household Department's general relationship with Xinjiang? The evidence shows that, when the court requested specific jade objects, the Imperial Household Department did indeed dispatch its own officials to Xinjiang to oversee the process of quarrying and transport the necessary raw materials. This was the case, for instance, in 1761 when the Qianlong emperor ordered his officials to organize the manufacture of a set of individual chimes (*teqing* 特磬) made from nephrite jade. Dekui 德魁 and several craftsmen, all from the Imperial Household Department, were sent to Xinjiang to carry out this task.[53] In addition, there were of course bondservants, such as Gao-pu, who were appointed as Xinjiang officials. But their involvement seems sporadic, particularly when compared to the crucial role that officials from the Imperial Household Department played in Jiangnan. Many of those serving as salt supervisors (*yanzheng* 鹽政) in Yangzhou and superintendents of the Imperial Textile Manufactories (*zhizao* 織造) in Suzhou, also acted as agents of the emperor in finding craftsmen to manufacture jade objects for court use.[54] During Qianlong reign, the Imperial Household Department had already established an extensive network to facilitate the carving of imperial jade objects in Jiangnan. However, no such similar network had been constructed in Xinjiang by 1778, besides appointing a few bondservants as officials. Was the court not prepared to build a Xinjiang network because the region had only been taken over by the Qing Empire less than 20 years prior? Or did the Imperial Household Department not consider it worth investing its own resources into this new territory because Xinjiang was a faraway frontier consisting mainly of deserts and mountains which were sparsely populated and poor in agricultural land? The uneven involvement of the Imperial Household Department in the two areas necessary to ensure the material outfitting of the court with jade objects – jade quarrying in Xinjiang and carving in Jiangnan – indicates that the department's motives,

53 Dekui's memorial on 12 Aug. 1761 (QL 26/7/13), *Manwen yuezhe bao*, 1883-15, 060-1083. For a detailed discussion of this jade chime project, see Yulian Wu, 'Chimes of Empire: The Construction of Jade Instruments and Territory in Eighteenth-Century China', *Late Imperial China*, 40 (2019): 43-86.

54 For some examples of bondservants' management of making court-requested objects in Yangzhou, including jade objects, see Yulian Wu, *Luxurious Networks: Salt Merchants, Status, and Statecraft in Eighteenth-Century China* (Stanford, CA.: Stanford University Press, 2017), chapter 2.

and possibly also its capabilities for engagement, in Xinjiang were quite different from those in Jiangnan.

About the Author

YULIAN WU is Assistant Professor of History at Michigan State University. She specialises in material culture, gender relations, and Manchu studies in Qing China. She is the author of *Luxurious Networks: Salt Merchants, Status, and Statecraft in Eighteenth-Century China*. Her current project explores jade production and consumption in Qing China.

Part III

Mobilizing Nature: Plants and Animals

The Qing emperor claimed rulership over everything in his dominion – animate and inanimate. The mobilization of plants and animals for imperial consumption, however, posed challenges to the palace machine that were of a different order from the manufacturing of imperial objects. Part Three investigates how the palace machine strived to adjust to the natural cycles of growth and decay as well as the uncertainties of disease and climate, while maintaining an idealized image of 'pristine nature'.[1] These chapters take an ecological approach to the Imperial Household Department's management of living nature, which often refused to be tamed. Lotus flowers bloomed and withered according to the seasons, and the sweetness of their roots and seeds depended on the quality of water, not on any imperial decree. However, as Siebert shows in Chapter Seven, growing and harvesting of the plant was segmented into rigid workflows, and the water spaces where lotuses grew in the imperial city and its vicinity were abstracted into monetary value. Even though the rent collected from this was tiny, it was duly integrated into the accounting machine year after year, from the late-17th century to the fall of the Qing in 1911.

The handling of unpredictable nature required not only abstraction but often involved commercial transaction, as Guan describes in Chapter Eight. She examines the delicate choreography required to secure a steady flow of fresh *materia medica* from all corners of the empire into the Imperial Dispensary, to safeguard the health of the emperor and his family. Some of these herbal and mineral substances had to be sourced from particular locations, to meet the expectations set by canonical works of pharmacology; some herbal medicines had a limited shelf life, and others would be unavailable when a harvest failed. When the routine channels of local tributes were exhausted, the court gave in and simply purchased the medicines it needed from private purveyors on the commercial market.

The handling of tribute elephants detailed by Yu in Chapter Nine reveals a third approach employed by the Imperial household Department to 'tame' living organisms: by literally disciplining them. Being herded on long treks from the empire's southern border to the capital, the animals that survived were trained to perform in imperial parades or put on display in the imperial zoo.

The Qing Imperial Household Department's efforts to tend to these living organisms went a long way to illuminate the imperial ambition of

1 For a recent study of the ideological invention of a 'pristine nature' in the Qing borderlands by means of regulating the foraging, trading and consumption of fur, mushrooms, and pearls, see Jonathan Schlesinger, *A World Trimmed with Fur: Wild Things, Pristine Places, and the Natural Fringes of Qing Rule* (Stanford, CA: Stanford University Press, 2017).

controlling nature in an economically efficient and ideologically efficacious way. Total control remained an unachievable dream due to the glitches and uncertainty inherent in looking after living organisms and in the workings of the palace machine itself. Taken together, the chapters in Part Three show how the processes of planning, action, and decision-making – indeed the entire palace machine that thrived on rationality and abstraction – would have had no chance of success unless they synchronized with the unpredictable rhythms of organic living matter and the inevitability of decay and oblivion. As such, the Qing palace machine, and the empire, were the co-productions of people, materials, and nature that together formed one 'body' or 'machine-body' (*jiti* 機體) – a 'self-contained structure' (*ti*) that moves according to a certain 'triggering mechanism' (*ji*).

The vignette by Akcetin that opens Part Three argues that the twin imperatives of control and of yielding to unexpected contingencies were both impulses that also structured the imperial warehouses. She uses the term 'decluttering' to describe the constant adjustments needed to balance these two imperatives. Thus, the vignette essay points to the discrepancy between realities and the palace machine's accounting apparatus that is a recurring theme in all three parts of this volume.

Vignette essay III

Decluttering

On the Classification of Objects at the Imperial Household Department

Elif Akcetin

Abstract

Drawing on the example of the Imperial Household Department, this essay offers a reflection on the cultural practices of Qing governance. It argues that a reading of the Qing state's mobilization of material resources through an economic lens reveals only part of the story. The classification of objects and the underlying material epistemologies did not merely represent a concern with calculating monetary value; they also served as an ordering mechanism through which the ruling elite visualized the subjects of the empire. The examples provided in the essay illustrate some of the ways in which the Qing state produced structural resources (such as systems of classification) to manage its imperial and colonial expansion.

Keywords: China, Qing Dynasty, material culture, strategies of governance, classification

The practice of decluttering encapsulates the spirit of 18th-century China. By 'decluttering' I refer to the state bureaucracy's preoccupation with simplifying the complexities and messiness of the world by filtering out elements deemed unnecessary for governance and fabricating standard types of people and objects, so as to rearrange nature and society 'into a legible and administratively more convenient format'.[1] The Qing ruling elites constructed a tidy, well-organized, decluttered field of vision, with the territories of the empire

1 James C. Scott, *Seeing Like a State: How Certain Schemes to Improve the Human Condition Have Failed* (New Haven, CT: Yale University Press, 1998), 3.

Siebert, Martina, Kai Jun Chen, and Dorothy Ko (eds), *Making the Palace Machine Work: Mobilizing People, Objects, and Nature in the Qing Empire*. Taylor & Francis Group, 2021
DOI: 10.5117/9789463720359_AKCETIN

neatly divided into categories defined by an intersection of social, economic, material, environmental, cultural, and behavioural patterns. For example, the peasant population was classified according to levels of poverty ('very poor', 'less poor', etc.) as part of the state's management of famine administration, and each region of the empire, from the provinces to the county level, was slotted into a specific economic and cultural model ('extravagant', 'poor and frugal', 'martial', etc.; these features also served to demarcate stereotypical ethnic identities, such as Manchus being 'frugal' and 'martial').[2] It is important to note that these administrative efforts to create a synoptic view of the territories relied on the incessant operations of appraising, categorizing, ordering, ranking and classifying, creating boundaries, hierarchies, and connections. One possible avenue of research into the history of Qing material culture is to analyse the rationale and cultural logic that governed these operations.

The Qing state's deep commitment to streamlining information and rationalizing governance during the 18th century was fuelled by the goal of enhancing its capacity for surveillance and intervention in the context of imperial expansion and the consolidation of power. The state's quest for systematizing knowledge also occurred in the broader cultural context of the early modern period (roughly 1500-1800), where there was a growing sense that the 'all under Heaven' (*tianxia* 天下) was expanding and that the things and people which made it up were moving at faster speed and in greater quantities. This consciousness emerged in a social and economic setting that Chinese historians have described in detail: some of the iconic developments of the period include the commercialization of agriculture, an increase in domestic production, a rapid growth in interregional and international trade, and a proliferation of consumer goods. Travel became more common and, with the unprecedented expansion of printing by the late Ming dynasty (1368-1644) and the decreasing price of printed materials, books (i.e. travel guides, administrative handbooks, vernacular literature) reached a larger number of readers – not only the literati – but 'ordinary' consumers of cities and towns as well.[3] With knowledge being more ac-

2 Pierre-Étienne Will, *Bureaucracy and Famine in Eighteenth-Century China*, translated by Elborg Forster (Stanford, CA: Stanford University Press, 1990), 97-102; Mark C. Elliott, *The Manchu Way: The Eight Banners and Ethnic Identity in Late Imperial China* (Stanford, CA: Stanford University Press, 2001), 276. Cultural and economic classifications commonly appeared in local gazetteers and in administrative handbooks such as, for example, The Comprehensive Directory of the Officials of the Great Qing (*Da Qing jinshen quanshu* 大清搢紳全書).

3 Timothy Brook, *The Confusions of Pleasure: Commerce and Culture in Ming China* (Berkeley, CA: University of California Press, 1998); Evelyn S. Rawski, *Education and Popular Literacy in Ch'ing China* (Ann Arbor, MI: University of Michigan Press, 1979).

cessible, more people than ever became aware of the existence of a realm beyond their own which consisted of a myriad of things and people. They were overcome by an urge to rein in this overwhelming world and the 'unruly artefacts'[4] within it – to understand it, organize it, and downsize it in order to make it manageable.[5]

During the Qing dynasty (1636/44-1911), the Imperial Household Department was one of the institutions that provided an outlet for the ruling elites to control all these unruly objects of the empire, to incorporate them within a state-defined regime of value and create a stable, coherent and legible view of the material realm (in fact, paralleling the above-mentioned ways by which the territories and people of the empire were integrated within an administrative grid). Given that the Imperial Household Department's main function was to generate the revenue required to cover the imperial family's daily consumption expenses and, in some cases, for financial emergencies in the provinces, the repetitive acts of categorizing, classifying, ranking and so forth that occurred under its roof represented a part of the everyday routines of assessing the quality and value of objects. However, such acts not only carried an economic dimension, but a cultural one as well, for they reflected a particular way of perceiving and ordering the different people of the empire. In this sense, the Imperial Household Department's classifications, as intimated through memorials, open a window for understanding the cultural practices of state-making and the mentalities of governance in Qing China.

The revenues that filled the Imperial Household Department's coffers came from different sources, including, but not limited to, rents from the imperial domains and shops; commodities obtained as tributes from tribal leaders, merchants, provincial officials, and foreign emissaries; and, frequently during the Qianlong reign, possessions confiscated from officials implicated in corruption or other malpractice. As one might imagine, there

4 I have borrowed the term 'unruly artefacts' from Kenneth M. George, 'Objects on the Loose: Ethnographic Encounters with Unruly Artefacts – A Foreword', *Ethnos: Journal of Anthropology* 64.2 (1999): 149-50. George describes unruly artefacts as follows: '...as things become unmoored or dislodged from their place of origin, manufacture, or intended use, they are inevitably snared in new hierarchies of value, exchange, and recognition' (p. 149).

5 For an example, see Li Shizhen's (1518-1593) case in Carla Nappi, *The Monkey and the Inkpot: Natural History and Its Transformations in Early Modern China* (Cambridge, MA: Harvard University Press, 2009) and that of Zhao Xuemin (fl. 1753-1803) in He Bian, *Know Your Remedies: Pharmacy and Culture in Early Modern China* (Princeton, NJ: Princeton University Press, 2020), chapter 6. Bian underscores the social diversification of knowledge (p. 155). In the context of consumer culture, see Elif Akcetin and Suraiya Faroqhi, eds., *Living the Good Life: Consumption in the Qing and Ottoman Empires of the Eighteenth Century* (Leiden: Brill, 2017).

was a constant movement of objects flowing in and out, and officials spent a great deal of their time classifying them, deliberating over whether they should be kept aside for the court's consumption, whether they were appropriate to be gifted to certain people, or whether they should be sold and converted into silver.[6] As discussed below, however, sometimes commodities that entered the storerooms remained there for a long time before finally being discovered and revalued by the officials in charge.

As part of both the political centralization and the rising consumerism of the 18th century, the Imperial Household Department underwent bureaucratization, just like other Qing institutions of the time: tasks were specialized, procedures were standardized, timelines were established, and a rationale for classification was developed in order to increase efficiency and maximize revenues. A significant example of this was fur, which was not only an important source of revenue for the court, but also an integral part of the identity of the Manchu rulers of China.[7] Once tribute and private sable furs[8] from the northeast ('Manchuria') were submitted to the palace, their movement was controlled at every step of the way: before they were transferred to the appropriate storage room they were inspected, graded and affixed with a seal (inscribed with grade information) by officials specifically assigned to the task.[9] Classifications were negotiable, and some

6 This description focuses on one aspect of the Imperial Household Department; for more detail on the social, economic and political dimensions of the institution, see Chang Te-Ch'ang, 'The Economic Role of the Imperial Household in the Ch'ing Dynasty', *The Journal of Asian Studies* 31.2 (1972): 243-73; Preston M. Torbert, *The Ch'ing Imperial Household Department: A Study of Its Organization and Principal Functions, 1662-1796* (Cambridge, MA: Harvard University Press, 1977); Qi Meiqin 祁美琴, *Qingdai neiwufu* 清代內務府 (Beijing: Zhongguo renmin daxue chubanshe, 1998); Lai Huimin 賴惠敏, *Qingdai de huangquan yu shijia* 清代的皇權與世家 (Beijing: Beijing daxue chubanshe, 2010), chapter 7; Lai Huimin, *Qianlong huangdi de hebao* 乾隆皇帝的荷包 (Taipei: Zhongyang yanjiuyuan jindaishi yanjiusuo, monograph no. 98, 2014).

7 Jonathan Schlesinger, *A World Trimmed with Fur: Wild Things, Pristine Places, and the Natural Fringes of Qing Rule* (Stanford, CA: Stanford University Press, 2017), especially chapter 1.

8 'Private sable' was acquired from merchants in the market. On tribute and private fur trade in the northeast, see Schlesinger, *A World Trimmed with Fur*, 135-6.

9 'Shouyong diaopi deng xiang 收用貂皮等項 (Collecting and Using Sable Furs and the Like)', in *Zongguan Neiwufu xianxing tiaoli*: *Guangchusi* 總管內務府現行條例：廣儲司, in *Jindai Zhongguo shiliao congkan* 近代中國史料叢刊 (Taipei: Wenhai chubanshe, 1972), 52-3. It is not clear exactly how the process of 'affixing a seal' worked, whether the seal was stamped on the fur pelt, whether it was embossed on a paper or silk slip which was then attached to the fur, or another method. The Regulations and Precedents describe the process as *qianya tuji* 鈐押圖記, *qianyong yinji* 鈐用印記, *qianyin tuji* 鈐印圖記, and *qian dengdi yinji* 鈐等第印記. A nonpermanent form of labelling must have been preferred, since a fur's grade was sometimes modified, in which case officials were required to 'change the seal' (*genghuan tuji* 更換圖記). See 'Shouyong diaopi deng xiang', 53, and Kungang 崑岡 et al., *Da Qing huidian shili* 大清會典

furs were upgraded to a higher rank when need arose; for example, when there was insufficient second- and third-grade sable available, the shortage was compensated for by elevating the rank of the best fourth-grade sable furs.[10] Although this kind of flexibility was allowed in the grading system, it was not extended to the circulation of undocumented fur pelts since, prior to relocation, each pelt had to be affixed with a seal, and the officials on the receiving end were required to check the seals and report if any fur lacked one.[11]

These careful valuations and classifications were necessary to delineate which furs would be gifted to whom. The rarest and most expensive furs were set aside for the emperor and his closest family members. On special celebratory occasions, for instance, imperial princes were rewarded with a number of fur pelts that was commensurate with the degree (and therefore proximity) of their relationship to the throne; the higher their rank in the kinship hierarchy, the more fur pelts they received.[12] Foreign emissaries who visited the court on a tribute mission received their share of furs as well, the furs' grade matching the emissaries' position of precedence in the imperial classification.

Another example of how material objects embodied a hierarchy of value can be seen in a memorial that officials presented to the court towards the end of 1739, after one of their inspection tours of the depots. When the officials had rummaged through the mountains of fur pelts, woollen fabrics, silks and satins, bed curtains, rhinoceros horns, ivory, crystals, various kinds of ram-horn gauze lanterns and other such objects in the Six Vaults of the Grand Storage Office (*Guangchusi Liuku* 廣儲司六庫), they had discovered old, moth-eaten, damaged fabrics which

事例, *juan* 1190, rep. in *Xuxiu Siku quanshu* (Shanghai: Shanghai guji chubanshe, 1995-2002), vol. 814, 459-460.

10 Officials were to select those fourth-grade sable furs which were 'almost new and of a beautiful colour' (*nianjin sehao* 年近色好) and change the seals accordingly. For a discussion on how the quality of a fur was evaluated, see Elif Akcetin, 'Consumption as Knowledge: Pawnbrokers in Qing China Appraise Furs', in Akcetin and Faroqhi, eds., *Living the Good Life*, 371, 378-9.

11 'Shouyong diaopi deng xiang', 53. This last example refers to the QL 39/12 (January 1775) regulation on the return, to their original store, of the leftover sable and black fox furs from the Imperial Wardrobe.

12 In other words, rank was converted into material value. See, for example, gifts distributed to imperial princes in 1661 (Shunzhi 18) on the occasion of the Kangxi emperor's ascension to the throne; E'rtai (Ortai) 鄂爾泰 et al., *Baqi tongzhi chuji* 八旗通志初集 (Changchun: Dongbei shifan daxue chubanshe, 1985), *juan* 53, 1013. In this instance, at least, the daughters of the emperor and of the imperial princes were granted mostly silks and satins, but no furs. It would be worthwhile to investigate whether the examples of gifting practices provided here represented a pattern.

had been sitting there since the Shunzhi (r. 1644-1661) and Kangxi (r. 1662-1722) reigns. After deliberation, they wrote their recommendation in a memorial: they had already identified the goods which were in a decent condition and therefore would remain in the stores. As for the rest of the objects, they had been in storage for years, and if they were left there any longer, they would be completely destroyed and unusable. Thus, the officials continued,

> We found it on record that Lu Jing had previously requested and been given permission to sell, at the Chongwen Gate (*Chongwenmen* 崇文門), the silks and satins, etc., in the stores of the Board of Revenue (*Hubu* 戶部), which were old, faded, stained and mildewed and which therefore could not be retained there for an extended period of time. We request that the unusable objects in the Grand Storage Office also be evaluated and sold at the Chongwen Gate. However, there are among those items 1,478 bolts of yellow silk fabrics with five-clawed and four-clawed dragon motifs (*huangse long mang bu duan sha* 黃色龍蟒補緞紗), which are for the exclusive use of the imperial family (*shangyong* 上用); it is, therefore, inappropriate to price and sell them off (*bubian zuojia bianmai* 不便作價變賣).
>
> We officials have thought about this matter extensively [and have prepared a suggestion for action]: ... According to precedent (*xiangli* 向例), when the Lamas come to pay tribute at court, only the Dalai Lama and the Jebtsundamba Khutuktu should be presented with return gifts (*zheshang* 折賞) of such *shangyong* silks; whereas the Khutuktu, Da-Lama and the like [i.e. Buddhist religious figures in a lower hierarchical position in the imperial classification] should be presented with return gifts of *guanyong* 官用 silks [*guanyong*, i.e. for officials' use].[13]
>
> Now, given that the five-clawed and four-clawed dragon silks are old and decayed, it is inappropriate to present them as gifts to the Dalai Lama and other such dignitaries. [We suggest] that they be used to compensate for a shortage (*zhuoliang dibu* 酌量抵補) in the *guanyong* silks and offered as return gifts to the Khutuktu, Da-Lamas, etc., instead, when the latter come to pay tribute at court.[14]

13 The titles designate the leaders and members of the Buddhist hierarchy in Tibet and Mongolia. See H. S. Brunnert and V. V. Hagelstrom, *Present Day Political Organization of China*, translated by A. Beltchenko and E. E. Moran (Shanghai: Kelly and Walsh, 1912), nos. 914a, 916, 916a, 918(4).

14 Imperial Household Department memorial to the emperor, *Zouzhe* 奏摺, *Guangchusi*, no. 05-0033-001, QL 4/11/6 (6 December 1739) (The First Historical Archives of China, Beijing). Translated freely.

The rest of the memorial proposes action for numerous other objects needing immediate attention, such as five patterned carpets from Russia (*Eluosi huatan* 鄂羅斯花毯), and 5,407 pieces of sable-calf fur (*diaolian* 貂臁). This is a significant document which gives us a glimpse into the material strategies of rulership. For instance, it was deemed inappropriate to sell *shangyong* cloths in the market, even when they were old and damaged. Not only was the emperor himself physically separated from the public, creating a sacred distance, but so were the textiles exclusive for his use, whatever their condition. At the same time, because the fabrics had been damaged, they were downgraded and considered most suitable for Tibetan and Mongolian religious figures on the lower echelons of the imperial hierarchy. In this way, material objects demarcated political space and boundaries.

What did these kinds of constant negotiations mean? To be sure, they reflected the imperial concern for efficiency in the allocation of resources. But we cannot ignore the cultural significance here; the obsession with defining objects and controlling their movement betrayed a desire to declutter and order nature and, by extension, the social and political landscape. In other words, the classification of goods at the Imperial Household Department was an exercise in defining social and political value, for the classifiers were appraising not only the objects of the empire, but their recipients (and makers and givers) as well. Hence, unruly and corrupt silks and furs, such as those which lacked a seal, which were unidentified and on the loose, or those which were fusty and decaying in the dark corner of a storage room, posed a danger to order; they had escaped imperial control and so needed to be reincorporated into the imperial regime of value.

About the Author

Elif Akcetin is Visiting Assistant Professor at Haverford College. She is a cultural historian whose research focuses on the art of governance in Qing China. She is the co-editor of *Living the Good Life: Consumption in the Qing and Ottoman Empires of the Eighteenth Century* (2017). Her monograph, currently in preparation, takes a corruption trial and investigation as an entry point to explore the intricate link between knowledge and power.

7 Growing and Organizing Lotus in Qing Imperial Spaces

Interlocking Cycles of Money and Nature

Martina Siebert

Abstract
The chapter explores the growing of lotus in water spaces under the control of the Imperial Household Department in and around Beijing. This seemingly minor organizational task with meagre financial returns was nevertheless regulated to the detail and established dependencies of tenants and officials, of working tools and paper trails, as well as between the flows of money and an unpredictable nature. Together they built a functional sub-part of the court's ideological project of presenting itself as economically efficient. The chapter argues that the undertaking was an ideologically efficacious spectacle visible in the bureaucratic process and in the lakes and moats around the Forbidden City which were beautifully covered with lotus plants.

Keywords: Qing Beijing, Westpark garden, lotus, bureaucracy, interconnectedness

Introduction

Lotus (*Nelumbo nucifera*) grows in lakes and rivers with its roots and rhizomes anchored horizontally in the mud and its huge leaves and flowers rising above water level. It is pleasing to the eye when in flower, and its edible seeds and roots, which are harvested in autumn, are a renowned Chinese delicacy. But lotus is even more than that: the beauty of its flowers and the splendid green of its leaves have inspired poets and painters, every part of the plant has its acknowledged efficacy in Chinese medicine,

Siebert, Martina, Kai Jun Chen, and Dorothy Ko (eds), *Making the Palace Machine Work: Mobilizing People, Objects, and Nature in the Qing Empire*. Taylor & Francis Group, 2021
DOI: 10.5117/9789463720359_CH07

Figure 7.1 View in northerly direction over the lotus-covered Middle Lake (*Zhonghai* 中海) with Beihai Bridge and White Stupa (*Baita* 白塔)

Ogawa Kazuma, *Shinkoku Pekin kōjō shashinchō* 清國北京皇城寫眞帖 (Tōkyō, 1906), vol. 2, plate 123. Courtesy of the Frances Loeb Library, Harvard University Graduate School of Design

and the imperturbable cleanliness of its petals and leaves has made it an emblem for the Buddhist enlightened mind. All of this literary, medical, and religious significance surely played a role when lotus became an occupant of water spaces under the administration of the Imperial Household Department (*Neiwufu* 內務府) during the Qing. This chapter examines the more worldly affairs of managing these lotus ponds and the sale of lotus roots (*ou* 藕) – or more correctly rhizomes – grown within the highly-controlled space of the imperial city itself (see Figure 7.1) and explores the workflows and knowledge that an array of people established around and in co-production with this plant. The various aspects connected with the plant and its designated spaces were codified into specific administrative regulations and the profits translated into 'lotus rent money'. Lotus plants, water spaces, officials and tenants, regulations, and the flow of silver and materials thus together constituted a small but clearly distinguishable, functional part of the Qing dynasty 'palace machine' and contributed to several of the pivotal aims of the larger machine, namely the production of money, prestige, and 'things'.

As a plant, lotus was a 'thing' that needed planting, care, and looking after. Lotus was one of the agricultural products grown in the luxurious, water-dominated space of the Westpark (*Xiyuan* 西苑) enclosure in the middle of imperial Beijing. After labourers had delivered all the roots requested by the Imperial Kitchens (*Shangshanjian* 尚膳監) from the Westpark lakes, the regulation demanded that superfluous lotus roots were sold for money; some more money came from renting out water areas dedicated to growing lotus under the administration of the Imperial Household Department in and around Beijing. The lotus from the Westpark moreover added a poetic atmosphere and religious meaning to the water spaces when in flower and demonstrated the court's air of austerity and economic zeal when the roots were sold for money, served to guests at festivities, or given as presents to honourable officials.[1] In short, lotus moved between different meanings, states, and functions while following its tracks defined on the one hand by its natural cycle of growth and decay and on the other hand by the rules and financial interests of the palace machine.[2]

The lotuses and the water bodies they grew in were administered by the Bureau of Imperial Gardens and Parks (*Fengchenyuan* 奉宸苑, hereafter Garden Bureau), one of the 'Seven Departments and Three Bureaus' (*Qisi Sanyuan* 七司三院) of the Imperial Household Department. Similar to other activities of the department and its subsections, responsibilities and workflows were codified in 'Regulations and Precedents' (*zeli* 則例), however, for lotus as a plant, the rhythm of actions such as collecting lotus rent money from tenants and selling lotus from Westpark were tied to the yearly routines of plant care and harvest. The lotus rent money in turn instigated further movements of the machine: small repairs in the Westpark area were financed with the money; it paid for replanting, digging waterways, and repairing sluices to stabilize the water level in the Westpark, for sharpening sickles, caulking boots, and to pay for the day labourers who swept the pathways, did seasonal weeding, cut the leaves and harvested the roots of the Westpark

1 See *Qing shilu* 清實錄 for the *renyin* 壬寅 day in the 7th month, year 20 of Kangxi's reign (3 September 1681), which mentions the empress' invitation to officials to attend a feast in Westpark, where they were served lotus and fish grown in the Taiye Lakes 太液池, i.e. the Westpark lakes. *Qing shilu, juan* 96, 26b (rep. Beijing: Zhonghua shuju, 1986).

2 Victoria Cha-Tsu Siu mentions lotus only once, as flourishing near the imperial apartments [of Changchun yuan 長春園], in her view 'appropriately enough, as this region contains major Buddhist temples' (see her *Gardens of a Chinese emperor: imperial creations of the Qianlong Era, 1736-1796* (Bethlehem, Pa.: Lehigh University Press, 2013), 59). Siu focuses on the aesthetic and emblematic aspects of Qianlong's 'Three Mountains and Five Gardens' project, rarely considering the web of shared duties that connected the gardens with each other and with their organizational heart, the Imperial Household Department.

lotus. To fulfil some of these tasks the Garden Bureau had to interlink with other departments or 'machine parts' of the Imperial Household Department, such as the Imperial Workshops (*Zaobanchu* 造辦處), Imperial Armoury (*Wubeiyuan* 武備院) and the Grand Storage Office (*Guangchusi* 廣儲司) or, for larger construction projects, with state agencies beyond the Department such as the Board of Work (*Gongbu* 工部). Viewed from the financial side, lotus rent money evolved into a tool that granted the Garden Bureau some self-sufficiency and managerial independency from the Department's central agencies. From 1681 on the Garden Bureau was no longer required to hand this rent money over to the Bureau of Grand Storage, but was allowed to keep it to handle smaller issues and acquire diverse items; in the mid-19th century the Garden Bureau was encouraged to lend out leftover lotus rent money and charge interest (*shengxi* 生息) on it.[3]

Thus, the example of growing and organizing lotus, a seemingly marginal part of the palace machine, reveals the complexity, interdependency, and adaptability of workflows and expertise in the machine. Furthermore, this telling case study foregrounds the situated processes by which knowledge is produced, codified, and controlled in the highly-regulated and seemingly self-contained organization of the Qing court.

Waters and Spaces

The space in focus in this chapter is the Three Lakes (*Sanhai* 三海) of the Westpark directly adjacent to the West of the Forbidden City (*Zijincheng* 紫禁城) in Qing-era Beijing. They are part of a man-made system of connected water reservoirs, lakes, and rivers that has evolved since the Yuan dynasty, with their source in the mountains to the northwest of Beijing.[4] The waters stretch like a chain of pearls from the Yuquan Mountains 玉泉山 in the northwest of Beijing right into the heart of the city. Beginning with the water reservoir of the Kunming Lake at the summer palace Yiheyuan 頤和園, the Gaoliang river 高粱河 brought the water to the lakes of the Zizhu Garden (*Zizhuyuan* 紫竹院, today south of the Beijing National Library), then further on to the Jishuitan 積水潭 water reservoir right inside the

3 See below for an account of the usage of 'lotus rent money'; for an archival find concerning the interest earned on lotus money and who the money was lent to, see footnote 25.

4 On the development of the Three Lakes' water space from the Yuan to the Ming and Qing times, see *Shichahai zhi* 什剎海志, ed. Shichahai yanjiuhui 什剎海研究会 (Beijing: Beijing chubanshe, 2003), 5-7.

northern wall of Beijing's Inner City (*Neicheng* 内城). Passing Shichahai 什刹海 – an exclusive living quarter for noblemen in the Qing dynasty – the water entered the Imperial City (*Huangcheng* 皇城) and finally reached its end in the form of the Three Lakes, i.e. the North, Middle, and South Lake, within the Westpark's imperial enclosure adjacent to the Forbidden City. To secure the water level in the Three Lakes, in the mid-Ming a canal was dug to directly connect Jishuitan with the Three Lakes, turning Shichahai into a dead-end sidearm of this water system that could be disconnected from the wider network in times of water shortage (see Map 2 at the beginning of this volume).

This chain of water areas connected quite differently-defined spaces. There was first the open, agricultural area northwest of Beijing which had imperial rice fields at the foot of the Yuquan Mountains but also fields tended by ordinary peasants. Secondly, the water passed through the areas of the Leshan travel palace (*Leshanyuan* 樂善院) and Zizhu Garden, to then, thirdly, reach the wall and moat of the Inner City,[5] passing by the city residents' dwellings. Most of those inhabitants were probably Manchu bannermen bonded to the Qing emperor in terms of both ethnic identity and labour duty.[6] The final area that connected into this water system was the walled Imperial City which was dedicated to administration offices and the delicate, imperial area of the Westpark enclosure.

Under Ming rule the Westpark was especially famous as the Jiajing emperor's (r. 1522-1566) refuge. He lived there for fifteen years, fashioning the park into a Daoist paradise stuffed with exotic animals and plants where he could perform his immortality rituals.[7] During the Qing the Westpark evolved into a park area that was used for less private activities. It served as a space for official rituals such as the annual imperial ploughing by the emperor (*yangeng* 演耕) and the plucking of mulberry leaves by the empress (*qincan* 親蠶), hosted feasts for honourable officials and imperial birthday parties, opened its gates to the most successful examinees for the

5 When the Qing took the city over from the Ming in 1644, this Inner City of Beijing (and other Chinese towns) was made into what foreigners used to called a 'Tartar city', restricted to Manchus and bannermen. The Han ethnic inhabitants of Beijing had to move out to the so-called Outer City (*Waicheng* 外城) – or 'Chinese city' – a rectangular city space attached to the south of the Inner City. For an account of the history of Beijing see, for example, Susan Naquin's *Peking: Temples and City Life, 1400-1900* (Berkeley: University of California Press, 2000).

6 On bannermen's special role in the Imperial Household Department and as a labour force, see Chapters One and Two in this volume.

7 Maggie Wan, 'Building an Immortal Land: The Ming Jiajing Emperor's West Park', *Asia Major* 22, no. 2 (2009): 65-99.

final round of the military palace examinations (*wu dianshi* 武殿試), and was the venue for imperial entertainments such as ice-skating games in winter (*bingxi* 冰嬉). The Pavilion of Purple Light (*Ziguangge* 紫光閣) on the western shore of the Westpark's Middle Lake was one of the halls used to welcome foreign ambassadors, its walls decorated with images of Qing military prowess. Between the South and Middle Lake lay the Garden of Abundant Beneficence (*Fengzeyuan* 豐澤園) with ten *mu* 畝 of rice acres, several mulberry trees, and a shed for raising silk worms that provided for the annual agricultural rituals and also enabled the Kangxi emperor to develop a new strain of early harvest rice. It was also here that the Qianlong emperor welcomed Mongolian princes and foreign delegates.[8]

When the Empress Dowager Cixi transformed the Westpark into her retirement residence in 1887 many of the more agricultural activities in the garden were suspended. The park instead acquired some icons of modern times: in 1888 Li Hongzhang 李鸿章 convinced Cixi and the Guangxu emperor to build a short railway line along the western shore that connected the southern end of the Middle Lake to the northern end of the North Lake – in a bid to gain further support for his railway projects. That same year the first electric light was installed in the Yiluan Palace 儀鸞殿, the Empress Dowager's residence and the train's final southern stop.[9] One of Cixi's first steps on moving into the Westpark was to further extend the area around the Middle Lake to the west – with the (intended) consequence that the Catholic church at Canchikou 蠶池口 had to be moved to Xishiku 西什庫.[10]

8 Liu Duying 劉杜英, 'Qing di qingeng Fengzeyuan 清帝亲耕豐澤園', *Zijincheng* = Forbidden City, no. 133 (2005): 124-125.

9 This 'entertainment train' was the second Chinese-initiated railway project – the first being the Tangshan-Tianjin coal line started in 1877. Two short-lived foreign railway projects preceded these activities, one in 1865 outside Xuanwumen Gate in Beijing and one between Shanghai and Wusong in 1877. See Liu Yan 劉燕, 'Qing mo Beijing xiyuan de yuyong tielu 清末北京西苑的御用鐵路', *Beijing dang'an*, no. 3 (2003): 56-57, and Yang Naiji 楊乃濟, 'Xiyuan tielu yu Guangxu chunian de xiulu dazhan 西苑鐵路與光緒初年的修路大論戰', *Gugong bowuyuan yuankan*, no. 4 (1982): 84-94. A slightly earlier twin project to the Westpark train was launched at Yiheyuan with a first segment finished in May 1888. See Wu Qian, 'Cixi zhuanyong de "youwan" tielu 慈禧專用的'遊玩'鐵路', *Wenshi bolan*, no. 11 (2013): 21-22; Yan Jiesheng 嚴介生, 'Yiheyuan Cixi zhuanyong tielu 頤和園慈禧專用鐵路 (Sightseeing Railway for Empress Dowager in Summer Palace)', *Wenhua jiaoliu*, no. 1 (2010): 72-74. We will meet this Yiheyuan railway again later in this chapter, as it repurposed some former lotus pond areas. On electric lighting at Westpark see, for example, Fan Liangshu 樊良樹, 'Beijing diyi diandeng shijian kao 北京第一盞電燈亮燈時間考', *Huabei dianli daxue xuebao* (Shehui kexue ban), no. 6 (2016): 19-22.

10 On the history of the various versions of the Beitang 北堂 (North Church) see Alan Richard Sweeten, *China's Old Churches The History, Architecture, and Legacy of Catholic Sacred Structures in Beijing, Tianjin, and Hebei Province* (Leiden: Brill, 2019), 113-130. A photograph shot in 1901

Figure 7.2 Mid 19th century illustration of Rainbow Bridge with a view of the White Stupa

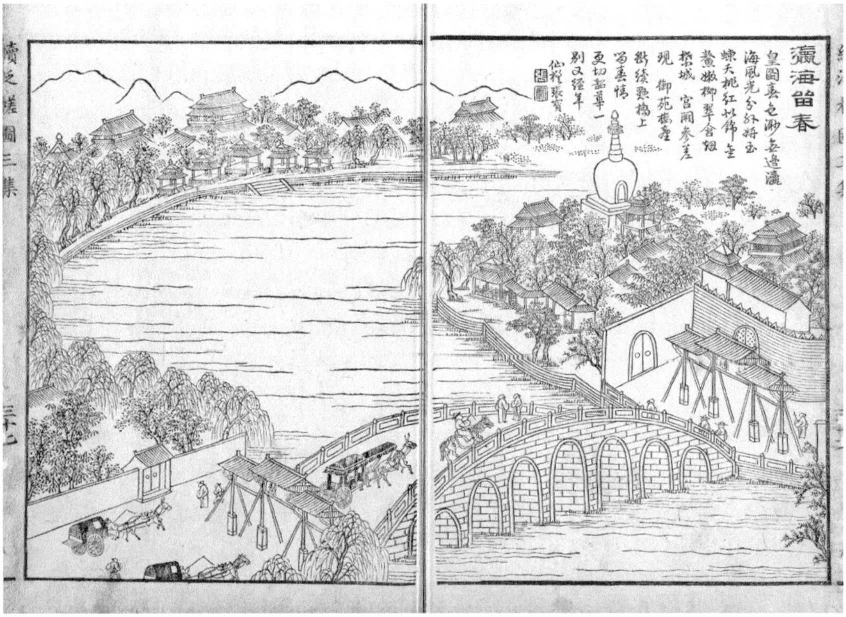

Source: Zhang Bao 張寶, *Xu Fancha tu san ji* 續泛槎圖三集 (Shanghai: Dianshizhai, 1831), 36-37. Interestingly, the lotus cover of the Middle Lake is omitted from this image. Courtesy of Bayerische Staatsbibliothek München, 4 L.sin. K 308-1/3, p.36b-37a

Different interests and activities overlapped in the space of the Westpark, so had to be negotiated and coordinated in order to make things work. The Westpark area was at the same time confined and open, exclusive and connected. When on their way into the Forbidden City – the innermost area of the 'onion layered' city of Beijing – to perform their daily duties or hand in reports or petitions, officials and commoners could take the Gold-Turtle Jade-Rainbow Bridge (*Jin'ao-Yudong qiao* 金鳌玉蝀桥, today's Beihai Bridge 北海大橋) which cut right through the Three Lakes between the North and Middle Lakes. A walk over this bridge thus allowed glimpses into the scenic landscape of the Westpark, including the White Stupa (*Baita* 白塔), which had been built for the Fifth Dalai Lama's visit in 1651 (Figure 7.2).[11]

shows the Canchikou church still standing at that time (see Laohu 老虎, 'Canchikou jiaotang chaichu shijian kao 蚕池口教堂拆除时间考', http://blog.sina.com.cn/s/blog_5ea5774b0100c1ea.html, accessed September 2020). The picture also shows that lotus leaves densely covered the North and Middle Lake to the right and left of the bridge (today's Beihai bridge) in the centre of the photo.

11 Naquin, *Peking*, 310-11.

Institutions and Rules

As mentioned above, the Garden Bureau was one of the 'Seven Departments and Three Bureaus' that comprised the Imperial Household Department, some of which were directly modelled on civic departments. For example, the Imperial Workshops – in charge of providing for the palace's needs of objects and repairs – was to some extent equivalent to the Board of Work, as the Court of Control and Punishment (*Shenxingsi* 慎刑司) was to the Board of Law (*Lübu* 律部). However, others, like the Garden Bureau, were unique to the Imperial Household. The Garden Bureau was established in the 23rd year of the Kangxi reign (1684) as one of the 'Three Bureaus' of the Imperial Household Department, along with the Imperial Stud (*Shangsiyuan* 上駟院) and the Imperial Armoury.[12] The Garden Bureau was not only in charge of the garden areas, with their plants and animals, palaces, temples and shrines, but also of the lakes, waterways and channels inside and outside of Beijing, their sluices and bridges, the boats and sledges used for recreation and for transport across the lakes in the imperial gardens.[13] Furthermore, the bureau administered the Imperial Rice Growing Estates (*Daotian chang* 稻田廠) in the Yuquan Mountains northwest of Beijing, as well as the Southern Park (*Nanyuan* 南苑), a hunting ground that was also used to raise deer, to grow vegetables and animal fodder. The Westpark, as a highly controlled, representationally and ritually important area, provides an especially dense example of the Garden Bureau's duties and areas of expertise. To be able to function properly it needed to link up with other departments in the exchange of materials, money, personnel, and expertise. This interdependence was particularly prominent in the case of water control and keeping up the functionality of canals and sluices. Water was both part of the Westpark's scenery and was crucial for growing lotus in the lakes and the other lotus ponds under the aegis of the Department.

Before the Garden Bureau was established as a separate office, the Westpark garden area had been directly supervised by the Central Administration

12 See *juan* 167 on the Garden Bureau, 166 on the Imperial Stud and 168 on the Armoury, in *Da Qing huidian zeli* (Qianlong ban) 大清會典則例 (乾隆版). I accessed this and the other imperial editions of the *Precedent Cases of Statutes of the Great Qing* using the database *Qingdai wubu huidian* 大清五部會典, Wang Guangyue 王光越 et al., eds. Beijing: Shutongwen Guji shujuku, 2007; accessed via CrossAsia, Feb. 2020).

13 Based on either rank or age, officials could receive special permits to use the boats in summer and sledges in winter to shortcut across the lakes. Permission was also needed to ride a horse or be carried by a sedan chair within the Westpark area. See for example the allowance given to Li Hongzhang 李鴻章 to use a two person sedan chair on 14 November 1898 within the Westpark in *Junqichu shangyudang* (fang ben) 軍機處上諭檔 (方本) (Grand Council record book of ordinary matters): *tiao* 条 3, *hehao* 盒號 1440, *cehao* 冊號 1 (The First Historical Archive of China).

of the Imperial Household Department (*Neiwufu zongguan* 內務府總管); before that, under the Ming, the area was under the administration of the Imperial Kitchens (*Shangshanjian* 尚膳監).[14] It is most probable that the Ming already grew lotus at the Three Lakes, but lotus does not seem to have been embedded into an elaborate, fine-tuned administration, let alone been organized and controlled as prestigious and monetary objects.

The *Precedent Cases* that accompanied the Qianlong, Jiaqing, and Guangxu editions of the *Collected Statutes of the Great Qing*, the *Da Qing huidian shili* 大清會典事例,[15] note down the evolving history of the 'mixed incomes' (*zazheng* 雜徵) of the Imperial Household Department, of which lotus sale and lotus field rent constituted one part. The Qianlong *Precedent Cases* record a decree issued by the Kangxi emperor in 1677. This commanded that the yearly tribute of lotus roots from the Three Lakes must be delivered to the Imperial Kitchens, any surplus must be sold for money, and an accounting report (*zouxiao* 奏銷) must be submitted in the 4th month of the following year. Moreover, the annual rent for growing lotus in the moat encircling the Forbidden City was fixed at 30 *liang* 兩 (taels) in silver.[16] This decree was the initiating event for a growing system of lotus fields and their renting out under the auspices of the Imperial Household Department. The chapters on the 'Mixed incomes of the Garden Bureau' (*Fengchenyuan zazheng* 奉宸苑雜徵) in the *Precedent Cases* for the Jiaqing and Guangxu Statutes (compiled in 1818 and 1899) further outline how the lotus areas were actively expanded under Qianlong, starting in the 9th year of his reign (1744). The *Precedent Cases* record a series of land surveys which provided new calculation bases for rent requests.[17] They also mention several 'diggings of new water ponds'

14 *Qinding Zongguan Neiwufu xianxing zeli er zhong, di er zhong: Fengchenyuan* 欽定總管內務府現行則例二種, 第二種: 奉宸苑 (hereafter *Fengchenyuan xianxing zeli*), compiled around 1850 (facsimiled in *Gugong zhenben congkan* 故宮珍本叢刊, Haikou, 2000), vol. 307, here page 159.

15 No such detailed *Precedent Cases* exist for the earlier two *Statutes* from the Kangxi and Yongzheng eras. Note that the additions to the *Statutes* from the Qianlong era were actually called *Regulations and Precedents* (*zeli* 則例) instead of *Precedent Cases* (*shili*). To avoid confusion with the other *zeli* they are here also called *Precedent Cases*.

16 *Da Qing huidian zeli* (Qianlong ban), *juan* 167, 15a. This information is repeated verbatim in the *Precedent Cases* for the Jiaqing and Guangxu *Statutes* (*Da Qing huidian shili* (Jiaqing ban), *juan* 902, 1a and (Guangxu ban), *juan* 1194, 15a). The Qianlong *Precedent Cases* add the information that the rent from the different lotus areas could directly be used by the Garden Bureau for replanting needs in the inner gardens. Any leftover budget had to be transferred to the Grand Storage Office; see *Da Qing huidian zeli* (Qianlong ban), *juan* 92, 15a.

17 The 1744 survey of the different types of land outside of the Desheng Gate 德勝門 in the north of the Inner City raised the total annual rent figure for this area from 45 taels to over 80 taels in silver. See *Da Qing huidian* (Jiaqing ban), *juan* 902, 4b (section 'Neiwufu 內務府' 18: 'Gardens Yuanyou 園囿', subsection 'Fengchenyuan zazheng 奉宸苑雜徵'); the text is copied

(*kaiwa shuipao* 開空水泡) or 'deepening' of existing ponds (*diwa* 低窪) to facilitate lotus planting.[18] But, as the chronological accounts in the Jiaqing and Guangxu *Precedent Cases* show, the assessment of fields and rent prospects were in constant flux and needed to be adapted to temporary or permanent environmental changes. In years when it was 'not possible to plant' (*bu kan zaizhong* 不堪栽種) the rent was reduced. But the rent could also be raised ad hoc when a harvest was especially abundant.[19] Reassessment of fields according to their environmental condition led to changes in rent classes – from the highest status of rice fields, to lotus or reed fields and, if the fields dried out, to arable dry land. If the land was not even arable, it was taxed at the same rate as land with buildings.[20] For instance, for Jiaqing 7 (1802) the *Precedent Cases* document the permanent silting up of a number of large lotus pond areas at Zhenguo Temple 鎮國寺 and Grass Bridge 草橋 to the southwest of Beijing, as well as of areas north of Beijing near Yuanmingyuan 圓明園, i.e. the Old Summer Palace.[21]

Where the chronological account of the *Precedent Cases* ends, the Garden Bureau's *Regulations* compiled in the mid-19th century, steps in (see footnote 14). Yet, the introduction to the *Regulation's* chapter on the

verbatim in the Guangxu edition of the *Precedent Cases* (*Da Qing huidian shili* (Guangxu ban), *juan* 1194, no pagination).

18 For example, in Qianlong 26 (1761) a water pond of 23.1 *mu* for growing lotus was dug at Di'an Gate 地安門, i.e. the northern gate of the Imperial City (*Da Qing huidian shili* (Jiaqing ban), *juan* 902, 7b); in Qianlong 29 (1764) 18.01 *mu* of water fields at Gengzhitu 耕織圖 at the west bank of the Kunming Lake, i.e. in the northwestern corner of the gardens surrounding the Yiheyuan summer palace, were deepened to make them fit for growing lotus (ibid., *juan* 902, 8b).

19 Assessed as 'not-plantable' were, for example, one-fourth of the lotus fields West of the Di'an Bridge in 1752 (ibid., *juan* 902, 6a). A rich (*feiyu* 肥腴) harvest was reaped in 1772 (Qianglong 37) and tenants of the lotus fields outside Desheng Gate and around Gaoliang Bridge 高亮橋 had to pay an extra 56.23 taels; only three years later (Qianlong 40) the same area was allowed a rent reduction of 34.61 taels because of a poor harvest (*qianshou* 歉收) (ibid., *juan* 902, 11b and 13a). On the larger economic construct and astuteness of the Imperial Household Department and its Grand Storage Office, see Lai Hui-min's book *Qianlong huangdi de hebao* 乾隆皇帝的荷包 (Taipei: Zhongyang yanjiuyuan jindaishi yanjiusuo, 2014).

20 From the land survey outside Desheng Gate in 1744 mentioned in footnote 17 the *Precedent Cases* provide the following list of field types: rice fields (*daotian* 稻田), dry land (*handi* 旱地), lotus ponds (*hehua chi* 荷花池) and reed fields (*pudi* 蒲地). The yearly rent per *mu* for rice fields was fixed at 0.42 taels, that of lotus ponds and reed fields at 0.3 taels and rent for dry land 0.25 taels (*Da Qing huidian shili* (Jiaqing ban), *juan* 902, 4b).

21 Ibid., *juan* 902, 14b. For Zhenguo Temple and Gras Bridge the *Precedent Cases* state that an area as big as 229.34 *mu* was silted up. The lotus ponds in those two areas had only been opened up about 30 years earlier, in 1775 and 1777, with a total size of 269.94 *mu* (ibid., *juan* 902, 13a). Subtracting the silted-up area from this we (nearly) arrive at the area size of 41.3 *mu* named in successive lotus rent reports and the Garden Bureau's *Regulations* (see Table 7.1).

'Official reporting about the amount of money received from renting out lotus areas and similar issues' (*Zouxiao hehua dizu deng xiang qianliang* 奏銷荷花地租等項錢糧), starts with the Kangxi decree of 1677 that had initiated the system of lotus rent money. It then provides a condensed list of the location, size, and rent of all lotus fields under the control of the Garden Bureau in the 14th year of Jiaqing reign (1814). After that date only a few areas producing lotus rent money were added, in 1837 and 1840 (i.e. numbers 8 and 16 in Table 7.1).[22] Table 7.1 below gives an overview of the size and rent of the listed areas as laid out in the *Garden Bureau Regulations*. Some areas – for example no. 2 – were actually used to grow reeds instead.

Whereas the *Precedent Cases* and *Garden Bureau Regulations* describe the regulatory framework of the lotus rent system, the annual 'Reports on the receipt of lotus rent with balances and disbursements' (*Zhengshou hehua dizu cunyong yinliang shumu* 徵收荷花地租存用銀兩數目; from here on *Lotus Reports*) allows for glimpses onto its actual workings. I was able to inspect seventeen examples of these *Lotus Reports* among the archival documents that the First Historical Archives of China in Beijing has identified as belonging to the Garden Bureau.[23] They cover the years 1820 (Jiaqing 25) to 1861 (Xianfeng 11); some documents were of sequential years, i.e. 1835 to 1839, and 1857 to 1858. Most of the ledgers (*qingdan* 清單) attached to these *Lotus Reports* are either lost or detached and may now be buried under the huge number of still-unprocessed Imperial Household Department documents at The First Historical Archives in Beijing. Nevertheless, the available ledgers already provide evidence to support what the *Garden Bureau Regulations* suggested, namely that the lotus rent money functioned as a separate moving part of the machine, producing and spending silver (or materials and working power that needed to be paid for with silver), whilst simultaneously accumulating a deposit (*cun* 存)[24] that allowed for a certain degree of independence. From 1850 on the deposit itself was put to work, since part of it was lent out to generate a return in interest.[25]

22 *Fengchengyuan xianxing zeli*, 176-178. The *Garden Bureau Regulations* cites the year 1814 as the basis for the number, size and rent of the lotus fields, but the latest year mentioned in the whole section on the Garden Bureau is 1850 (see, for example, ibid., 179).

23 This section of Imperial Household Department documents is currently still being newly processed by The First Historical Archives of China and is thus unavailable to users. Once they have finished re-cataloguing (and scanning), these documents most probably also have new shelfmarks. Below I will use their 'old' system of which some already had numbers stating *xinzheng* 新整 'newly arranged'.

24 On accounting, see Chapter Three in this volume.

25 *Fengchenyuan xianxing zeli*, 179. See also the 'Account book of repairs, lotus rent and interests' (*Suixiu hezu shengxi qianliang zhang* 歲修荷租生息錢糧賬), GX27/1 (Feb. 1901), at

Table 7.1 Sources of lotus income according to *Garden Bureau Regulations* as of 1840

		Sources of 'lotus rent money'	area *mu* 畝	rent in silver *liang yin* 兩銀	rent per *mu*
water areas	1	Forbidden City moat 紫禁城護城河	288.7	129.915	0.4500
	2	Jinhe reed area 金河蒲草地	520.3	208.12	0.4000
	3	East and West of Leshanyuan 樂善園東西墻外荷花地	79.6	39.8	0.5000
	4	Leshanyuan/White Stone Bridge/ Zizhuyuan lotus area 樂善園白石橋紫竹院荷花地	69.9		
	5	Small Stonebridge lotus field 小石橋荷花地	49	127.755	0.5032
	6	Shichahai lotus field 什剎海荷花地	135		
	7	Gaoliang Bridge lotus field 高亮橋荷花地	84.4	42.2	0.5000
	8	Small lotus pond at Gaoliang Bridge 高亮橋小河泡荷花地 (1840)	30	13.5	0.4500
	9	Diaoyutai. SE lotus field 釣魚臺西南荷花地	393.26	176.967	0.4500
	10	Huxinlou lotus field 湖心樓荷花地	29.13	16.735	0.5745
	10	Zhenguosi lotus field 鎮國寺荷花地	41.3	18.585	0.4500
	12	Jishuitan lotus field 積水潭荷花地	135	81	0.6000
	13	Di'an Bridge lotus field 地安橋荷花地	224.14	134.484	0.6000
	14	Deshengmen-Longwang jiaolou lotus field 德勝門龍王堂角樓荷花地	336.65	134.66	0.4000
	15	Diaoyutai lotus field 釣魚台荷花地	743.89	223.167	0.3000
	16	Leshanyuan lotus field 樂善園荷花地 (1840)	51	[22.95][26]	[0.4500]
sum			**3211.27**[27]	**1369.838**	

Neiwufu Fengchenyuan archives (hereafter NWF-FCY) of The First Historical Archives of China, Beijing, NWF-FCY, no. 131, 13-15. This document lists the income and expenditure of 'interest' from lotus money lending for the year Daoguang 26 (1846). The main borrowers of the money seem to have been imperial relatives and the salt administration of Lianghuai. The latter paid 260 taels interest in silver (*li yin* 利銀) for the thirteen-month period from the 10th month of Daoguang 25 to the 9th month of Daoguang 26, i.e. twelve regular and one intercalary months.

26 The *Garden Bureau Regulations* give the size of both areas – lotus fields and dry land – but only the sum of the rent, i.e. 95.85 tael. Considering an average rent for the lotus fields of 0.45 per *mu* leaves 72.9 tael for the dry land area with a rent of roughly 0.35 per *mu*.

27 The standard number given in the reports between 1820 and 1852 is 3164.61 *mu* with a rent of 1333.388 tael. The closest one gets to this number is by excluding the lotus ponds at Leshanyuan

	Sources of 'lotus rent money'	**area** ***mu* 畝**	**rent in silver** ***liang yin* 兩銀**	**rent per** ***mu***
dry land areas	17 Leshanyuan reed area 樂山園葦草地	210.11	[72.9]	[0.3469]
	18 Leshanyuan residential area 樂山園房基旱地	56	16.8120	0.3002
	19 Leshanyuan additional residential area 樂山園房基旱地 (1837)	154	46.2	0.3000
	20 Jiao'erbao Lishui Bridge dry land 茭兒堡利水橋旱地	206.52	20.6530	0.1000
	21 Grass Bridge reed area 草橋 葦地 (1840)	87	26.1	0.3000
other	22 Guangyuan sluice area with two empty houses 廣源閘臨河空房二間		2.4000	
	23 Sale of superfluous lotus roots from Three Lakes (Westpark) 三海瀛臺剩餘蓮藕鬻買		104.848	
	Total	**3924.9** ***mu*** (261.8 ha)	**1659.751 tael** (61.4 kg)	

Source: *Fengchenyuan xianxing zeli*, 176-178

Regularities and Realities

The existing *Lotus Reports* narrowly reflect the rules set out by the *Garden Bureau Regulations*. As 'tables in prose' they inscribe the changing – and unchanging – numbers of balances and disbursements, area sizes, and rent money into a highly formalized textual grid. Every report started with the same statement of the total size and rent of the various lotus fields under the Garden Bureau's management, namely 3,164.61 *mu* of water or land, for which the bureau had 'recruited tenants to grow lotus and reeds' (*zhao tianhu cheng zhong hehua pucao* 招佃戶承種荷花蒲草) and to pay a rent total of 1,333.388 taels in silver every year. A standard component of the reports' opening passage was also the sale of Westpark lotus, that was accounted for at the fixed amount of 104.848 taels (i.e. about 3.88 kilograms of silver) in all reports I was able to consult. *Lotus Reports* from subsequent years show how the money left over from a previous year was carried over into the next year's account and how expenses differed each year. Into this steady rhythm of the 'reporting

(no. 16) and Gaoliangqiao (no. 8) which were both added in 1840. Then the numbers are 3130.27 *mu* and 1333.388 tael.

machine's' clicking and stamping, though, occasionally there were unexpected squeaks of reality.

It is often only in these situations when certain parts of the machine become visible. The *Lotus Reports* for the years Daoguang 18 and 19 (roughly 1838 and 1839) document a minor 'hiccup' in the system:

Table 7.2 Lotus money balance of the years 1838 and 1839[28]

	Carried forward from previous year	Lotus rent	Lotus rent plus other rents/sales income	Expenses	Left over in current year
DG18 (1838)	3,724.737	1,138.766 (i.e. the standard 1,333.388 taels reduced by 194.622 taels because of an unusually dry spring and summer)	1,425.729	1,255.462	3,895.004
DG19 (1839)	3,895.004	1,431.199 (in addition to the annual rent of 1,333.388 taels, half of the rent reduction from the previous year, i.e. 97.311 taels, had to be repaid)	1,757.062	1,000.676	4,645.39

The leniency granted because of an expected bad harvest in cases such as that of 1838 often only changed the tenants' rent into instalment payments, it did not cancel it. Only in cases where it was 'not possible to plant' was an actual exemption granted.[29] Sometimes these instalments overlapped. For instance, the report for Daoguang 28 (1848) shows that 113.27 taels were received as the last instalment of a three-year agreement, together with 318.817 taels that was the first payment of a newly-started three-year instalment cycle.[30] Thus, while the wheel of the annual rent demands kept on turning at a steady speed, the actual payments progressed in their own bumpy way.

The ledgers (*qingdan*) attached to these *Lotus reports* recorded what expenses were paid with the lotus rent money. The items listed in all the

28 NWF-FCY *xinzheng* 新整, no. 4590, contains two *Lotus Reports*, one dated DG18/12/13 (27 Jan. 1839), one DG19/11/25 (30 Dec. 1839).

29 See the examples above quoted from the *Precedent Cases* (footnote 19) and the document discussed below, where the tenant Wang Long was exempted from paying his rent for the year Guangxu 16 (1890) because his fields had silted up.

30 NWF-FCY 新整, no. 4590, *Lotus Report* of DG28 (1848).

ledgers I inspected are more or less identical. Only their amount varies. The year Daoguang 18 (1838) is a typical example:[31]

- offerings at the Temple of Sericulture (*Can tan* 蠶壇): 6.2 taels;
- replanting of various withered trees, bamboos and shrubs in the Westpark area (Ningshou Palace, Silkworm Altar etc.). Some replanting was covered by selecting suitable pines from in front of the Pavilion of Purple Light, but an additional eighty trees (pine varieties, crab apple, mulberry, lilac, etc.) and 192 peony shrubs had to be bought on the market: 333.7 taels;
- acquisition of 735 bamboo plants sized from 1.1 to 1.4 *zhang*: 60.3 taels;
- 300 clay pots for tending chrysanthemum plants: 6 taels;
- 1,200 bushels of straw to build windshields for vegetables and covers for earth pits for winter flowers: 26.4 taels;
- 700 catties (*jin* 斤) of cattail stalks to build sun covers for the vegetable gardens at South Lake: 1.556 taels;
- 748.11 catties of white flour glue for the spring and autumn window paper repairs at Jingshan 景山, Three Lakes, etc.: 19.106 taels;
- repairing the lanterns (*chuodeng* 戳燈) at Jingshan, Three Lakes, etc.: 55.2 taels;
- provisions for the overseer, assistant overseer and gardeners at Zhaigong 齋宮 (in Tiantan park): 365 taels;
- writing up the 'inventories of things on display' (*chenshe dang'an* 陳設檔案)[32] : 30 taels;
- salary for scribes: 52 taels;

Sum of the above: 1,255.462 taels.

Similar to the *Lotus Reports*, the ledgers were bureaucratically-formalized versions of realities, in which these items may have functioned as stand-ins and 'financial placeholders' for other necessary expenses. These reports and ledgers thus only show the tip of an iceberg of the numerous small actions and messy day-to-day businesses that was obscured from the central administration's view and remains hidden from our sight today.

The Garden Bureau's archival material contains a small number of more detailed ledgers that allow some insights into the distortions of everyday management that were concealed beneath that smooth surface of official reporting. There are first two ledgers for the years Daoguang 11 (1831) and

31 Ibid., *Lotus Report* of DG18 (1838).

32 These 'inventories of things on display' had to be compiled and checked with the actual displayed objects every second year by two officials. They were produced in two copies, one of which was stored at the site and the other in the Yamen (*Fengchenyuan xianxing zeli*, 184).

27 (1847), reporting the flow in and out of the Garden Bureau's finances.[33] Both are neatly written and seem to be clean copies for archival storage produced from the actual on-the-spot information, but they still differ significantly from the seventeen standardized reporting ledgers quoted above. The 1831 document lists expenses sorted by month, while the 1847 document records both lotus rent silver flowing in and disbursements for various 'small matter', i.e. silver flowing out again, giving the exact date of each transaction. Moreover, both ledgers name items beyond those found in the 'standard' reporting ledgers, for example buckets (*tong* 桶), iron vessels (*tieqi* 鐵器), and steamed buns (*bobo* 餑餑) used to feed the goldfishes in the South Lake. They also show how tasks were distributed according to the actual climate of a year: 100 taels in silver were paid for 'felling and weeding' (*genzha* 根楂) in the 1st, 4th and 9th months of 1831 but in the 3rd and 6th months of 1847. The 1847 ledger also records receipts of lotus rent money. Altogether 600 taels in rent arrived at the end of the 8th and early 9th months; on the 10th day of the 10th month a belated rent payments (*zhaojiao yin* 找交銀) in the amount of 975.773 taels was received.[34]

Another important pair of archival documents pertaining to the lotus rent system are the documents NWF-FCY no. 144-1 and no. 144-2, whose information complement each other.[35] Document no. 144-1 gives a detailed account of lotus rent money received between the 9th and 12th months of the years 13 and 14 of Guangxu's reign (i.e. roughly from mid- to late-October 1887 and 1888 to the ends of January 1888 and 1889 respectively). Document no. 144-2 reviews changes of tenants and lotus areas between 1878 and 1900. Together they show that the lotus rent system was based on many small-scale negotiations and agreements. From no. 144-1 we learn how tenants' payments of rent and debt trickled in over the year. Besides two larger payments of 'jointly delivered lotus rent money of individual tenants' (*ge tian gongjiao* 各佃共交) – i.e. 233.5 and 620.728 taels in 1887 and 399.1 and 599.626 taels in 1888 – numerous belated payments arrived at various times up until the end of year. Some of them were again 'joint' payments, but others were individual, revealing the tenant's name (see Figure 7.3). Combining this information

33 NWF-FCY no. 126: '*Hehua dizu yinliang bu* 荷花地租銀兩簿 (Account book of lotus rent silver)', GX11(1838); NWF-FCY no. 131.

34 NWF-FCY no. 126, 5-9, NWF-FCY no. 131, 1-7.

35 NWF-FCY no. 144-1: '*Shou hehua dizu yinliang bu* 收荷花地租銀兩簿 (Account book of the receipts of lotus rent silver)', GX13/9 (mid-October 1887); NWF-FCY no. 144-2: '*Zhengshou hehua dizu yin weicao qianliang ce* 徵收荷花地租銀葦草錢糧冊 (Booklet on the receipts of lotus rent silver and reed payment)', GX29/8/1 (21 Sept. 1903).

Figure 7.3 Ledger on lotus rent and debt payments

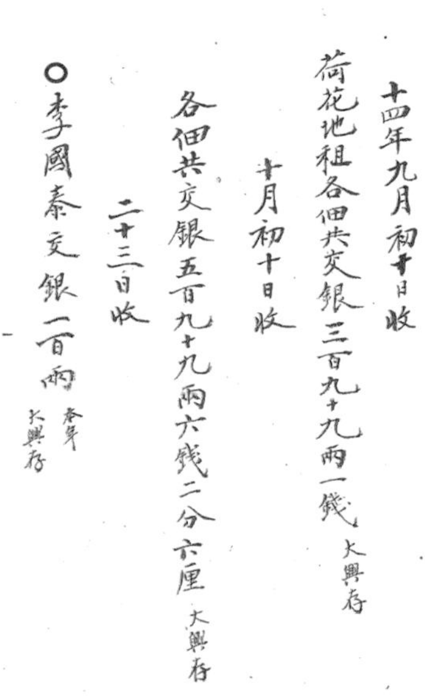
十四年九月初十日收
荷花地租各佃共交銀三百九十九兩一錢 大興存
十月初十日收
各佃共交銀五百九十九兩六錢二分六厘 大興存
二十三日收
○李國泰交銀一百兩 大興存

First page of the document NWF-FCY 144-1 showing various jointly or individually arriving lotus rent and debt payments from tenants from the 10th to 12th month of Guangxu 14 (1888); the circle marks Li Guotai (NWF-FCY no. 144-1, 3; The First Historical Archives of China, Beijing)

Figure 7.4 Register on rented lotus areas and changes in tenancy

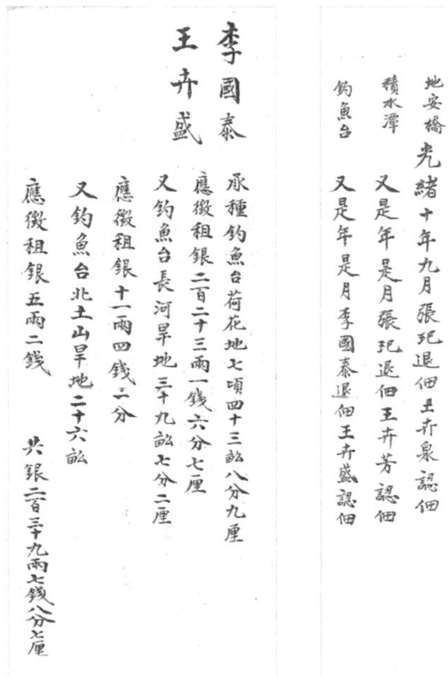
李國泰
王卉盛
承種釣魚台荷花地七頃四十三畝八分九厘
應徵租銀二百二十三兩一錢六分七厘
又釣魚台長河旱地三十九畝七分二厘
應徵租銀十一兩四錢二分
又釣魚台北土山旱地二十六畝
應徵租銀五兩二錢
共銀二百三十九兩七錢八分七厘

地安橋 光緒十年九月張玘退佃王卉泉認佃
積水潭 又是年是月張玘退佃王卉芳認佃
釣魚台 又是年是月李國泰退佃王卉盛認佃

To the left, the entry in the 1903 register naming tenants and the size and rent of their lease, here Li Guotai and his successor Wang Huisheng as tenants of the Diaoyutai area. To the right, the respective entry from the perspective of the lotus rent areas showing that among several changes in GX10/06 (Oct. 1884) Diaoyutai tenantship had moved from Li to Wang (NWF-FCY 144-2, 7, 23; The First Historical Archives of China, Beijing).

with that in document no. 144-2 exposes the long-term dependence that tenants had on the Garden Bureau – and therewith indirectly the Imperial Household Department – due to pending lease payments. For example, according to document no. 144-1, tenant Li Guotai 李國泰 handed in 100 taels of silver on the 23rd of the 10th month in Guangxu 14 (26 Nov. 1888). The first part of document no. 144-2 lists Li Guotai as the tenant of 743.89 *mu* of lotus fields in Diaoyutai (Table 7.1, no. 15) with a yearly rent of 223.167 taels. The last page of document no. 144-2 records the rent system from the angle of the lotus areas, showing that Li Guotai's lease of the Diaoyutai fields had actually already ended in 1884 (*tui tian* 退佃), when Wang Huisheng 王卉盛 succeeded him as the tenant (*ren tian* 認佃). So, Li's 100 taels payment in 1888 was at least four years overdue (see Figure 7.4).

The lotus rent debts and payments were detailed with 'tickets' (*piao* 票). This is confirmed by a note scribbled under tenant Wang Long's 王隆 allotment of the Zhenguo Temple lotus field (Table 7.1, no. 11). The note copies part of a memorial dated May 1903 in which the emperor was asked to exempt Wang Long from paying that year's 'rent ticket' (*zupiao* 租票), the 24th since he had become tenant of the area near Zhenguo Temple in Guangxu 4 (1878). Wang had settled, literally 'returned', all his previous tickets (*huipiao* 回票) except that of 1890 when the fields had suffered from flooding and silting up so Wang had one year's rent waived. No reductions were granted in 1900 when his harvest was affected by calamities due to war, however. In 1903, the year the note was written, seven of the 41.3 *mu* at Zhenguo Temple had been occupied by the Yiheyuan railway 頤和園鐵路 (see footnote 9) since the third month of Guangxu 29 (April 1903), so the memorial recommends exempting him from one year's rent as compensation.[36]

In respect to the size of the lotus fields and their rent, the design of the system appears not to have changed much between what was described in the 1814 *Garden Bureau Regulations* and the 1903 register. Lotus rent money remained a stable building block in the Garden Bureau's financial activities. That this stability was more one of expectation than reality becomes particularly clear in the case of the so-called 'sale of superfluous lotus roots from the Three Lakes' (item no. 23 in Table 7.1). One would expect this to be an especially vulnerable and flexible amount of money because harvests, as well as market prices, tend to differ between years.[37] But the existing *Lotus Reports* unshakably report the amount delivered each year as 104.848 taels in silver. Within the seventeen inspected reports only that of 1861 made an exception. In this no 'profits' were expected because the Beijing canals have dried up and the lotus sprouts in the Three Lakes had all withered. The memorial instructing this exemption underlined that it expected conditions to return to normal the following year, when the silver would be due as usual.[38]

NWF-FCY document no. 126 on the lotus money balances of 1831 shows that it was not only tenants outside the palace who accumulated debts. Official garden deputies (*yuancheng* 苑丞) who were responsible for the Westpark lakes were not always able – or willing – to hand-in the expected

36 NWF-FCY no. 144-2, 9. Concerning how the ticket system was used to keep track of materials and costs between institutions of the Imperial Household Department, see Chapter Three in this volume.

37 That the Neiwufu did not consider itself to be dependent on changing market prices, but tended to fix prices, for example for medical materials, is shown in Chapter Eight in this volume.

38 NWF-FCY 新整 no. 4592: *Lotus Reports*, XF11/11/12 (= 12 Dec. 1861).

'profits'. Document no. 126 contains an entry for 791.432 taels in silver of 'lotus root money paid by Three Lakes, etc.' (*Sanhai deng chu qianjiao lianou yin* 三海等處欠交蓮藕銀). A large portion of this sum was belated payment of the expected income from the 'sales of superfluous Westpark lotus roots' by garden deputies of the North, Middle, and South Lakes. Again, some payments were delayed for many years. For example, garden deputy Guang'en 廣恩 was 25 years late submitting 80 taels of 'lotus root money' for the years 1807 and 1808 when he was in charge of the North Lake.[39] It is difficult to know whether these annual receipts of fixed profits were intended to leave room for the garden deputies to make some extra money on the side through the actual sales. But it seems obvious that these 'sales profits' functioned as a form of 'rent'.[40]

In the early 19th century it had become routine for officials to be dispatched to inspect and assess any damage in lotus growth and decide on an amortization schedule. The archives document a case of Qianlong 46 (1781) that might have triggered the installation of this routine. When the Garden Bureau officials had forwarded a request to continue the previous year's reduction in lotus rent due to bad weather, the request was swiftly granted by the Imperial Household's central administration. But doubts were raised by a competing official within the central administration,[41] the weather archive was checked and it was found that, indeed, no weather irregularities for the Beijing area had been recorded for 1781. Thus, any need for a rent reduction had to be the result of a 'carelessness in management' (*bu yongxin jingli* 不用心經理) – on the side of the tenants for taking insufficient care of their plants, and on the side of the reporting officials of the Garden Bureau for filing a false request. As a result, the tenants and officials had to share the debt and pay back the rent reduction. The tenants were allowed to do this over a span of two years, while the officials had to repay it within the same year. In a second round of investigations high officials of the Imperial Household Department were accused of not having thoroughly researched the truthfulness of the claims of the reporting officials when they had granted the request, so a fine of six months' salary was imposed. As a consequence, and to prevent future requests for reductions on false

39 NWF-FCY no. 126, 11-12.

40 See also the epigraph opening my conclusion, that points to how this Westpark 'rent' might have been instead meant to allow for an extra income on the side of those in the lower echelons.

41 A personal feud might have played a part in this case: on the side of the investigators we find Heshen 和珅 (1750-1799), mostly known for his misuse of the system, on the side of the convicted is Jin Jian 金簡 (?-1794), the 'good' official who was, for example, in charge of the project to produce a set of moveable wooden type for printing, the *juzhen* 聚珍.

pretences, on-site controls by officials were installed as necessary routine precaution.[42]

Another example, reported in the meticulous style of Qianlong era reporting, provides a glimpse into the security problems arising from the huge garden space of the Westpark with its many levels of employees.[43] In the 9th month of the year 1771, the final report on a case of 'Stealing of fish by the commoner Xingfu 興福 and others inside Westpark' and the penalties imposed on the eunuchs and officials involved was issued. The document records that a group of commoners had frequently sneaked into the walled area of the Westpark at night to catch fish in the Middle Lake, until a certain Yang San 楊三 had been caught. A thorough investigation eventually brought to light the identities of his collaborators in Westpark, as well as the higher officials who should have exposed this clandestine undertaking much earlier, so needed to be punished as well. In passing, we learn from the document that 'in the 5th, 6th, 7th and 8th months, Yang San could not continue his fishing because the lake was densely covered with lotus leaves [which obstructed his fishing net]; after the 10th month the lake was frozen'.[44]

The paper trails regarding the regulatory administration of the imperial lotus business outlined above come in three levels of granularity: first, the *Regulations and Precedents* which froze into rules the once-administrative decisions triggered by a specific event; second, the annual reports in which the ups and downs, failures and problems appear in their solved, finalized form of cash flows; and finally, the small, day-to-day communication and bookkeeping archived in registers and inventories. Cases of misconduct and their investigations are often the only way to gain insights into the actual practices and lived realities inside the Westpark. Luckily, the Empress Dowager Cixi's fascination for photography provides us with a late glimpse of Westpark lotus and her personal use of it (see Figure 7.5).[45]

42 *Qianlong chao Neiwufu zouxiao dang* 乾隆朝內務府奏銷檔, vol. 367, 112-115 (rep. *Qing gong Neiwufu zoushaodang* 清宮內務府奏銷檔. Beijing: Gugong chubanshe, 2014, vol. 131).

43 'Gate control' (*menjin* 門禁) was an important issue, especially after 1887 when Cixi moved into the Westpark. See the Imperial memorial: 'On gate control and guard shifts at all gates of the Three Lakes' (Sanhai ge menjin ji shouwei zhangcheng 三海各門禁及守衛章程) of Guangxu 13/06/06 (26 July 1887) by Yixuan 奕譞 et al., which stipulated who could enter through which gate, needing what kind of authentication, and how the gates should be guarded, etc. See *Junqichu shangyudang* 軍機處上諭檔: 条 2, 盒號 1394, 冊號 4 (The First Historical Archives of China, Beijing).

44 The document was filed by the Shenxingsi 慎刑司 in QL36/09/14 (21 Oct. 1771), NWF no. 5-291-4: 'Zou wei minren Xingfu deng zai xiyuan nei tou yu 奏為民人興福等在西苑內偷魚'.

45 For a MIT *Visual Cultures* web-essay on Cixi and photography, see David Hogge, 'The Empress Dowager and the Camera. Photographing Cixi, 1903-1904'. Massachusetts Institute of

Figure 7.5 The Empress Dowager Cixi and attendants on Middle Lake

The photo shows Empress Dowager Cixi and her entourage on the imperial barge on Middle Lake impersonating Buddhist deities and their assistants. The shape of the Chinese character for 'longevity' (*shou* 壽) was added to the photo to simulate auspicious smoke rising from the incense burner. The photo was taken between 1903 and 1905 and became well-known by its inclusion into Princess Der Ling's *Two Years in the Forbidden City* (facing page 90). Courtesy of Freer Gallery of Art and Arthur M. Sackler Gallery (Archives Purchase, 1966, Xunling, FSA_A.13_SC-GR-243)

Interlocking Cycles

The growing and organizing of lotus was interlinked with numerous other areas of expertise, administrative units, and responsibilities, all of which were constrained in one way or another to the natural cycle of lotus growth and decay. The archival resources at our disposal translate all the different cycles, actions and relations into monetary equivalents in the form of income (rent, sale, or interest), payment or fine.

As mentioned earlier, until 1681 the Garden Bureau had to submit the income from its lotus rent and sale to the Grand Storage Office; from 1681 on the Garden Bureau was allowed to keep the lotus rent money for 'acquiring miscellaneous items' (*banmai lingxing wujian* 辦買零星物件). The ledger

Technology, 2011 Visualizing Cultures (http://vcarchive.com/mitvc/empress_dowager/index.html; accessed September 2020).

of Daoguang 18 (1839) translated above shows which items were eligible to be paid for with lotus rent money. Since 1850, in cases where funds were insufficient, the Garden Bureau was asked to invest the money and wait until enough interest (*shengxi* 生息) has accumulated; in the case of bigger tasks, the Ministry of Work would provide financial and logistic assistance.[46] Boats and sledges were part of the Westpark equipment used for pleasure rides and to shortcut across the lake by traversing the water or ice. Up until 1842, annual checks and necessary repairs were the responsibility of the Imperial Workshops; from that date the Garden Bureau was in charge and had to cover all expenses with lotus rent money. Other responsibilities were not transferrable: yellow cloth rugs (*huangbu wadan* 黃布挖單) used for wrapping the Westpark lotus roots before sending them to the Imperial Kitchens had to be ordered from the Grand Storage Office;[47] the acquisition of items such as felt, carpets, metal-pieces, and nails, remained in the hands of the Imperial Armoury, along with sharpening the sickles used in the Westpark.[48] Sickles, an everyday tool used to cut weeds and lotus leaves, vividly illustrate how work processes interconnected across bureaus and how objects were often assemblages of distributed responsibilities as well as of new and recycled materials. For instance, whereas the Imperial Armoury was responsible for sharpening sickle blades, replacing their handles had to be done by the Imperial Workshops, preferably by recycling the Garden Bureau's discarded poles used to steer the flat-bottomed boats across the Westpark lakes.[49]

Not all interlinks are readily evident. Whereas the *Garden Bureau Regulations* clarify the collaboration among the various offices involved in sickle maintenance, two major annual tasks related to the Westpark lotus remain almost invisible. These are, firstly, cutting and removing tonnes of lotus leaves and harvesting the flowers' seedpods, and secondly, harvesting the roots buried in the muddy lake beds, which was facilitated by 'trampling' on the lake ground to loosen it so that the roots could be pulled out.[50]

46 *Fengchenyuan xianxing zeli*, 176 and 179.

47 Ibid., 219 and 226.

48 Ibid., 179 and 185.

49 Ibid., 226.

50 When Xu Ke 徐珂 wrote in the early years of the 20th century, Erzha 二牐 ('Double Sluice') in the eastern half of Beijing was the main scenic spot to go for pleasure trips on boats. In his time Kunming Lake did not allow private boats and he claimed that Shichahai only had boats for harvesting lotus, which Xu called 'small lotus trampling boats' (*ta ou xiaochuan* 踏藕小船). Xu Ke, *Qingbai leichao* 清稗類鈔 (Beijing: Zhonghua shuju, 1984, 3rd printing 2003), vol. 1, 133 (*mingsheng lei* 名勝類: Erzha 二牐).

Moreover, lotus roots needed to be replanted every year in early spring. Again, it is by way of a case of 'carelessness and neglect of duty' that we can gain an insight into how this planting was managed and how gardens may have shared seedlings. When, in summer Qianlong 37 (1772), lotus plants were not developing as expected at the lakefront of Kunming Lake and other nearby ponds, an extra requirement of 5,525 catties of lotus tubers (*yang* 秧) for the subsequent year's replanting was calculated. The reporting officials estimated that 1,025 catties could be taken from the river at Xichunyuan 熙春園 (today's Qinghua Garden 清華園), the remaining 4,500 catties had to be bought on the market at 20 taels of silver per 1,000 catties. Another 4.158 taels had to be added to that to pay for the planters.[51] Two almost identical reports from 1812 and 1874 add a few more puzzle pieces to the picture, showing how people and materials moved the lotus machine and were simultaneously moved by it. Before winter, potted flowers and vines from the Westpark were transferred into storage pits (*gao* 窖) warmed with horse dung, the roots of bamboo plants in the gardens were covered to shelter them from the cold, weeds were picked and the grass cut for the final time that year and, last but not least, lotus leaves and algae at the Three Lakes were cut and removed. Fifty short-term hired labourers (*sula* 蘇拉) were assigned to move 221 baskets of horse dung from the Imperial Stables to the Westpark; 80 *sula* supervised by three foremen were needed to move the plants to the vaults; and hundreds of *sula* were involved in earthening up the bamboo: 160 *sula* plus 16 foremen at the Pavilion of Purple Light, 260 *sula* plus 26 foremen at Yong'ansi 永安寺 and 600 *sula* plus 60 foremen at Jingshan Park. An even more substantial number of labourers was needed for the water-bound work of cleaning the lakes of algae and dead lotus leaves. Here the reports list 30 *sula* per day for each of the lakes for a period of fifteen days, i.e. 1,350 *sula* working days.[52] Another report issued in September 1902 records the orally-transmitted urgent request to 'fish out and cut the lotus leaves and stalks at the Three Lakes without delay' (*ganjin jiang Sanhai he ye geng dalao jiejing* 趕緊將三海荷葉梗打撈潔淨), for which the department was seeking allowance to

51 NWF no. 05-0299-052: 'Chayi Qingyiyuan nei Gengzhitu deng chu zaizhong ouyang xishao zhi dachen He [erjing'e] deng fafeng shi 查議清漪園內耕織圖等處栽種藕秧稀少之大臣和[爾經額]等罰俸事 (Reviewing the request of the high officials He[erjing'e] etc. for punishment because of the meagre development of planted lotus tubers at Qingyiyuan, Gengzhitu and other places)', QL37/7/27 (25 Aug. 1772). The reporting officials voluntarily offered to refund the money to cover the additional costs. They further suggested punishing the garden deputies and foremen. But Qianlong found the self-denunciating high officials guilty of 'carelessness' (*shuhu* 疏忽), so eventually they were punished as well and three months of their salary was seized (ibid.).

52 NWF-FCY no. 4587 (JQ17/8 = Sept. 1812) and NWF-FCY no. 4593 (TZ13/8/12 = 22 Sept. 1874).

hire fifty additional labourers for a month to enter the premises every day and perform the task. The cost was computed as 2.5 *diao* 吊 (strings of copper cash; each string equals 1,000 *wen* of copper cash strung in ten bundles of 100 *wen* each) for food and salary per worker per day, plus expenses for tools such as sickles, baskets, ropes etc., adding up to a total sum of 3,698 *diao*.[53]

Rare evidence of actual sales of Three Lakes' produce is provided in a report about income and expenses of the Middle Lake between the 5th and 8th months of the year Guangxu 21 (end of May to mid-October 1895):[54]

- 3,040 catties of red-flower lotus roots (*honghua ou* 紅花藕) sold for 300 *wen* 文 per catty, in total 912 *diao*;
- 652 bushels (*ba* 把) of lotus seedpods (*lian peng* 蓮蓬) sold for 500 *wen* per bushel, in total 326 *diao*;
- 13,600 catties of lotus leaves (*heye* 荷葉) for 6 *diao* per 100 catties, in total 816 *diao*.

The document also names another agricultural product from the Middle Lake, 'chicken heads' (*jitou* 雞頭), or Euryale ferox or fox nuts, worth 4 *diao* per 100 pieces (*ge* 箇), with a total sale of 14,400 pieces in 1895. But the turning cogs of income and expenses barely matched each other. The expenses listed in the same report for water-bound starch-containing plants bought on the market – namely roots of white-flowered lotus, lotus seedpods, fox nuts and water caltrop (*lingjiao* 菱角) – added up to a total of 17,755 *diao*, more than six times the amount of the reported income of 2,630 *diao*. Moreover, whereas the palace sold its own lotus seedpods for 500 *wen* per bushel, it bought them on the market for 600.[55] This discrepancy appears especially stunning when one considers how consistently economically-minded the Imperial Household Department claimed to be. It instead reveals that the Westpark lotus served 'economic' functions beyond purely monetary ones.

Conclusion

> We only need a small number of things: oil for our hair, rouge, powder, incense and paper. These have to be given in a fixed amount to us and

53 NWF no. 03-6656-173: 'Wei Boa qingli Sanhai hemian yongguo qianwen shumu shi chengwen 為報清理三海河面用過錢文數目事呈文 (Report on the costs of cleansing the water surfaces of Three Lakes and rivers)', GX28/8/25 (26 Sept. 1902).

54 NWF-FCY no. 4606: 'Zhonghai chengbao shi qingdan 中海呈報事清單 (Ledger on reported tasks at Middle Lake)', GX21 (1895).

55 Ibid.

> our maids; of other costs that need to be paid there are of course brooms, baskets, etc. and fodder for the animals in the garden. All this can be covered by the income from selling the produce of the garden. This will save us 400 taels per year and the older maids we entrust with the business will also be able to earn some extra money for themselves.

The above quote comes from the 56th chapter of the *Dream of the Red Chamber* (Honglou meng 紅樓夢) in which the two highbrow teenage ladies Jia Tanchun 賈探春 and Xue Baichai 薛寶釵 discussed how to make economic use of the gardens that surrounded their mansions by assigning one of their older maids to manage each marketable plant species as its 'garden official'. According to the historian Yang Naiji, Cao Xueqin 曹雪芹 (fl. mid-18th century), author of the *Dream of the Red Chamber* and member of a high-ranking bannerman family, took the organization of the imperial gardens under the Garden Bureau of the Imperial Household Department as a blueprint for Jia Tanchun's management plan.[56] The anticipated profit from the sale of superfluous Westpark lotus roots under the direction of the garden deputies indeed appears to be an astonishingly exact model for the teenage ladies' plan.

In his study of Ming gardens, Craig Clunas identified three phases in the development of the Chinese garden. Originally an agronomic symbol of reclusive living and bucolic self-sufficiency, by the mid-Ming gardens had become experimental playgrounds for well-connected gentlemen specialists to collect and nurture plant varieties and exotic species. In the late-Ming, an age of rapacious commercial transactions, garden owners had to mask the economic value of the fruits grown in order to safeguard the social distinction between 'scholar' and 'merchant'. Gardens thus mainly featured as aesthetic entities to be evaluated according to the 'more slippery mechanism of taste'.[57]

In Westpark under the Qing Imperial Household Department, the seemingly conflicting identities of an economically-productive place and aesthetically tasteful space were ambiguous rather than mutually exclusive. Economic productivity was one of the values that the Qing sought to display publicly, making it one emblem of the place and space of the Westpark. Constructing one of the first working railway projects under Chinese aegis on the shores of Three Lakes was surely another example of this. Neither

56 Yang Naiji 楊乃濟, 'Tanchun li jia zhi suo ben 探春理家之所本', *Hongloumeng xuekan*, no. 4 (1981): 39-40.

57 Craig Clunas, *Fruitful Sites* (London: Reaktion Books, 1996), 77-78, 89-94.

of these activities was self-contained. The Westpark's Three Lakes and the 'lotus money' they generated were at the heart of a system that extended along the ecology of the waterways to the northwestern suburbs beyond the outer city walls of Beijing, economically interlinked with the market for goods and workers and that administratively connected several departments of the Imperial Household with each other.

In early 1906 the reconfigured Board of Work, then renamed the Ministry of Agriculture, Work, and Commerce (*Nong-gong-shang bu* 農工商部), requested the transmission of a large area at the Leshan travel palace into its jurisdiction in order to install an Agricultural Experiment Field (*Nongshi shiyan chang* 農事試驗場) – which later became the Beijing Zoo (see Chapter Nine). As a result, all the tenants in the area had to move out and their due rent was deleted from the books. Further rent deletions in the area were expected in the course of the construction of railways by the 'Railway company' (*Tielu gongsi* 鐵路公司).[58]

With the end of the Qing era approaching the palace machine began being dismantled and its parts exploited for other uses. It was not only former lotus fields that changed their purpose, ultimately in the 20th century the Westpark space itself was transformed. The lakesides of the Middle and North Lake were turned into a public park in 1925, while the southern part of the garden became Zhongnanhai, the highly-guarded area and residence of the Headquarters of the Communist Party of China and the Chinese president.

Figures 7.1 and 7.2 at the beginning of this chapter illustrate a double perspective that can also be detected in the archival sources. Whereas the photograph of 1900 shows Middle Lake completely covered with lotus, a sea of leaves, the woodblock print from 1830, taken from an almost identical viewing angle, instead depicts the lakes as an empty water space. Similarly, money trails and regulations as well as incidents and accidents on the ground reveal many aspects of the organizing of lotus in Qing imperial spaces and how lotus played its various roles in the 'palace machine'. Meanwhile, the messy business of cutting lotus leaves and pulling their roots out of the muddy lake bed remain almost invisible.

The presence of lotus in the Three Lakes partly defined the space of the Westpark, even financing some of the minor tasks required for its upkeep. Lotus was economically and visually meaningful. When, in 1923, the Swedish art historian Osvald Sirén took photographs of the Imperial City and gardens,

58 NWF no. 05-1049-068: 'Yingzheng huhua dizu weicao qianliang bing bennian yongyu yinliang shumu 應徵荷花地租葦草錢糧並本年用遇銀兩數目 (Expected income from lotus rent and reed money and this year's expenses)', GX32/12/18 (31 Jan. 1907).

the last Emperor Puyi was about to move out of the Forbidden City.[59] The photographs Sirén took of the Westpark lakes show that, by then, the area had fallen out of its agricultural use and into disarray; the end of Westpark lotus can be viewed as a symbol of the end of the Qing palace machine, along with the demise of the dynastic order itself.

About the Author

MARTINA SIEBERT works as subject librarian at the East Asia department of the Staatsbibliothek zu Berlin and as independent scholar. She researches and publishes on the history of Chinese exploration into nature and technology with a focus on the styles and agendas of presenting and organizing that knowledge in writing.

59 Osvald Sirén, *Imperial Palaces of Peking*. Paris: Van Oest, 1926.

8 The Medicine Supply System of the Qing Court

Xueling Guan

Abstract

To secure the supply of medicinal products for the Qing palace, the court took a series of measures to mobilize several institutions to participate in the operation. The operation resulted in a comprehensive, flexible, and multivariate supply system. Although many offices were involved in the system, all of them were under imperial control. The central government's strong vertical control over the regional societies facilitated the smooth operation of the system. The collaboration of various institutions ensured the functioning of medicinal projects for the Qing court.

Keywords: medicinal products, supply, court, Qing dynasty

During the Qing dynasty, pharmacotherapy – the consumption of prepared medicines – was the main form of medical intervention used in the imperial court.[1] To ensure the uninterrupted supply of the best medical materials for the benefit of the emperor's health, an entire medicine supply chain was established, mobilizing a host of institutions from the local to the central to participate in the processes of collection, procurement, preparation, dispensation, and disposal of medicinal ingredients and products. This chapter investigates the operations of this supply system and infers from it the Qing court's conceptions and attitudes towards materiality, efficacy, and profitability. Since both the imperial body and the medicinal substances administering to it were living entities, they were both in a state of flux and equally liable to decay. Thus, in order to harmonize the

1 The term 'medicine' in this article refers to Chinese medicinal ingredients, Chinese ready-made medicine, and Western medicines.

Siebert, Martina, Kai Jun Chen, and Dorothy Ko (eds), *Making the Palace Machine Work: Mobilizing People, Objects, and Nature in the Qing Empire*. Taylor & Francis Group, 2021
DOI: 10.5117/9789463720359_CH08

two vulnerable systems of bodies and medicines, the supply chain and its attendant accounting system needed to function with regularity and flexibility – as a reliable, predictable machine on the one hand which, on the other hand, could be adjusted to changing situations by ad hoc tinkering as required.

Collecting Medicinal Stuff: The Local Tributes

Traditional Chinese medicine put special value on the place-based authenticity or *au terroir* quality (*daodi* 道地) of medical materials. Tributary medical substances not only indicated imperial control of these regions, but also symbolically flaunted it. It was believed that only habitats of certain regions could produce the best kinds of medical materials, thereby determining the quality of the manufactured medicine, and eventually affecting its healing capability. The Qing dynasty continued its predecessor's practice of requiring local tributes from disparate corners of the empire, to ensure the supply of high-quality medicines for the court. As stated in the *Collected Statutes of the Ming* (Ming huidian 會典), 'all the provinces [have to] submit their annual collection in the original form of medicinal materials (*bense*本色) together with the converted silver cash (*zhese* 折色)'.[2]

This statute carried several messages: firstly, the provinces that produced medical materials were obliged to render them to the court as local tributes. This guaranteed the variety and *au terroir* authenticity of medicines. Secondly, the medicines could be rendered to the court in two alternative forms: in their 'original form' (*bense*) – as the medical material itself – or in a 'substitute form' (*zhese*), that is, the silver cash converted from the value of medical materials. Thirdly, the statute makes evident that, whereas each province had a fixed tribute quota, if the quota of medical materials could not be fulfilled due to poor harvest or other reasons, the shortfall had to be paid for in the substitute form of silver cash. This silver would then be used to purchase the materials that the Imperial Dispensary (*Yuyao fang* 御藥房) needed from the market. This system, with an internal backup measurement (or dual intake tracks), was designed to maximize supply, quality and flexibility by supplementing the age-old tribute system with market mechanisms. The former afforded regularity whereas the latter

2 *Da Qing huidian* (Yongzheng ban) 大清會典 (雍正版), eds. Yunlu 允祿 et al., *juan* 248, in *Jindai Zhongguo shiliao congkan sanbian* 近代中國史料叢刊三編 (Taipei: Wenhai chubanshe, 1985), vol. 790, 15731.

provided flexibility. The subliminal message seems to be that the emperor's health was everybody's business.

Renhe County 仁和縣 in Zhejiang Province was one of the empire's major producing areas for herbal medicinal materials. Its county gazetteer, printed in 1687, provides a detailed summary of the actual data showing how the local administration implemented the state tribute quotas between the 2nd year of Shunzhi (1645) and the 13th year of Kangxi (1674), i.e. in the early years of the Qing dynasty.[3] For instance, in 1645, the amount of submitted medicinal materials that was converted into silver cash reached twenty-seven percent. In the following year, the entire annual tribute was collected in the form of silver cash. Then in 1647, the collected medicinal materials consisted of the following substances: gold leaf, silver leaf, *maimendong* (麥門冬, Ophiopogon root, which soothes the stomach), *xusuizi* (續隨子, caper seeds, a cathartic medicine), *qianjincao* (千金草, Fortune Eupatorium herb, which cures heatstroke), *xiangbaizhi* (香白芷, Dahurian Angelica root, which cures headache and toothache), Chinese chestnut shell (*li ke* 栗殼, which calms nausea and coughing), and *mituoseng* (密陀僧, the oxidized lead ore litharge, which alleviates sores, eczema, and ulcers). Some of the products did not actually originate from Renhe but were imported from Wuhu 蕪湖, about 280 kilometres away from Renhe. In general, the court collected medical materials from their original habitats. But as time went on and some materials no longer grew in those areas, the court still maintained the previous rules and kept assigning a quota to them. In order to fulfil these quota, local officials would collect money from the community and purchase the respective medical materials from the habitats they were still available in.

The situations in 1648 and 1649 were the same as in 1646, when the entire tributes were converted into silver cash. The total amount of cash – including delivery expenses – was 68 *liang* (taels), 6 *qian*, and 4 *fen*. Between 1653 and 1661, both the assortment of materials collected and the amount of cash paid remained consistent. For those years the gazetteer also gives the weight of each collected medicinal: 64 catties (*jin* 斤) of *baizhu* (白朮, the root of white Atractylodes), 50 catties of *maimendong*, 500 sheets (*tie* 貼) of gold leaf, 100 catties of *xiangbaizhi*, 50 catties of *qianjincao*, and 20 catties of *xusuizi* – all

3 Zhao Shi'an 趙世安 (Qing) et al., *Renhe xianzhi* 仁和縣志, 28 *juan*, printed in the 26th year of Kangxi reign (1687) (rep. in *Zhongguo difangzhi jicheng: Zhejiang fuxianzhi ji* 中國地方志集成: 浙江府縣志輯, vol. 5), *juan* 6: tu gong 土貢 (local tributes), 28b-31a. See also the digital copy at Harvard-Yenching Library: http://nrs.harvard.edu/urn-3:FHCL:14361176, accessed September 2020.

considerable amounts. The value of the materials paid in substitute cash was in total 37.11 taels. Delivery expenses had to be added to both: for the medicinal materials this amounted to 3.89 taels, and for the cash to 13.98 taels of silver. In the first two years of the Kangxi reign (1662-63) the same amounts of medicinal and silver are listed. From the third to the 6th year of Kangxi (1664-67), all of the medicines were again collected in the 'converted form' (*zhese*); in 1668 some were delivered in the original form, others in the substitute form. Medical materials delivered in their original form included 60 catties of *baizhu*, 50 catties of *maimendong* and 50 sheets of gold leaf, while the equivalent of 50 catties of *qianjincao*, 20 catties of *xusuizi*, was paid in the substitute silver cash. Of *xiangbaizhi* 50 catties were delivered in their original form and the equivalent of 50 catties in substitute form.

From 1669 to 1674 all the medicinal materials were delivered in both forms, in kind and as substitute cash in a ratio of roughly one to two. Thus, 22 catties of *baizhu* were delivered in kind and 43 in cash. For *xiangbaizhi* the numbers were 33 and 66 catties, for *maimendong* 16 and 33, for *xusuizi* six and 13, for gold leaf 16 and 30 sheets. The delivery expenses followed this ratio: for in-kind material it amounted to 1.21 catties and for cash to 2.67 catties (see Table 8.1).

Table 8.1 Annual tribute of medicinal material from Renhe County

Year(s)	Original form		Converted form	
	Medical materials	**Quantity**	**Medical materials**	**Quantity/Value**
1653-1661	*baizhu*	64 catties		37,11 taels silver
	maimendong	50 catties		
	xiangbaizhi	100 catties		
	xusuizi	20 catties		
	qianjincao	50 catties		
	gold leaf	500 sheets		
1668	*baizu*	60 catties	*qianjincao*	50 catties
	maimendong	50 catties	*xusuizi*	20 catties
	xiangbaizhi	50 catties	*xiangbaizhi*	50 catties
	gold leaf	50 sheets		
1669-1674	*baizhu*	22 catties	*baizhu*	43 catties
	maimendong	16 catties	*maimendong*	33 catties
	xiangbaizhi	33 catties	*xiangbaizhi*	66 catties
	xusuizi	6 catties	*xusuizi*	13 catties
	gold leaf	16 sheets	*qianjincao*	50 catties
			gold leaf	30 sheets

Source: Zhao, *Renhe xianzhi*, *juan* 6, 28b-31a

In the period from the 4th year of the Kangxi era (1665) until the compilation of the Renhe gazetteer report twenty-two years later, the actual tribute of these medicinal materials including delivery expenses was reduced to 21.32 taels in silver. The alternation between the two tracks for collecting medicine within twenty-nine years in early Qing reveal flexibility and adaptability as an important aspect of the transforming palace machine.

Once collected, the transport of these tributary medicines also required careful management. After the prefectures and counties levied medical materials from their people, the materials and substitute cash had to be delivered to the designated central agency in the capital. The Qing statutes stipulated that 'as was codified at the beginning of the Shunzi reign, for all deliveries of original kinds of materials and substitute cash sent by the provinces the regional guardian (*shoudao* 守道) and provincial administrative commissioner (*buzhengsi* 布政司) shall assign honest and efficient officers (*lian gan* 廉干) to file and examine the documents, and dispatch labour for water and land transportation in order to deliver the materials to the capital before the deadline date'.[4] The attention paid to time management and the conflation of officials' moral quality and managerial skills are extremely similar to the case of the Gao-pu jade transport analysed in Chapter Six.

Upon arriving at the capital, the handover and inspection also had to follow a fixed procedure. The materials were first delivered to the Medicinal Material Warehouse (*Shengyao ku* 生藥庫) of the Imperial Academy of Medicine (*Taiyiyuan* 太醫院). The officials in the warehouse would inspect the quality of the delivered materials, count the quantities, and note the results down in documents. The warehouse was only responsible for evaluating and – in early Qing times – also for safekeeping the medical materials; it had no executive power over their use. That responsibility lay with the Board of Rites (*Libu* 禮部), which granted the reception and consumption of medical materials. From the 3rd year of Kangxi (1664) onwards a third institution became involved in the administration of medical materials. Due to administrative changes the materials were no longer stored in the Imperial Academy of Medicine, but in the Board of Revenue (*Hubu* 户部). Nevertheless, power over their delivery was still held by the Board of Rites. Thus, when the Imperial Dispensary needed any medical raw materials, the Imperial Academy of Medicine needed to file an application to the Board of Rites, which in turn would forward the request to the Board of Revenue, whereupon the materials could be picked up from the Medicinal Material

4 *Da Qing huidian* (Yongzheng ban), *juan* 47 (*Jindai Zhongguo shiliao congkan sanbian*, vol. 768), 2745.

Warehouse.[5] This collaboration between departments from both the inner court and the civil government mirrored the border-crossing trajectory of the medical substances.

Harnessing the Market: Medicine Merchants and the Court

Connecting the palace machine to the market mechanism, some of the medical materials needed by the court were purchased by commissioners, a job entrusted to prominent medicine merchants. According to the archives examined so far, the main medicine merchants or shops commissioned were the Wanquan House (*Wanquan tang* 萬全堂) and Tongren House (*Tongren tang* 同仁堂) in Beijing.[6] This arrangement followed a proposal by Liu Shengfang 劉聲芳, director of the Imperial Academy of Medicine, who stipulated that, from the 2nd month of the 1st year of the Yongzheng reign onwards (March 1723), Wanquan House and Tongren House should be in charge of purchasing medical materials, on an annually rotating basis in which each shop would be exclusively responsible for one year. In contrast to the tributary medicinal materials, these purchased items did not have to be inspected by the Imperial Academy of Medicine, but were delivered directly to the Imperial Household Department (*Neiwufu* 內務府), to be inspected and received by the Department's director. The Wanquan House or Tongren House was required to initially provide the medicinal materials at their own expense. Then, in the 8th month of each year, the Imperial Dispensary would report its actual yearly usage of medicines, in names and quantities, to the Imperial Academy of Medicine, which, together with the Ministry of Revenue, would inspect and review the market price of the medical materials, to then finally issue corresponding payments to the relevant merchant. In the 9th year of Yongzheng (1731), Lin Zucheng 林祖成, vice director of the Imperial Academy of Medicine, proposed setting the price permanently, in line with its market price in the year 1731, no matter how the market price might change in the future.[7] His proposal became an established rule for commissioned

5 *Da Qing huidian shili* (Guangxu ban) 大清會典事例 (光緒版), ed. by Kungang 崑岡 et al., 1899 (Beijing: Zhonghua shuju, reprint 1991), vol. 1105, 10.

6 After the middle of Qianlong's reign, Tongren House became the only merchant commissioned to purchase medicines for the court.

7 *Zou'an* 奏案, no. 05-0032-006, 'Caiban yaowei zhi weiyuan Yue Fengming deng cheng wei wu li dianban qing jietang banli yi an 采辦藥味之委員樂鳳鳴等呈為無力墊辦請借帑辦理一案 (On the case of medicine commissioners such as Yue Fengming who were unable to pre-pay for the medicines themselves and have requested an advance payment from the court)', QL 4/10/3

purchases for a long time subsequently. For the merchants this meant not only having to deliver the materials first and receive payment later, they also lost any influence over the pricing but had to take whatever was dictated as being the current price by the administrative unit; for any adjustments to the rising market price, they had to submit a request to the emperor himself, who would then decide whether or not to permit any price increase in their invoice.

The rules and procedures established to interlink the palace machine's cogs with the merchants' world were time and time again challenged by obstacles arising from the rules themselves and from the needs of the time. As the example above has shown, the rules were conceived to the detriment of the merchants. Being required to pay in advance for imperial purchases – a *de facto* no-interest loan to the emperor – created a capital crunch for the medicine houses. Especially when a natural disaster or epidemic drove up the prices of medicines, the merchants would suffer financial loss. Furthermore, since the payment to merchants was not based on the current market price, but in line with an artificial price set years earlier, the merchants were effectively made to shoulder the risks of price fluctuations by themselves. Maintaining an adequate supply of capital to navigate these uncertainties became one of the merchants' heaviest burdens. Therefore, on many occasions the merchants of Tongren House asked for adjustments to the standard procedures. They either petitioned to be relieved of their commission job, requested advance payments, or asked to be allowed to increase the set prices on their invoices.[8] Motivated in part by trust in the Tongren House and in part by a desire to maintain a regular supply of medicine for the court, the emperor permitted deviations from his own rule many times, by agreeing to Tongren House's entreaties. During the Yongzheng reign (1723-1736), for example, as much as 40,000 taels of silver was issued as advance payments.[9] In the 9th year of Qianlong (1744) an increase of one-third of the set price was permitted; in addition, in a number of the following years advance payments of 3,000 taels of silver were issued each year. In the 7th month of the 16th year of Daoguang (August 1836), the Tongren House received a loan of 1,000 taels of silver from the Bullion Vaults of the Grand Storage Office (*Guangchusi Yinku* 廣儲司銀庫), under the Imperial Household Department.[10] In sum, the court archives show that, from the early 18th century to the end of the Qing dynasty in 1912, the Tongren House

(3 November 1739). The *Zou'an* (official memorials submitted to the throne) cited in this chapter are all stored in The First Historical Archives of China, Beijing. The numbers refer to their shelfmarks.

8 *Beijing Tongrentang shi* 北京同仁堂史, ed. Beijing Tongrentang Group Co., Ltd. et al. (Beijing: Renmin ribao chubanshe, 1993), 22.

9 *Zou'an*, no. 05-0032-006.

10 *Beijing Tongrentang shi*, 20.

Figure 8.1 Medicinal bottle offered by the Tongren House to the court

Source: The Palace Museum, Beijing

merchants consistently petitioned for advance payments and price increases, with the emperor usually granting these requests. A complementary and interdependent relationship had evidently been established between the merchants of Tongren House and the Qing imperial court. The court needed Tongren House to purchase the medicinal material through its commercial network and their experts to guarantee the quality of these purchases, whereas the medicine shop capitalized on the prestige thus derived to further develop and gradually attain a monopolistic position in the industry (see Figure 8.1, which shows a bottle containing musk, *shexiang* 麝香, offered by the Tongren House to the court in the early 20th century).

Local Production and Court Monopoly: The Case of Donkey-Hide Gelatine

In parallel to these professional, commercial routes for acquiring medicinal materials, their systematic collection directly from their local sources remained an important activity of the imperial court. Some of these materials never reached the open market – at least not in the same quality – but were exclusively produced for the court. *Ejiao* (阿膠, donkey-hide gelatine from Dong'e county 東阿縣 in Shandong) provides an excellent example of such a local product, which the court safeguarded its exclusive access to. This traditional medicine was made from water and donkey skin, stewed with crystal sugar, yellow rice wine and herbal oil. It was used as a tonic to replenish a person's blood and nourish their *yin*. The quality of *ejiao* was highly dependent on two things – the raw materials used, especially the water quality, and the artisans' manufacturing skills. Jean-Baptiste Du Halde (1674-1743) gave an extensive account of the material and production of *ejiao* in his *Description de l'Empire de la Chine*.[11] The water used to make *ejiao* for the court had unique qualities. It came from a deep well close to the town of Dong'e County. 'Its Water is extremely clear, and more ponderous than common Water; if it be mix'd with foul Water, it refines it instantly, by precipitating the Filth to the Bottom of the Vessel'.[12] Since this water could filter impurities, the *ejiao* made from it was very clean. The local government assigned guards to protect and regulate access to the well. In addition, most of the year the well was closed with a seal of the local office, only being opened during the set season of making *ejiao* for the emperor. Political power ensured exclusive access to natural resources – in this case, the superb water and the best donkey skins. For the other main raw material healthy donkeys with strong muscles and shiny black coats were selected and skinned. *Ejiao* production was a very complex process with exacting standards. The manufacturing procedure comprised more than ten stages, including choosing and washing the materials, melting the skin, extracting, cutting and shaping, and weathering. The finest materials and artisanship were required to make top-quality *ejiao*. Hence, the local government directed the best local artisans to focus on making *ejiao* for the emperor.[13]

Storage and Recycling: When Medicine Became Stale

The data from Renhe county above have shown the large body of medicinal materials that the Qing court acquired every year via the tribute system. As

11 Du Halde's source of information on *ejiao* (Ngo-kyau) was a letter which Pater Parennin sent from Beijing, together with the actual ingredients, to the French Academy of Sciences in May 1723. See *Lettres édifiantes et curieuses écrites des missions étrangères* (Toulouse: Sens et Gaude, 1811), vol. 19: 'Memoire de la Chine', 242-261, here 253-254.

12 Jean-Baptiste Du Halde, *Description of the Empire of China and Chinese-Tartary* (trans. from the French, London: J. Watts, 1741), vol. 2, 230.

13 *Lettre edifiante et curieuses*, vol. 19, 242-261.

the efficacy of medicinal materials perishes over time, dealing with stale or out-of-date drugs was a standard issue faced by the Imperial Dispensary. The historical documents show that the Qing court adopted different approaches to deal with medicinals in various conditions. For instance, in the leap month after the 6th month of the 5th year of Qianlong (July/August 1740), the Imperial Dispensary checked the stored medical materials and found that fifty-three kinds had already been stored for a long period of time and had become stale. Thirty-five kinds of these out-of-date medicines were considered still usable, including cinnabar (*zhusha* 朱砂), orpiment (*cihuang* 雌黄) and ginseng cream. Of the other eighteen kinds some had become infested by worms so were unusable while for others – for example toluene (*duoermendina* 多而門第納) sourced from Western countries, root of silverweed cinquefoil (*yanshouguo* 延壽果) and *wannianjian* (萬年劍, identification unclear) – the doctors were no longer knowledgeable about their uses and effects.[14] Prince Hongzhou 弘昼, who was in charge of the Imperial Dispensary at that time, proposed that the medicines that were considered still usable should immediately be delivered to the Imperial Dispensary and that new ones would only be allowed to be distributed after those old ones had been used up. This suggestion evinced an awareness that can be described as conservationist and ecological, encapsulated in the Chinese saying, 'extracting the maximum use from a material' (*wu jin qi yong* 物盡其用). He ordered that the worm-damaged medicines and those of unknown usage should be disposed of immediately. Prince Hongzhou further demanded strict compliance with the rule that, for any unfamiliar medicines that entered the storage from then on, the Imperial Dispensary was responsible for inquiring about their usages and effects and keeping a record of these for future reference.[15]

Regularly checking the storage was actually an established routine for the Imperial Dispensary (see also the vignette essay introducing part three of this volume). Nevertheless, many more cases like this are documented in

14 The phenomenon of medicines with unknown effects and usage in the court store rooms mainly applied to materials from foreign countries or vassal states. When arriving at the palace as a gift or tribute, Western missionaries and doctors from the Imperial Academy of Medicine were assigned to examine them and the Imperial Dispensary was asked to note down the main effects in their records and stick yellow tags with the names onto the medicines' packaging. But over time, these documents got lost and therefore people no longer knew what ailments these medicines could be used to treat.

15 *Zou'an*, no. 05-0038-023: 'Heqinwang zouwei Yaofang tianshe zhushi deng wutiao shi 和亲王奏為藥房添設主事等五條事 (Prince Heqin requesting the Imperial Dispensary to add the position of a secretary and other together five issues)', QL 5/r6/13 (5 August 1740).

the archival materials, including the following example. In the 10th month of the 28th year of Qianlong (November 1763), the Imperial Dispensary detected forty-one kinds of outdated medicines in storage. Again they were categorized into three groups: those 'stored for years but with no known usages or effects', those which 'might be still usable', and those 'mildewed, rotten and no longer applicable'. For those of the last group, namely Cordyceps sinensis (*dongchong xiacao* 冬蟲夏草) and silverweed cinquefoil (*yanshouguo*), the office asked for approval and, only after the emperor had agreed, disposed of them. Among those considered 'still usable', fourteen kinds were kept for future use, but the rest was sent to the Chongwen Gate 崇文門 Tax Bureau to set a price for them and then handed over to the Tongren House merchant Zhang Shiji 張世基 as a trade-in for other medical materials.

In this shrewd exchange, no doubt facilitated by imperial power, old stock was effectively deployed as payment for new purchases. The Qianlong emperor ruled that, 'hereafter all the medical materials coming to the court from the outside, except for a reasonable amount that needs to be kept as a reserve for use by the court, the rest should be constantly accounted for [possibly to serve as a trade-in]'.[16] This shows the mercantilist mindset of the Qing court, whereby whatever was not consumed or kept in reserve should be traded back to the market while still fresh enough. Accurate reckoning and control of the movement of medicinal materials in and out of the court was key to maximizing profit. Later on, in the 4th month of Qianlong's 40th year (May 1775), the rule was extended to 'those medical materials collected and stored that are never actually used or which have only few potencies'. Again, the Chongwen Gate Tax Bureau fixed a price, for which it was handed over to Zhang Shiji to sell, and the expected profit was traded in for other medical materials currently needed by the court.[17]

In addition to Chinese medicines, numerous Western medical materials poured into the Qing court, especially during the reigns of the Kangxi, Yongzheng, and Qianlong emperors. The Dew Room (*Lufang* 露房) in Wuying Palace (*Wuying dian* 武英殿), a small storage room originally used for perfumes, was the main locale for storing these Western medicines in the Qing court, even though they were seldom used. When staff were sent to check the store room in the summer of the 19th year of Jiaqing (1814), in preparation for renovating

16 '嗣後凡有外進各項藥味內酌量存留備用外，其餘隨時酌量奏明，照此辦理'. See *Qinding Zongguan Neiwufu zeli er zhong* 欽定總管內務府則例二種, rep. in *Gugong zhenben congkan* 故宮珍本叢刊 (Haikou: Hainan chubanshe, 2000), vol. 306, 379-380.

17 Ibid.

the Dew Room, they found that '[the room] was crammed with bottles full of cloves (*dingxiang* 丁香), nutmegs (*doukou* 豆蔻), cinnabar oil (*rougui you* 肉桂油), etc. The oil has thickened into a paste-like substance, so hard that even a spoon cannot scoop it out'.[18] This finding shows that the Western materials had been left untouched there for a very long time. Even though they had lost all their medical properties, these materials were still valued as a curiosity. In the end, the Jiaqing emperor gave some of these materials to selected officials, and handed the rest over to the Imperial Workshops.[19]

As already mentioned, the Qing court had a system for collecting new and fresh medicinal materials and disposing of unused or stale and out-of-date ones – ideally by exchanging them for new materials. This promotion of movement and circulation of medical material by way of 'dispensing with the old to make way for the new', can be said to be the working principle of the court's drug management. Although the ways of treating aged materials varied from case to case and each decision had to await imperial approval, there were generally three options: either be economical and keep what was still usable for immediate use and dispose of what had gone off; trade in aging medicinal materials for fresh ingredients via professional merchants; or rededicate them as gifts to officials.

Making Medicine in the Palace Workshops

The Imperial Dispensary, as one of the offices having specialists responsible for making medicine, was ordered to collaborate with the labour force in several other related departments in the palace machine. This collaboration was not only a form of shared expertise but also a protection mechanism to prevent mistakes. In the case of donkey-skin gelatine outlined above, local manufacturing was needed to take advantage of the freshness of raw materials, the exceptional quality of the locale's spring water, and the specialist skills of artisans in Dong'e county. In many other cases, ingredients were assembled from various parts of the empire to be processed within the palace compound by the Imperial Dispensary and the Imperial Workshops into consumables in the form of balls, powders, creams, and pills.

The Imperial Dispensary was in charge of manufacturing and processing medicines in the court. Since its establishment in the 10th year of Shunzhi

18 Yao Yuanzhi 姚元之 (1783-1852), *Zhuyeting zaji* 竹葉亭雜記 (rep. Beijing: Zhonghua shuju, 1982), vol. 1, 21.
19 Ibid.

(1653), the Imperial Dispensary's major responsibilities had always included maintaining the supply and overall management of concocting medicines for the court. The job of actually making the medicine was undertaken by so-called grinding artisans (*nianyao sula* 碾藥蘇拉) and compounding doctors (*heyao yisheng* 合藥醫生), some of whom were recruited from the common people rather than the court elite. The medical officers at the Imperial Academy of Medicine participated in the processes as well. According to the chapter concerning the Imperial Dispensary in the *Current Regulations and Precedents from the Imperial Household Department* (Zongguan Neiwufu xianxing zeli: Yuyaofang 總管內務府現行則例: 御藥房), 'for all the processing and compounding works for balls, powders, creams and pills in the Imperial Dispensary, the Imperial Academy of Medicine is required to assign medical officers to collaborate with our own officers to supervise the works'.[20] In other words, in a system of mutual control and support, officials from the Imperial Dispensary and the Imperial Academy of Medicine would in all cases co-supervise the manufacturing processes of medicines together. This rule was proposed and then permitted in the 6th month of the 5th year of Qianlong (June or July 1740). The arrangement was made to prevent mistakes, but it also demonstrates the meticulousness and prudence of the imperial court towards the making of imperial medicine. Liquid medicine was no exception to this rule. To prepare liquid medicines for the use of the emperor and the empress, that process was similarly put under the supervision of both institutions, but the Imperial Dispensary would send a eunuch to function as cooperating partner to the imperial doctor from the Imperial Academy of Medicine. Two servings of medicine were produced from each decocting process. They were cooked as one but then placed into two separate vessels. The medicine decoction from one vessel would be tested first by the imperial doctor from the Imperial Academy of Medicine and then by the imperial eunuch sent by the Dispensary. If no negative effect was detected, the other vessel would be presented to the emperor. In reality, this system of mutual control was not always used, and in several cases officials from the Imperial Academy of Medicine did not personally take part in the decocting process. Sometimes they just submitted the prescription via the official channels and asked the Imperial Dispensary to create the medicine according to a recipe.[21]

The Imperial Workshops specialized in manufacturing various household items and representational accoutrements of the palace

20 *Qinding Zongguan Neiwufu zeli er zhong*, vol. 306, 390.

21 *Da Qing huidian shili* (Guangxu ban), vol. 1105.

Figure 8.2 Various patterns for medical tablets (*dingzi yao* 錠子藥)

Source: The Palace Museum, Beijing

and imperial family under the management of the Imperial Household Department. The court archives show that, for a considerable period, the Imperial Workshops and the Imperial Dispensary shared responsibility for manufacturing medicines. The Imperial Workshops' main task in this was producing medical tablets (*dingzi yao* 錠子藥), which was undertaken by the Workshop for Producing Medicine (*Zhiyao zuofang* 製藥作坊; Figure 8.2 shows a selection of patterns used on these tablets), one of the many specialist workshops it supervised. Now and then other workshops were also involved.

The Imperial Dispensary was an institution that specialized in the manufacture of medicines and was stuffed full of experts such as doctors and their collaborating artisans. But this was not the case for the medicine-specific workshop in the Imperial Workshop. The archival documents show that the Imperial Workshop seems to have been a general manufactory which needed to borrow the necessary expertise from the Imperial Dispensary since it filed a document requesting doctors to be transferred into the workshop in order to accomplish their task every time they were assigned one. After the medicines had been manufactured, the doctors were transferred back to the Dispensary. Nevertheless, the emperor sometimes worried about the quality of medicines. When Emperor Qianlong once inquired about the manufacture of medicine at the Imperial Workshop, Shuwen 舒文, chief supervisor of the Imperial Household Department (*Neiwufu dachen* 内务府大臣), replied:

> Every year we follow the established prescription, summon doctors from the Imperial Dispensary to process, compound and manufacture the demanded medicines, and we assign an additional doctor and a vice director (*yuanwailang* 員外郎) to supervise the work... Since Your Majesty has today inquired about the processing and compounding of the medicines in the Imperial Workshops this year, your servant [I] will personally, together with the doctors, lead the personnel to work dedicatedly and meticulously in processing and compounding.[22]

This sentence suggests that in certain cases, when medicines were produced by workers who might lack expertise, or generally to prevent cheating and mistakes, two mid-level officers were assigned to monitor the manufacture and take responsibility if anything went wrong.

Conclusion

The supply chain of medicinal materials in the Qing imperial court was conceived as a fully integrated and flexible whole. As one of the many supply systems in other parts of the palace machine, it fed on diverse sources. The various specialized offices involved in procuring medicine for the court were integrated into a chain of actions, each office in charge of one stage of the supply chain – from collection and delivery, to reception and inspection, storage, manufacture, and the final stage of disposal. Responsibility for the medicinal materials and all other duties connected to them moved from stage to stage and actor to actor and thereby connected the different administrative spaces. Starting far from the court at the level of local prefectures, county governments and provincial administrative commissioners, the medicinals moved into the sphere of the central agencies at the heart of the Qing empire, such as the Board of Revenue, the Board of Rites and the Imperial Academy of Medicine, to eventually arrive in the palace under the administration of the Imperial Household Department, with its Imperial Dispensary and the Imperial Workshops. They all served to

22 *Qing gong Neiwufu Zaobanchu dang'an zonghui* 清宮內務府造辦處檔案總匯, ed. by Chinese University of Hong Kong and The First Historical Archives of China (Beijing: Renmin chubanshe, 2005), vol. 50, 28.

maintain the supply of medicinal materials to the court and for the bodies of the emperor and his family, performing their jobs in a linked chain of actions and responsibilities, thereby monitoring and coordinating with one another.

The flexibility of the system came to the fore when conflicts arose between the established regulations, precedents or conventions and the practical cases. In these cases expediency became the problem-solving principle. The quota of medical materials for each province did not remain static but changed year by year in terms of whether the tributes were collected in their original form, converted into cash, or as a mix of both. The rule of acquiring medicinal materials from special court purveyors for fixed prices and on a commission basis was moderated by issuing merchants advance payments and granting temporary adjustments of the price of medical materials, in line with market fluctuations.

The system dealt with medicinal materials of diversified origins, which included local tributes collected from each province, commodities purchased through commissioned merchants, and concoctions manufactured in the medical and artisanal workshops in court. It is noteworthy that the court did not perceive any inherent difference between raw materials – be they plants, animal parts, or minerals – and prepared consumables from its own dispensary or from Western countries. All were considered 'medicine', i.e. *yao* 藥.

Therefore, although the medicine supply system of the Qing court might at first glance appear to be simply a matter of the court procuring medicinal materials, in fact it involved multilayers of institutions, ranging from local governments as providers of raw materials and specialized regional products, to commercial houses trading in medicinals and concoctions, further on to state ministries and institutions at the court which not only controlled, produced, stored and re-sold medicines, to finally prescribe, produce and administer medicines to the emperor and his family. Although each institution held different positions in an extended network, they all served the shared duty of supplying medicines to the court. Each of them performed their own routine duties within the seeming self-sufficient palace machine (*zizhuan* 自轉), which were often unrelated to medicine, and interlocked and worked collaboratively with public actors (*gongzhuan* 公轉) when drawn into the supply chain of provisioning imperial health.

Although the processes involved many institutions and a large number of personnel, ultimately it was the emperor himself who controlled the system and made decisions on activities regarding collecting, purchasing, manufacturing, and disposing of medicines. While the respective agencies carried out the specific tasks, all the important issues were reported to the emperor, who was the final arbiter (and direct benefactor of the medicines). In other words, a *chaine operatoire* was established, extending out from the central court to the local fields, with the imperial body firmly at its centre. In the final analysis, the vast capacities and knowledge of both the political system of the empire and the manufacturing network of the medical institutions and artisanal workshops in the palace compounds, as well as commercial actors, were commandeered to guarantee a continuous supply of medicines for the court. The collaboration between those institutions in a regulated but responsive chain ensured the timely and adequate supply of the best medicine in the empire to enhance and secure the health of the emperor and his family.

Translated from Chinese by Yuanxie Shi

About the Author

Xueling Guan 关雪玲 is a researcher (equivalent to curator) at the Palace Museum in Beijing. She specialises in the history of medicine of the Qing court on which she has published numerous articles and monographs such as *Qingdai gongting yixue yu yixue wenwu* 清代宮廷醫學與醫學文物 (The Court Medicine of the Qing Dynasty and Historical Medicinal Artifacts).

9 When There Is Peace, There Are Elephants

Hui-chun Yu

Abstract

The Forbidden City, the symbolic and actual centre of power of the Qing Empire, was crowded with animals. This chapter focuses on elephants, a crucial part of imperial ceremonial processions. As performers in various imperial ritual ceremonies from the early Qing to almost the end of the dynasty, elephants exemplified how the palace machine of performative emperorship domesticated and imperialized wild animals, made them subject to imperial rule and metaphoric paragons of imperial virtue.

Keywords: animals, Qing Empire, elephant performance, zoological garden

The Forbidden City, the symbolic and actual centre of power of the Qing Empire, was crowded with animals: not just as emblems carved on stones but real, embodied, living animals. They were kept in tiger rings, horse or elephant stables, dog kennels, eagle cages, and institutions such as the Imperial Birdcage (*Bainiaofang* 百鳥房), and the Imperial Bestiary (*Baishoufang* 百獸房). The Qing court employed these live animals to perform imperial roles in three contexts: state ritual ceremonies (*si* 祀), military force (*rong* 戎), and the tribute system (*chaogong* 朝貢). Whereas a great variety of animal species played their part in the tribute system and as diplomatic gifts from afar to Qing emperors, some species were assigned special roles: elephants and tigers were mainly used in ritual ceremonies; horses, dogs, and eagles for military force. Animals used as raw materials for luxury objects were even more pervasive: from ivory used in home furnishings, to rhino horn drinking cups, to precious furs worn by officials, to name just a few examples. Furthermore, visitors, staff and officials, as well as the emperor and his family were surrounded by representations of

---

Siebert, Martina, Kai Jun Chen, and Dorothy Ko (eds), *Making the Palace Machine Work: Mobilizing People, Objects, and Nature in the Qing Empire*. Taylor & Francis Group, 2021
DOI: 10.5117/9789463720359_CH09

animals made from stone or metal, such as bronze lions and jade elephants, embroidered animals on rank badges on official robes, and animal-like costumes on the court theatre stage. All of these animals – dead or alive, real or representative, actual or imaginary – were gradually 'imperialized' into political actors by the 'palace machine' which provided the framework and incentive to acquire, possess, domesticate, train, transform, rank and integrate them into the political performances of the Qing monarchy. I here use the term 'imperialized' instead of 'imperial' based on the argument that turning animals into political actors was a long-term process of instruction, not a pre-existing, innate condition. In the same way, a person could be born as a Qing prince but only 'become' a Qing emperor by performing emperorship through the 'stylized repetition of acts' in various imperial ceremonies that were considered characteristic of this role.[1] Performance, as stylized repetition of acts in this case, does not mean to pretend as what one is not but to make the imperialization work.

This chapter examines what mechanism or operative chain was needed to transform these diverse animals into functioning components of the Qing imperial court, and why it was thought necessary to 'imperialize' them, turning them into subjects of Qing imperial rule. To do this, the chapter considers the administrative system of the Qing monarchy in the Forbidden City as a complex, multi-functional palace machine that executed performative emperorship together with animals. This palace machine of performative emperorship consisted of at least three main parts conducting various administrative tasks and policy-making duties: the Imperial Household Department (*Neiwufu* 內務府), which created and staffed imperial space, the Court of the Imperial Clan (*Zongrenfu* 宗人府), which managed imperial personnel, and the Imperial Procession Guard (*Luanyi wei* 鑾儀衛), which paraded imperial ritualization. In order to produce efficacious emperorship, this palace machine had to coordinate with administrative systems external to the court, sometimes even with administrative systems external to the empire, in order to cast and choreograph humans, animals and objects into imperial ceremonial performers.

This chapter focuses on imperial elephants. As performers in various imperial ritual ceremonies from the early Qing to almost the end of the

1 Here I invoke Judith Butler's claim that 'gender is in no way a stable identity... rather, it is an identity tenuously constituted in time – an identity instituted through a *stylised repetition of acts*' [my emphasis]. See Judith Butler, 'Performative Acts and Gender Constitution: An Essay in Phenomenology and Feminist Theory', *Theatre Journal*, vol. 40, no. 4 (Dec. 1988): 519. On Emperor Qianlong's ritualizing power, see Angela Zito, *Of Body and Brush: Grand Sacrifice as Text/Performance in Eighteenth-Century China* (Chicago: University of Chicago Press, 1997).

dynasty, elephants exemplified how the palace machine of performative emperorship domesticated and imperialized wild animals, made them subject to imperial rule and metaphoric paragons of imperial virtue. Of all the live animals kept by the Qing monarchy administration system, only elephants and horses progressed through ranks according to a logic that copied that which promoted human officials. Their status increased with age and they were rewarded for being obedient. Both animal species have a long life expectancy and had the opportunity to be physically very close to the emperor during state ceremonies.[2] Through the prominent roles assigned to them, these animals were transformed into an extension of the emperor's will and body. Whereas horses were essential to the armed forces of the Manchu empire, elephants were designated emblems of the emperor's divine virtue. The presence of elephants was a sign that the mandate of Heaven still rested firmly in his hands. As the saying goes, 'when there is peace, there are elephants' (*taiping you xiang* 太平有象).

Elephants had been a crucial part of imperial ceremonial processions ever since the Song Dynasty. Why elephants, of all animals? In traditional Chinese political ideology, elephants were regarded as sentient beings capable of assisting sage kings; for instance, it was said that, 'in the faraway land of Cangwu 蒼梧 where the legendary sage king Shun (舜, reigned 3rd millennium B.C.E.) was buried, there were often elephants ploughing for him'. This implied that, 'wherever elephants reside, the land is fertile'.[3] Elephants serving the emperor thus became a symbol for a prosperous land and for its leader's good governance.

Since an elephant's life expectancy ranges from sixty to eighty years, it is reasonable to assume that some imperial elephants kept in Beijing might have served several emperors through the Ming and Qing dynasties. The Ming monarchy instituted the Imperial Elephant Stables (*Xiangfang* 象房) located near Xuanwu Gate 宣武門 in the southwest corner of Beijing's Inner City wall. The stable was under the administration of the Brocade-clad Guard (*Jinyi wei* 錦衣衛), an elite corps of eunuch guards who were also in charge of palace security and surveillance in the empire. After the Manchu had conquered the Ming, the Qing monarchy took over those elephants and their

2 He Xinhua 何新華, 'Qingdai dongnanya gong xiang maoyi yanjiu 清代東南亞貢象貿易研究', *Dongnanya yanjiu = Journal of Southeast Asian Studies*, no. 1 (2011): 82.

3 The original text reads '舜葬蒼梧之野，有群象常為之耕。……象之所在，其土必豐', see for example Chen Dazhang 陳大章, *Shi chuan mingwu jilan* 詩傳名物集覽, *juan* 3, 33a (*Siku quanshu* 四庫全書 edition. Reprint. Taipei: Taiwan shangwu yinshuguan, 2008).

Figure 9.1 Treasure and leading elephant of the Grand Imperial Procession as depicted in the *Illustrations to the Statutes of the Great Qing* (Da Qing huidian tu 大清會典圖)

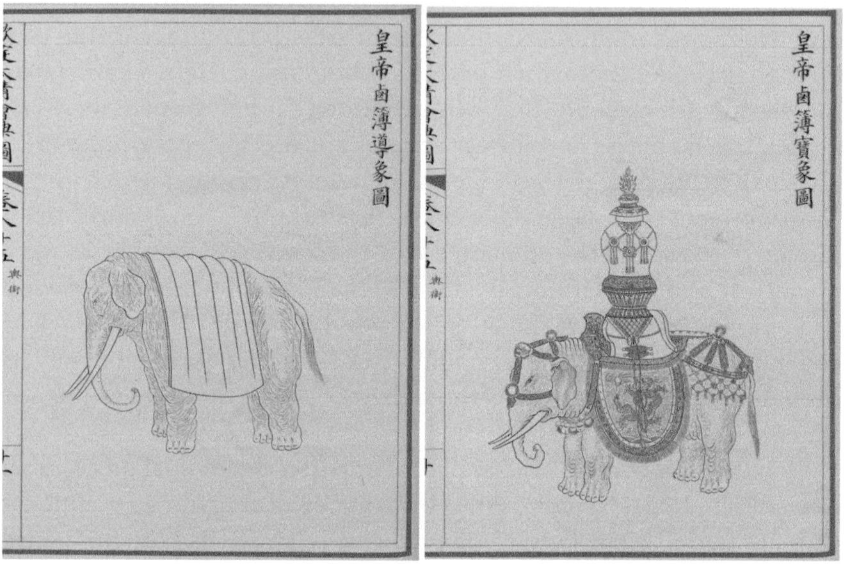

Source: Kungang et al., *Da Qing huidian tu*, *juan* 85, 10a, 11a. (Staatsbibliothek zu Berlin, Pekinger Slg. 4; http://resolver.staatsbibliothek-berlin.de/SBB000258D300031018 and SBB000258D300031020, accessed August 2020. CC BY-NC-SA 4.0)

stables but transferred their management to the Imperial Procession Guard.[4] The move emphasizes how crucial elephants were for Qing rituals – being indispensable for the set-up of the 'grand imperial procession' (*dajia lubu* 大駕鹵簿) and the 'ceremonial imperial procession' (*fajia lubu* 法駕鹵簿). The former arrangement was an integral part of all central imperial rites performed by the emperor himself, such as worshipping the Heaven and Earth, praying for harvests and praying for rain, while the latter was the procession set-up used for smaller-scale rituals, such as the daily gathering of officials at the imperial Outer Court (*chaohui* 朝會), and worshipping the Manchu ancestors at the Imperial Ancestral Temple (*Taimiao* 太廟). Nine elephants, five 'treasure elephants' (*bao xiang* 寶象) and four 'leading elephants' (*dao xiang* 導象) were on duty in the grand imperial procession (see Figure 9.1).

In order to sustain elephants as ceremonial guards in imperial processions, the monarchy had to inject a large amount of human resources and materials

4 Yu Minzhong 于敏中, *Qinding rixia jiuwen kao* 欽定日下舊聞, *juan* 72, 8a (*Siku quanshu* edition. Reprint. Taipei: Taiwan shangwu yinshuguan, 2008).

into a well-designed system that executed a series of logistical tasks: a steady source of elephant tributes needed to be maintained, elephants needed to be tamed and trained, someone had to guide them through the ritual performances, and so forth. One function of the palace machine of performative emperorship was not only to make those wild elephants become capable of performing specific acts according to ceremonial scripts, but also to tame those vassal states, such as Annam (Vietnam), Myanmar (Burma), and Cambodia, from which elephant tributes usually came. In other words, the entity of an imperial elephant was more than just its own physical body, it was also an extension of the Qing empire.

The chapter next demonstrates how imperial elephants helped to manifest the authority of the monarchy, then examines what happened to them once the power of the dynastic palace machine had dwindled.

Calling the Loyal Imperial Elephants

Similar to human subjects, elephants were deemed to qualify for appointment as imperial guards based on their seniority and loyalty – that is to say, whether they had been consistently obedient over a long time or not.[5] It was assumed that only elephants who were motivated to take on such a post and had shown their loyalty to the emperor were suitable. Lin Lu 林璐, a literatus in the 17th century, reported in his anecdote 'On Elephants' (*Xiang ji* 象記) that imperial elephants were recruited from among those animals who volunteered themselves instead of those that were taken by force. 'When recruiting elephants', Lin Lu explained, 'the territory keeper (*shoutuzhe* 守土者) goes into the mountains and announces that the imperial court is in need of guardsmen and any elephant answering the call will receive an official post. The territory keeper will only lead those elephants down the mountains who kneel down and lower their heads as if giving their assent (*nuo* 諾); he should not capture the animals by force'.[6]

Elephants were thus regarded as extreme loyalists. One anecdote mentioned in Yao Yuanzhi's 姚元之 (1783-1852) *Miscellaneous Notes from the Bamboo Leaf Pavilion* (Zhuyeting zaji 竹葉亭雜記) reflects on how people thought of elephants as 'sentient beings' (*lingwu* 靈物): 'In autumn 1717, while

5 Lin Lu 林璐, 'Xiang ji 象記 (On Elephants)', in *Yu chu xin zhi* 虞初新志, ed. Zhang Chao 張潮, *juan* 7, 63 (1700 edition; rep. *Xuxiu Siku quanshu* 續修四庫全書, vol. 1783. Shanghai: Shanghai guji chubanshe, 1995-2002).

6 Ibid.

marching with its fellow elephants, a sick imperial elephant stumbled and fell on Beijing's West Chang'an Street. This elephant first struggled to get up to its knees, then kowtowed three times to the north and three times to the west before collapsing. It died instantly. The kowtow to the north was to show its gratitude to the emperor; with the kowtow to the west it sent its regards to where it had been born'.[7]

In addition, imperial elephants of the former Ming dynasty, like human officials, were seen as 'remnant subjects' (*yimin* 遺民). It was said that, when the Ming rebel Li Zicheng 李自成 (1606-1645) stormed into the Forbidden City, all the imperial elephants lamented the dynasty's fall by howling and crying as Li's troops marched pass the Elephant Stable.[8]

The above-mentioned Lin Lu also wrote the following fictional anecdote about an elephant making a political stand during the Ming-Qing conflict:

> The Chongzhen 崇禎 emperor [r. 1627-1644] of the Ming Dynasty once asked for elephant tributes from the south. One of the elephants [in the herd] indicated that it was willing to come [to Beijing]. However, the Ming dynasty collapsed before this elephant was sent to the capital. Years later in our grand [Qing] dynasty, the same herd of elephants was summoned to serve as tributes to the new monarchy. Most elephants [of the herd] showed up except the one [that had previously agreed to serve the Ming]. A few days later, it did come but only to take its wives home. After all, the other elephants were still ready to go to Beijing. The territory keeper explained to this elephant that the new Son of Heaven was so blessed that the entire universe was at his disposal and asked how it dared to refuse to answer his call. The elephant once again refused to come. When the territory keeper set up a cannon to threaten the elephant the elephant slowly moved forward into the range of the cannon and waited. What brave and loyal behaviour that was![9]

Elephants were generally seen as loyal adherents who would rather die than betray their lord – just like the male elephant in the story above. But some were seen as turncoats (*erchen* 貳臣, literally 'officials who serve twice'), who willingly accepted the change of Heaven's mandate and swiftly transferred

7 Yao Yuanzhi 姚元之, *Zhuyeting zaji* 竹葉亭雜記, *juan* 2, 51 (1893 edition; rep. *Xuxiu Siku quanshu* 續修四庫全書, vol. 1139. Shanghai: Shanghai guji chubanshe, 1995-2002).

8 Yao Zhiyin 姚之駰, *Yuan Ming shi leichao* 元明事類鈔, *juan* 38, 7 (*Siku quanshu* edition. Reprint. Taipei: Taiwan shangwu yinshuguan, 2008).

9 Lin, 'Xiang ji', 63.

their allegiance to the new Son of Heaven – like the female elephants in the story. During the Ming-Qing conflict, being a 'turncoat' could be as precarious as being a loyalist remnant subject. Just like human officials, imperial elephants were portrayed as struggling to maintain their loyalty during the drawn-out and violent dynastic transition.

The depiction of such loyalty conflicts gradually ceased, as the stability of Manchu emperorship increased. Zhang Jiliang's 張際亮 (1799-1843) poem on the 'Elephants' Bathing Trip' (*Xi xiang xing* 洗象行) describes 'two elephants recruited from the previous dynasty that are too old to have any teeth left. But these two elephants are very docile and good at pulling the chariot, so they still follow the Son of Heaven on his tours of inspection'.[10]

Elephants were also sign of, and evidence of, Qing victory over a region. In 1646 Hong Chengchou 洪承疇 (1593-1665), an official in the Board of War,[11] reported to Qing Emperor Shunzhi 順治 (r. 1643-1661) that two tribute elephants from Yunnan 雲南 had been escorted by soldiers to Jiangning 江寧 (today's Nanjing 南京). To Hong, 'this proved that both Yunnan and Guizhou 貴州 have successfully been incorporated into the territory of the Qing Empire'.[12]

To survive the Ming-Qing transition, officials had to either position themselves as loyal remnants or turncoats; a third option was to betray both sides by trying to establish new dynasties of their own. Whichever option they chose, elephants functioned as pledge of loyalty and subservience. In 1655, in order to show their subservience and allegiance to the Qing Empire, Shang Kexi 尚可喜 (1604-1676), the governor-general of Guangdong, and Geng Jimao 耿繼茂 (?-1671), the governor-general of Fujian, endeavoured to deliver fourteen elephants from Guangzhou 廣州 to the imperial capital.[13] We do not know if Emperor Shunzhi 順治 actually received the animals or not. In any case, those elephant tributes were a political statement of

10 Zhang Jiliang 張際亮, 'Xi xiang xing 洗象行 (Elephants' Bathing Trip)', in *Wanqingyi shi hui* 晚晴簃詩匯, ed. Xu Shichang 徐世昌, *juan* 138, 3138 (1929 edition; rep. *Xuxiu Siku quanshu* 續修四庫全書, vol. 1629-33. Shanghai: Shanghai guji chubanshe, 1995-2002).

11 Hong Chengchou was one of the more successful 'turncoats'. He had originally sworn his allegiance to the Ming Chongzhen emperor but later defected to the Qing.

12 'Zhaofu Jiangnan daxueshi wei jie jin xiang shi shi 招撫江南大學士為解進象隻事 (Solutions from Jiangnan Scholars on Elephant Tributes)', in *Ming-Qing shiliao* 明清史料, ed. Zhongyang yanjiuyuan lishi yanjiusuo 中央研究院歷史語言研究所 (Taipei: Zhongyang yanjiuyuan Lishi yuyan yanjiusuo 1957-1958), *jia bian* 甲編, no. 6: 505.

13 'Pingnanwang wei jiesong xiang zhi shi 平南王為解送象隻事 (King Pingnan and Elephant Tributes)', *Zhongyang yanjiuyuan lishi yuyan yanjiusuo cang Ming-Qing shiliao* 中央研究院歷史語言研究所藏明清史料 (Ming-Qing historical materials stored at the Institute of History and Philology, Academia Sinica, Taipei), no. 150407-001.

peaceful surrender, although Shang and Geng later joined Wu Sangui 吳三桂 (1612-1678) to revolt against the Manchu emperor in 1668.

In sum, during the Ming-Qing transition tribute elephants and the loyal services they provided to the monarchy functioned as embodiments of the loyalty of the vassal states those elephants came from. As sentient beings who were considered to have volunteered for lifelong service and loyalty, elephants were welcomed as emblems of peace and stability in the realm.

Elephant Tributes for the Son of Heaven

Symbolism aside, transporting and looking after the animals on their way to Beijing posed considerable logistical and practical challenges. Thus, whether the delivery of elephant tributes was smooth or rough indicated the degree of stability of the Qing state in a tangible manner. The administrative system handling elephant tributes operated rather well during the 18th century. Thus, at certain times there were more than enough elephants at the disposal of the Qing Imperial Guards. At other times, however, when the elephant tribute system showed difficulties or problems, this was probably because the empire itself was in trouble. In 1853, for instance, the outburst of the Taiping Rebellion hindered the elephant tributes from arriving at the imperial capital. The elephants could not proceed north, so were temporarily kept in Yunnan.[14]

In order to guarantee tribute elephants and their trainers a safe journey from the southern edges of the empire to the imperial capital, a series of administrative systems – from the empire's borders to the court – needed to be activated. On the over 3,000-kilometre long journey the proper procedure for numerous tasks needed to be ensured, from crossing borders with the correct papers at hand, to food and accommodation at the various stations. In 1793 for instance, when the Gurkhas delivered elephants and horses to the Qing as tributes after the Sino-Gurkha War, General Fuk'anggan 福康安 (1754-1796), who was the governor-general of Guangdong and Guangxi (*Liangguang zongdu* 兩廣總督) at the time, devised a detailed security plan for the trek from Gurkha through Tibet to Beijing. Fuk'anggan personally accompanied the excursion to Tibet and frequently exchanged memorials with Emperor Qianlong to confirm the progress of the trip along stations

14 Wu Zhenyu 吳振棫, 'Hao gong xiang xing you xu 後貢象行有序 (Preface to Elephant Tributes)', in idem., *Huayiguan shi chao* 花宜館詩鈔, *juan* 15, 165 (1865 edition; rep. *Xuxiu Siku quanshu* 續修四庫全書, vol. 1521. Shanghai: Shanghai guji chubanshe, 1995-2002).

on its way to Beijing. Hosting a tribute excursion on its way to the capital could be a logistical challenge and financial risk for a place. But sometimes it could also have its profitable sides, as when the emperor instructed that both the Dalai Lama and the Panchen Lama in Tibet should receive an elephant when the expedition passed through their territory, and that all the tribal chiefs involved in this tribute campaign should be awarded as well.[15] The painting *Elephant and Horse Tributes from the Gurkha Campaign* (Guo'erke jin xiang ma tujuan 廓爾喀進象馬圖卷) by Prince Hongwu 弘旿 (1743-1811) depicts the conditions that elephants and horses might have experienced along these tribute journeys and also shows how they attracted large groups of commoners and children as onlookers.[16] Not only did these tribute campaigns take elephants to the capital, on their long journeys to the capital the elephants also served as a display of the omnipresence and omnipotence of imperial power. Ordinary people who might never have the chance to travel to the imperial capital themselves could at least imagine that, even though the emperor was as far away as the sky, he was able to exercise his power to tame those giant elephants as well as the faraway states they came from. Furthermore, the tribute campaigns also served as periodic tests to ascertain if there were any loopholes within the empire's airtight bureaucracy.

Tribute journeys were an endurance test for the administrative system. They were full of potential difficulties and problems and thereby challenged the stability, but also the flexibility, of several of the empire's internal administrative mechanisms. On 17th August 1875, a news report in *Shenbao* 申報 stated that a diplomatic corps from Burma with seven elephants were on their way to the capital.[17] They were scheduled to travel through several provinces – from Guizhou, Hunan 湖南, Hubei 湖北 and Henan 河南, before finally reaching Zhili 直隸. After arriving in Hubei, they took a substitute, slightly Western route via the Jingzhou 荊州 prefecture since the scheduled receiver in Wuhan 武漢 could not manage to host them. Therefore, a temporary elephant ring was set up from bamboo stalks outside

15 'Fukang'an deng zou Kuoerka jingong xiangzhi mapi zaitu xingzou qingxing zhe 福康安等奏廓爾喀進貢象隻馬匹在途行走情形各摺 (Memorials from General Fuk'anggan regarding Elephant Tributes from Gurkha), in *Huangchao xu wenxian tongkao* 皇朝續文獻通考, ed. Liu Jincao 劉錦藻, *juan* 62 (rep. *Xuxiu Siku quanshu* 續修四庫全書. Shanghai: Shanghai guji chubanshe, 1995-2002).

16 See the reproduction of the painting on the website of the Palace Museum in Beijing, https://digicol.dpm.org.cn/cultural/detail?id=092965fe51b644fcb84d98ca88be5630, accessed June 2020.

17 All *Shenbao*'s news reports about the Qing court were transcribed from the official reports of *Jingbao* 京報 or *Dibao* 邸報, the daily news published by the court.

the south gate of the town of Jiangling 江陵, which attracted 'children, women and idlers from the city' (*hong tong hong nü yiji youshou haoxian zhi bei* 黃童紅女以及遊手好閒之輩). Jiangling's local authority failed to control the excited crowd that had rushed over to see the elephants, and some impetuous people even climbed the bamboo wall to have a better look. When some of them were hurt by an elephant which hit them with its trunk, people started panicking and knocked down parts of the bamboo wall. Two elephants nearly escaped. Over thirty people died or were injured in the incident, mostly because they trampled upon each other.[18] The news report did not relate whether any officials were punished for this tragic event, but there was clearly a need to prevent further accidents. To better guard and facilitate the onward journey of the diplomatic corps, the inspector-general of Henan Province, Liu Qi 劉齊 ordered the prefectural magistrate Chen Yuanhao 陳元誥, who was to host the corps at their next station, to receive them in person at the border between Hubei and Henan.[19]

When the supply of imperial elephants became insufficient during the late Qing, the Imperial Household Department put provincial governors directly in charge of purchasing elephants and sending them to the capital. Whether the elephants came from domestic purchase or foreign tributes, the administrative and financial burden of their hosting and delivery had to be borne by the provincial governors in charge.[20] In addition, if anything went wrong, such as the Jiangling tragedy in 1875 recounted above, it would result in further administrative trouble and the chance of punishment. Therefore, local chiefs generally did their best to either deter or postpone any such delegations or tribute missions from passing through their county.[21] Meanwhile, the empire was also dependent on the internal situation in Burma and Vietnam for its purchase of elephants from those countries. The absence of political stability in Burma or Vietnam would lead to an absence of elephants.

The number of elephants kept varied during different reigns of the Qing period. The changing number suggested the scale, and probably depreciation, of the palace machine in this regard. According to the *Archive of the Imperial Procession Guard* (Luanyiwei quanzong dang 鑾儀衛全宗檔), the Imperial Elephant Stables housed between seventeen and 30 animals in the Qianlong

18 'Miandian gongshi gaidou 緬甸公使改道 (Burma Envoy Changes His Way)', *Shenbao* 申報, 17 Aug. 1875.

19 'Guangxu yuannian bayue ershi ri Jingbao quanlu 光緒元年八月二十日京報全錄 (News from the Capital, Guangxu 1, 20th day of the 8th month), *Shenbao* 申報, 30 Sept. 1875.

20 Mao Xianmin 毛憲民, 'Qingdai Luanyi wei xunxiangsuo yang xiang 清代鑾儀衛馴象所養象', *Zijincheng = Forbidden City*, no. 3 (1991): 17.

21 'Dijing jishi 帝京紀事 (Notes from the Imperial Capital)', *Shenbao* 申報, 12 Oct. 1892.

reign (1736-1795) and between sixteen and 35 in the Jiaqing 嘉慶 reign (1795-1820). From then on the number decreased. In the Daoguang 道光 reign (1821-1850) there were between eleven and fifteen elephants, and in 1866, the Xianfeng 咸豐 reign, only six remained.[22] The number went further down until there were only four imperial elephants left to perform ritual duties in 1883,[23] and, by 1893, only one young elephant was in the stable.[24]

Imperial elephants were not just role-bearers and emblems, they also were physical bodies that needed care and fodder. Each elephant consumed at least 160 *jin* 斤 (catty, about 94.4 kilograms) of straw every day, some of which came from the imperial fields managed by the *Fengchenyuan* 奉宸苑, (the Bureau of Imperial Gardens and Parks, henceforth 'Garden Bureau'), and the rest was purchased on the market. When in 1887, for instance, the imperial fields could not supply enough straw to feed the four imperial elephants, the Garden Bureau had to pay 129 taels of silver to buy an additional 227,200 catties of straw, i.e. 32,457 bundles in total.[25] In fact, it cost more to feed imperial elephants than carnivorous beasts.[26] As early as in the Yuan dynasty (1271-1368), the poet Ai Xingfu 艾性夫 (n.d.) had criticized the waste of imperial resources in keeping elephants for what he considered to be a system that presented a false appearance of peace and prosperity. Ai argued that what people really needed was not elephants, but good harvests.[27]

Elephant Training and Performances

The Imperial Elephant Stable was located outside of the Xuanwu Gate 宣武門 at the southwest corner of the Inner City wall of Beijing (see Map 2 at the

22 Mao Xianmin, 'Qingdai Luanyi wei xunxiangsuo yang xiang', 17.

23 'Jingdu jinxin 京都近信 (Recent news from the Imperial Capital)', *Shenbao* 申報, 7 Sept. 1883.

24 'Fengbi zanhao 楓陛簪毫 (Notes in Autumn)', *Shenbao* 申報, 29 Oct. 1893.

25 'She caochang 設草廠 (Establishment of Imperial Rice Straw Hall)', in *Da Qing huidian* shili (Guangxu ban) 大清會典事例 (光緒版), ed. by Kungang 崑岡 et al., *juan* 21, 1864 (rep. in *Xuxiu Siku quanshu* 續修四庫全書, vol. 798-814. Shanghai: Shanghai guji chubanshe, 1995-2002).

26 According to an entry in an accounting report (*zouxiao dang* 奏銷檔) of 1740, a mature tiger consumed around five catties of meat every day. If one catty of meat cost 40 copper coins (*wen* 文), then the court spent at least 73,000 *wen* of copper coins a year to raise one tiger, or 73 tales of silver a year. See 'Ying gou shi ge yong yin 鷹狗食鴿用銀', in *Qianlong chao zouxiao dang* 乾隆朝奏銷檔案, microfilm no. 353-354, Academia Sinica, Taiwan.

27 Ai Xingfu 艾性夫, 'Annan gongxiang 安南貢象 (Elephant Tributes from Annam)', in idem, *Sheng yu* 剩語, *juan* 2, 25 (*Siku quanshu* edition. Reprint. Taipei: Taiwan shangwu yinshuguan, 2008).

beginning of this volume, no. 8). The gate was open to officials who were granted the opportunity to visit the imperial palaces. On their way they could watch the elephants being fed or, by paying the trainers some small change, request the elephants to perform some simple tricks. For instance, Ji-Won Park 朴趾源 (1737-1805), a Korean envoy who visited Beijing in the 18th century, reported visiting the Imperial Elephant Stable where he watched the animals following their trainers' orders to perform tricks such as kneeling, kowtowing and then howling, waving their trunks, making sounds like flutes and pipes, and stumping their feet.[28] In addition to charging visitors to watch the elephants perform, the trainers at the Imperial Stable also made profits by engaging in other businesses, for example, selling elephant dung. In 1893 when there was only one young elephant left in the stable, the market price of elephant dung was sky-high.[29] When an imperial elephant died, its skin and bones were removed and stored at the Imperial Armoury (*Wubeiyuan* 武備院) in case the Imperial Dispensary (*Yuyaofang* 御藥房) needed them as medicinal ingredients (on the Imperial Dispensary see Chapter Eight).[30] Ivory was another valuable material for making luxury decorative objects.

Twelve full-time professionals shared the task of elephant-training in the Training Office (*Xunxiangsuo* 馴象所).[31] In preparation for every imperial procession, the Training Office had to carry out several rehearsals to ensure that the elephants were familiar with all the steps involved. In order to prepare for the ceremony of Worshipping the Earth (*jiaosi* 郊祀), for instance, in 1887 'the Imperial Possession Guard official ordered the keeper of the Elephant Training Office to take the elephants to the Imperial Chariot Garage (*Luanjiaku* 鑾駕庫) on the 26th day of the 4th leap month of the lunar calendar so that they could have a drill with the full set of five chariots and all other necessary equipment'.[32] The Qing grand imperial procession demanded five 'treasure elephants' (*bao xiang* 寶象) carrying bronze vases and four 'leading elephants' (*dao xiang* 導象) to walk in the procession (see Figure 9.1). In other words, nine elephants had to take the field together.

28 Ji-Won Park 朴趾源, 'Huangtu ji lue 黃圖紀略 (Notes on the Forbidden City)', in *Rehe riji* 熱河日記, *juan* 5, 327 (rep. Shanghai: Shanghai shudian, 1997). A similar performance at the elephant stable is given in a record from the Russian diplomatic corps visiting the imperial capital in 1692. See Chen Qiyuan 陳其元, 'Pin meng riji 聘盟日記 (Diary on A Diplomatic Trip)', in idem, *Yongxianzhai biji* 庸閒齋筆記, *juan* 5, 56-62 (1874 edition; rep. *Xuxiu Siku quanshu* 續修四庫全書, vol. 1142. Shanghai: Shanghai guji chubanshe, 1995-2002).

29 'Feng bi zan hao 楓陛簪毫 (Notes in Autumn)', *Shenbao* 申報, 29 Oct. 1893.

30 'Chun ming ji shi 春明紀事 (Notes in Spring)', *Shenbao* 申報, 5 June 1886.

31 Liu Jincao, *Huangchao wenxian tongkao*, *juan* 86: 5.

32 'Jing chen za lu 京塵雜錄 (Some Notes on Life in Imperial Capital)', *Shenbao* 申報, 24 June 1887.

During the late Qing, however, the number of imperial elephants in the procession was gradually reduced, due to a lack of available animals. In 1888, with only four elephants left in the stable, only one elephant walked in each procession of the ceremony of Worshipping the Earth.[33]

Some aspects of caring for the elephants were also entertaining for commoners in Beijing. For instance, the Elephant Bathing Day (*xixiang jie* 洗象節) was an annual event hosted by the Imperial Elephant Stable on the 6th day of the 6th month in the lunar calendar. It was a favourite summer festival for Beijing citizens, when all the elephants were led from the stable to the moat outside the Xuanwu Gate to bathe themselves in turn. One report in the newspaper *Shenbao* noted that, 'on that day the site would be packed with people from all over the place and looked like it was crowded with enormous ants'.[34]

Elephants' Impromptu Performances

Glitches occurred in the colossal palace machine as it tested the limits imposed by natural constraints. Qing imperial elephants were originally born and bred in countries with a warm climate all year round. When delivered to the north and faced with a severely cold winter and the regimen of long and harsh training to fulfil their imperial procession duties, some of these elephants turned into giant, lethal beasts.

In 1882, one particular elephant was responsible for three serious incidents. The first happened on the 1st day of the 7th month of the lunar calendar, when a trainer was killed by the elephant's huge tusks when he failed to control the animal properly while passing through the city at Eastern Chang'an Street on the way to a ceremony held at the Imperial Ancestral Temple.[35] About three months later, that same elephant became out of control again and stepped on and killed a street vendor near the Western Chang'an Gate.[36] Just a few days later the elephant escaped from the stable at night and trampled an old lady to death while running through the narrow alleyways of Beijing. The residents in the area neighbouring the Imperial Elephant Stable understandably panicked, and more than twenty

33 'Jiao si gong ji 郊祀恭紀 (Records of Earth worship)', *Shenbao* 申報, 10 Jan. 1888.

34 'Yu yuan chan zheng 御苑蟬聲 (The Sound of Cicada in Imperial Garden)', *Shenbao* 申報, 31 July 1893.

35 'Jingdu zawen 京都雜聞 (Some News on Life in Imperial Capital)', *Shenbao* 申報, 3 Sept. 1882.

36 'Dumen jinshi 都門近事 (Current Events in Imperial Capital)', *Shenbao* 申報, 19 Nov. 1882.

trainers from the Elephant Training Office were dispatched to track down the escaped elephant, eventually managing to capture the animal.[37] Due to these three incidents, on the next year's Elephant Bathing Day, only the other three elephants were allowed to have their bath in the moat. However, one of those, a female elephant, became angry when a trainer forced her to move forward. The creature 'rolled and waved its trunk, shook its head and roared loudly'. Most of the onlookers were so frightened that they instantly ran away, leaving their own belongings behind. In order to avoid any potential accident, the elephant trainers did not allow that elephant to have her annual bath either.[38] Furthermore, the *Shenbao* reports between 1886 and 1887 suggest that from then on, when rituals required imperial elephant processions, experts would first determine whether the elephants were tame and tractable enough to perform their duties; if not, they were made to stay in their stable to avoid any possible trouble.[39] The *Shenbao*'s last report on this issue appeared at the end of 1887: 'One of the imperial elephants has been suffering from a bad temper from time to time. Some days ago it became so angry that it attempted to break the gate bolt and escape; fortunately, a gate keeper arrived and stopped it before the incident got out of hand'.[40] This event led to the final decision to no longer parade elephants in imperial processions.

According to the memoirs of Lady-in-Waiting Der Ling 德齡 (1885-1944), in 1903 the Empress Dowager Cixi had once commanded a Russian circus which was performing in the Imperial Capital at the time to set up tents in the radish field beside the Kunming Lake 昆明湖 of the Yiheyuan 頤和園, the new summer palace northwest of Beijing. Around 200 imperial officials and royal family members then watched a performance. Empress Dowager Cixi had insisted that no dangerous beasts, such as lions and tigers, were allowed to enter the palace or take part in the show, but she did not object to elephants. She was so impressed by a baby elephant that it was later given to her by the ringmaster as a gift.[41] The empress dowager accepted the gift, but none of the palace staff knew how to make the elephant perform as it had done in the circus show, so it was sent to the Imperial Elephant Stable to live with the other elephants. If Der Ling's account was true, it would

37 'Xuanxiang fu 馴象復 (Elephant Training)', *Shenbao* 申報, 26 Nov. 1882.

38 'Jingdu jin xin 京都近信 (Current News in Imperial Capital)', *Shenbao* 申報, 7 Sept. 1883.

39 'Nian xia jiwen 輦下紀聞 (Notes on His Majesty)', *Shenbao* 申報, 28 June 1887.

40 'Jin tai qiu xu 金臺秋旭 (Autumn Morning in Imperial Capital)', *Shenbao* 申報, 1 Nov. 1887.

41 Der Ling, *Two Years in the Forbidden City* (London: T. Fisher, 1912). Thanks to Prof. Christine Moll-Murata for bringing my attention to Der Ling's memoirs. On Der Ling's and other palace ladies' memories see Moll-Murata's Chapter One.

mean that elephants were still being kept in the stable in 1903. Be that as it may, the Qing Imperial Elephant Stable did not acquire any new animals and was eventually closed down after all the elephants had died.[42]

The End of Imperial Elephant Performances

As elephants gradually lost their role in Qing imperial ceremonies and rituals, because of their diminishing numbers and the fear of incidents such as those recounted above, circuses presenting shows co-performed by human beings surfaced as a new form of animal performance. Elephants and other exotic animals gained a new role in the scheme of Western commercial performances and exhibitions, either in the form of circuses or in small-scale private zoos. In the beginning both emphasized the entertaining function of animals. By the end of the 19th century, there was already an international trade in animals to feature in circuses and profit-making private zoos. The earliest mention of European circuses performing in China in the *Shenbao* newspaper can be traced back to 1874.[43] On 14th June 1882, G. Chiarini's Royal Italian Circus and Performing Animals arrived in Shanghai, after touring big coastal cities in Asia. To promote their show, which included tigers, monkeys, kangaroos, bears and others, they launched a series of advertisements in *Shenbao*.[44] The spectacle was so popular that this circus visited Shanghai again in 1886 and 1887. The illustrated supplement to the *Shenbao*, the *Dianshizhai Pictorial* (Dianshizhai huabao 點石齋畫報), depicted the densely packed Western and Chinese spectators at the Chiarini performances and the scenery of caged wild animals displayed outside the show (see Figures 9.2a and b). In 1889, some of the older, retired animals from this circus, including tigers, lions, bears and monkeys, were sold to the Yangshupu Garden (Yangshupu da huayuan 楊樹浦大花園) in Shanghai.[45] The Yangshupu Garden was a privately-owned, Western-style botanical and zoological garden which opened to the public in 1889 and displayed wild

42 'Xiang fang 象房 (Imperial Elephant Stable)', *Guanhua zhuyin zimu bao* 官話注音字母報 (Official Language Newspaper), vol. 24 (1918): 27 (rep. in *Zhongwen qikan quanwen shuju ku* 中文期刊全文数据库, 1911-1949, edited by Quanguo baokan suoyin, Shanghai).

43 'Xu lu xiguo maxi 續錄西國馬戲 (Continued Records on Western Circus)', *Shenbao* 申報, 1 July 1874.

44 'Zhi shou qian 鷙獸起岸 (Birds and Beasts Arrive Ashore)', *Shenbao* 申報, 15 June 1882. Advertisements where launched on the last pages of the *Shenbao* in June and July of 1886 and 1887 and between late April and mid July in 1889.

45 'Zhong shou de suo 眾獸得所 (All Beasts Settle Down)', *Shenbao* 申報, 9 July 1889.

Figure 9.2a-b 'Western Show [i.e. G. Chiarini's Royal Italian Circus and Performing Animals] Arrived Again (in Shanghai)' (*xi xi chong lai* 西戲重來)

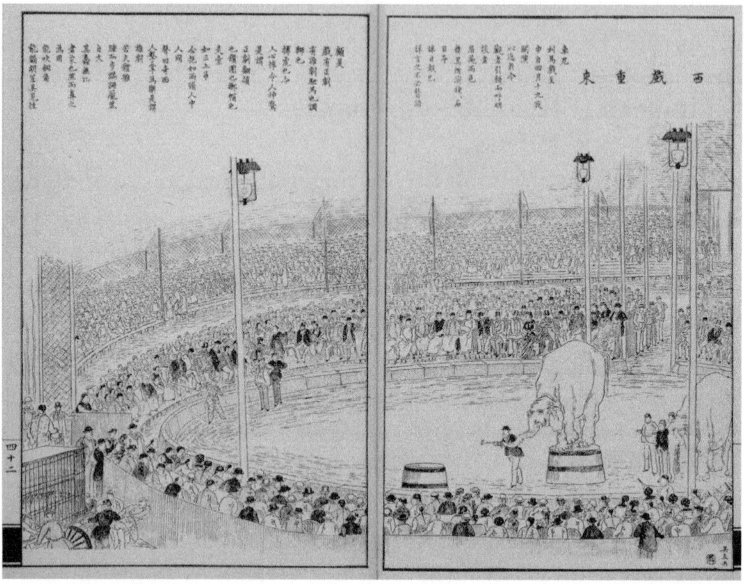

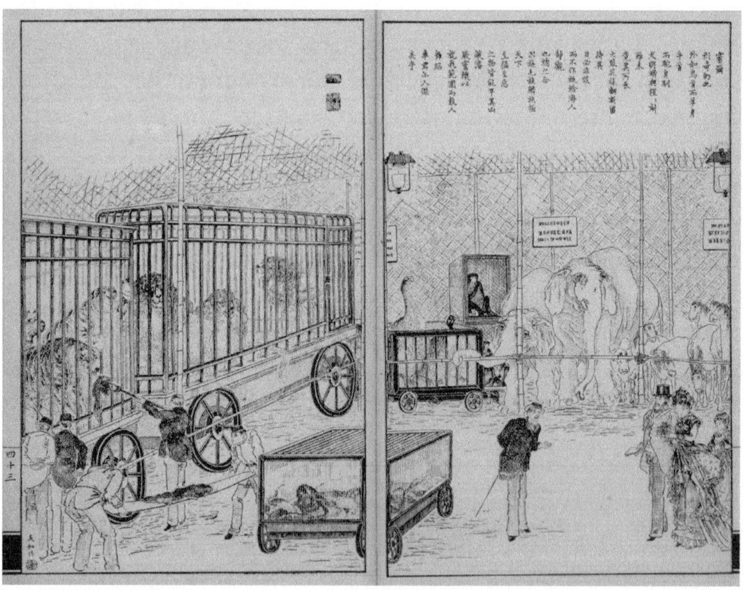

The illustrations were made by artist Wu Youru 吳友如 (1840-1894). *Dianshizhai huabao* 點石齋畫報, Fifth collection, part 1 (*di quji shang* 第五集上), 42-43, 1898 edition, Shanghai: [Shenbao guan]. Part of 'Shanghai's Lens on the New(s)', *Visualizing Cultures*, Massachusets Institute of Technology, 2015. https://visualizingcultures.mit.edu/dianshizhai/dsz_gal.html, accessed September 2020.

animals such as tigers, leopards, lions, elephants and turkeys in cages.[46] However, due to the venue's poor management, the owner soon found himself unable to afford to feed the larger animals, and sold the lions and tigers to another circus in 1892.[47]

In 1907 the first state-run zoological garden in Beijing opened to the public.[48] While animals had lost their roles in Qing imperial performances, these new forms of entertainment shows now facilitated their appearances – in the case of the Beijing zoo, charging twenty copper coins for every admission ticket.[49] These animals differed from the elephants described above, since they were neither captured nor tribute animals, but goods bought on the international market. Many of the first group of exotic animals in the zoo, including one elephant, were purchased from Germany in 1906 by Duanfang 端方 (1861-1911), who was serving as the governor-general of Jiangsu and Zhejiang (*Liang Jiang zongdu* 兩江總督) at that time; the other domestic animals were collected from local governments all over the country.[50]

In this transition from Qing imperial stables to Beijing public zoological garden, one may discern a change in the mechanism through which the managers of the Qing empire domesticated its subjects – human and animal alike. It might also indicate an accelerated modern turn of the palace machine. Following Empress Dowager Cixi's original plan, the public zoological garden was a section of the Agricultural Experiment Field of the Ministry of Agriculture, Industry, and Commerce (*Nong-gong-shang bu nongshi shiyan chang* 農工商部農事實驗場, henceforth Agricultural Experiment Field), about three kilometres outside Beijing's Xizhi Gate (*Xizhimen* 西直門). The Agricultural Experiment Field occupied 71 hectares structured into four sections: the zoological garden (*dongwuyuan* 動物園), the botanical garden (*zhiwuyuan* 植物園), the agricultural experiment laboratories (*shiyanchang* 試驗場), and the Agricultural High School (*Chudeng nongxuetang* 初等農學堂).[51] The entire Agricultural Experiment Field was officially opened to the public on 16th June 1908, named the 'Park of Ten Thousand Living

46 'You dahuayuan ji 遊大花園記 (A Visit to the Grand Garden)', *Shenbao* 申報, 14 Sept. 1889.

47 'Shihu yi zhu 獅虎易主 (Lions and Tigers have a New Master)', *Shenbao* 申報, 18 May 1892.

48 Yu Hui-chun 余慧君, 'Cong huangji lingyou dao wanshengyuan – Da Qing diguo de dongwu shoucang yu zhanshi 從皇家靈囿到萬生園——大清帝國的動物收藏與展示 (From Imperial Menageries to Public Zoological Garden: Captive Wild Animals at the Qing Court)', *Xin shixue = New History*, vol. 29, no. 1 (March 2018): 1-57.

49 *Dagong bao* 大公報 (Tianjin edition), no. 1804 (20 July 1907).

50 *Dagong bao* 大公報 (Tianjin edition), no. 1759 (5 June 1907).

51 Yang Xiaoyan 楊小燕, *Beijing dongwuyuan zhi* 北京動物園志 (Beijing: Zhongguo linye chubanshe, 2002), 96-99.

Beings' (*Wanshengyuan* 萬生園) by its visitors.[52] It was equipped with many additional diverting facilities including restaurants, cafés, a photo studio and travel lodges for the royal family. In 1955, it became the Beijing Zoo (*Beijing Dongwuyuan* 北京動物園).

The Agricultural Experiment Field was Cixi's last eye-catching political accomplishment, and one which the empress dowager herself visited quite often. The land had formerly belonged to the Leshan travel palace (*Leshanyuan* 樂善園) and been managed by the Imperial Household Department as lotus fields rented to tenants (see Chapter Seven). With this new identity and function, the Agricultural Experiment Field was placed under the administration of the Ministry of Agriculture, Industry, and Commerce. This brand new ministry of Qing central government, established in 1906, resulted from the political reform movement that Empress Dowager Cixi started in 1905, the 'movement of preparation for constitutionalism' (*yubei lixian yundong* 預備立憲運動). Cixi appointed Duanfang 端方 (1861-1911) and four other high-ranking officials to visit Japan, the US and many European states to scrutinize their constitutional systems. Leaving China on 11th December 1905, it took six months to complete the entire trip and these five officials brought back many Western ideas to reshape the Qing Empire into a modern state with a constitutional monarchy.[53] Duanfang advised Cixi that it was necessary to build libraries, museums, zoos, parks, and other public facilities for civic education.[54] Without hesitation, Cixi immediately added the zoo to the blueprint of the Beijing Agricultural Experiment Field, as one aspect of the state's 'preparation for constitutional change'.

52 Some literatures used a different character for *sheng*, i.e. 牲 (animal, including sacrificial animals) instead of 生 (living being), thus naming the Agricultural Experiment Field Wansheng 萬牲園 ('Park of Ten Thousand Animals'). If we consider the Agricultural Experiment Field as a whole, and the zoological garden as just a portion of it, Wansheng yuan written as 'living being' 生 seems more accurate than written as 'animal' 牲. However, since the zoological garden was open to the public one year earlier than the rest of the field, it is quite possible that tourists used the latter term, i.e. 萬牲園 at first, and then changed to the other one, i.e. 萬生園, afterwards. Whichever *sheng* was used in its name, in Qing official documents its appellation is 'Agricultural Experiment Field of the Ministry of Agriculture, Industry and Commerce'. The Agricultural Experiment Field was the sixth state-run experiment field in the late Qing. See 'Shangbu zouqing bo guandi xingban nongshi shiyanchang zhe 商部奏請撥官地興辦農事試驗場摺', *Dongfang zazhi* 東方雜誌, vol. 3, no. 6 (1906): 112-113 (rep. *Zhongwen qikan quanwen shuju ku* 中文期刊全文数据库).

53 For the details of Duanfang's trip, see Dai Hongci 戴鴻慈, *Chushi jiu guo riji* 出使九國日記 (Beijing: Diyi shuju, 1906).

54 'Dai Duan liang dachen zou geguo daomin shanfa qing cidi juban zhe 戴端兩大臣奏陳各國導民善法請次第舉辦摺', *Shengjing shibao* 盛京時報, 7 Oct. 1906.

Responding to requests for political reform, the Qing palace machine of performative emperorship gradually transferred some of its power and driving force into arms of the Qing State central government, such as the new Ministry of Agriculture, Industry, and Commerce. Formerly the Qing palace machine had exclusively been able to create imperialized bodies – transforming humans into emperors and tribute animals into elephant officials. Despite allowing for reforms the court still expected institutions such as the Agricultural Experiment Field to keep the image of a prospering empire alive and nurture the imperialized body of the whole Qing palace machine, of which animals, human subjects, the imperial family and emperor were constituent parts. It was considered vital to maintain the myth that the Manchu empire was sustained through the mandate of Heaven, whilst changing the mechanism by which this was accomplished. Instead of transforming animals into imperialized bodies, the animals in the Beijing Agricultural Experiment Field were displayed as a nation state's collection, aiming to satisfy the public gaze and thirst for knowledge. They introduced the civic body to an alternative state-making myth, that of a state of China with a constitutional Manchu monarchy. In this new political context, elephants and all other living beings in the Agricultural Experiment Field became simultaneously scientific subjects and objects for citizens' everyday amusement. In contrast to the elephants that had performed in imperial processions, zoo animals were not expected to be loyal or tame. They were meant to be just wild animals, showcasing the abundance and variety of various species. These creatures were presented as animals, who 'should not be endowed with [such human qualities as] morality, fierceness, stupidity, and sentience' (*wuliang* 無良、*wumen* 無猛、*wuchun* 無蠢、*wuling* 無靈), as Qiu Fengjia 丘逢甲 (1864-1912) commented after visiting the zoological garden.[55]

Along with the abolition of imperial elephants and their guardsmen in imperial processions, natural animals were no longer undergoing the transformation into imperialized bodies to become metaphors of loyalty and virtue or displays of imperial power. Nonetheless, the animals at the Beijing Agricultural Experiment Field were also fully integrated into the political mechanisms and metaphors of the Qing monarchy. The change from elephants and imperial guards leading the emperor's procession to the Temple of Heaven or Temple of Earth to pray for peace and prosperousness,

55 Qiu Fengjia 丘逢甲, 'Wan sheng yuan/Garden of ten thousand animals 萬牲園', in idem, *Lingyun hairilou shichao* 嶺雲海日樓詩鈔, *juan* 11, 132 (1937 edition; rep. *Xuxiu Siku quanshu* 續修四庫全書, vol. 1576. Shanghai: Shanghai guji chubanshe, 1995-2002).

to elephants caged in the public zoological garden of the Agricultural Experiment Field reflected the transfer from a palace machine powering performative emperorship to a performative statecraft operated by a state administration system. Empress Dowager Cixi's Agricultural Experiment Field was intended to demonstrate the appealing stance of 'breeding and nurturing all lives' (*shengyang wanwu* 生養萬物) in order to 'love and nourish all ten thousand human subjects' (*ziyang wanmin* 滋養萬民). This was to be achieved not through praying and ploughing or impressive performances – as in the centuries of imperial rule in the past – but through a new political theatre in which various animals, plants, agricultural professionals and students (as future professionals), were staged to perform Cixi's 'preparation for constitutionalism', a modern state-making enterprise leading up to the 1911 revolution and the birth of the Republic of China, a dramatic but anticipated change. However, it took another three decades to fully bring the Qing palace machine to a close, when Puyi abdicated as Emperor of Manchukuo in 1945.

About the Author

HUI-CHUN YU is Associate Professor, Department of Arts and Creative Industries, National Dong Hwa University, Taiwan. Her researches focus primarily on the history of institutional collectibles in early modern China, including imperial art collections, museum collections, animals and plants in parks and zoological gardens. By examining how politicians and institutions collected and displayed collectables, she explores the complicated politics of subject-making from the Qing Empire to early Republican China.

Coda

Dorothy Ko, Kai Jun Chen, and Martina Siebert

'Men and women of the eighteenth century – before the West arrived in force – were already creating the conditions of China's modern society', historian Philip Kuhn wrote in his classic study of the Qianlong reign, when the Qing empire was the most productive and populous in the globe.[1] Examining a sorcery scare that engulfed the empire, from the emperor, his bondservants and bureaucrats on down to the itinerant monks and beggars in remote provinces, Kuhn discerned under the sheen of prosperity brewing ethnic animosity, awareness of resource scarcity, fear of precarity, as well as the erosion of social mores that gave life stability and meaning. The 'conditions of China's modern society' that Kuhn spoke of were the dark forces of disquiet that removed people from traditional moorings and availed them to be reshaped into modern individual selves and subjects.

In examining the longer history of this modern becoming by tracing the inner workings of the Qing palace machine, the contributors of the present volume have identified other conditions of China's modern society: the principles of rational planning, accountability, and economic rationality. The Imperial Household Department, an enormous apparatus staffed by over 22,000 people in the mid-18th century, was fashioned to function with machine-like regularity, guided by a set of transparent rules that aim at producing mechanical objectivity. As structure and as process, this remarkable institution is the 'ghost in the machine', the intelligence and agency that is ideally free from human intervention that may introduce errors or biases. The economization of costs, not maximization of profit, was an overarching concern of the Department. The seventeenth-century Italian author Giovanni Botero praised the Chinese empire for its rich natural resources and effective management which produced bountiful surpluses

1 Philip A. Kuhn, *Soulstealers: The Chinese Sorcery Scare of 1768* (Cambridge, Ma.: Harvard University Press, 1992), 1.

Siebert, Martina, Kai Jun Chen, and Dorothy Ko (eds), *Making the Palace Machine Work: Mobilizing People, Objects, and Nature in the Qing Empire.* Taylor & Francis Group, 2021
DOI: 10.5117/9789463720359_CODA

to fuel overseas trade and growing population.[2] The late Ming and Qing empires appear more mercantilist than those in Europe that bore the name.

Indeed, one of the most striking findings of this book is the remarkable extent to which the Qing court, the bureaucratic state, and the market were interwoven in logic of practice as in logistics. This is evident in the accounting and fiscal systems, whereby the accounting and auditing ledgers of commercial firms were adopted by the Imperial Household Department, itself a microcosm of the fiscal state. The intimacies between the court, the state, and the market are even more salient when one follows the money and the goods, from jade boulders conveyed from newly conquered territories of Xinjiang, the fresh medicine and provisions procured for residents in the palace, to the myriad tools, machine parts, and skills required in the Imperial Workshops that were purchased from the market. The private (court) and public (state) spheres interpenetrated, and the official (court and state) could not have functioned without the (private commercial) market. This characteristic of the late-Ming and Qing early modern empires deserves further research.[3]

If the conditions of modernity identified by Kuhn are rife with tensions, the chapters of this book reveal even more tensions if not paradoxes. The modern reader may find a blatant contradiction between a core operating principle of the palace machine – rational management of resources – and its predilection for conspicuous production/consumption of luxuries and spectacles. The contributors to this book, however, suggest that this contradiction is more apparent than real. The cultivation of the reality *effect* of the performance of order and rules, of social and financial accountability, and of public extravagances was a persistent working principle of the Qing palace machine that accounted in no small part to its success. Such spectacles as imperial parades of elephants and gilded temple roofs are but an extension of the performance of rationality in the accounting and manufacturing procedures. How else can rationality, order, and splendour exist in the world if they were not duly manufactured, displayed, and performed? From this perspective, dutiful adherence to regulations or precise reckoning of the price of a nail to six decimal points are not empty gestures but the

2 Giovanni Botero, *The Reason of State and the Greatness of Cities* (1606), cited in Nathaniel Wolloch, *Nature in the History of Economic Thought: How natural resources became an economic concept* (Abingdon and New York: Routledge, 2017), 9.

3 Historian Liu Zhiwei 劉志偉 argues that the market became a de facto agent in governance of the empire after the Ming 'Single Whip' reforms in the sixteenth century, whereby corvée labour and tax in kind were converted into silver payments. Liu, *Gongfu tizhi yu shichang* 貢賦體制與市場: 明清社會經濟史論稿 (Beijing: Zhonghua shuju, 2019).

very foundation of engagement and action. This insight may illuminate other paradoxes that haunt the workings of the Qing palace machine: the contrast between the exacting precision in the accounting ledgers and the fuzziness in the actual handling of resources; the simultaneous desires for stable structures and motion, the insistence on the virtues of both cost control and waste, as well as the asymmetries between the efficacies of words and things.

Lest the reader leaves with the impression that the Imperial Household Department, and the Qing palace machine it managed, was a clockwork-like entity, one only needs to recall the sick elephants, shards of misfired pots, expired medicines, as well as other artefacts of failure and malfunctioning documented in the chapters. For all the modern, rational principles that undergird its design, the Qing palace machine is better imagined as a machine-body (*ji ti* 機體), a Chinese term which describes equally well the physical presence of a machine, an institution, and a living body. A machine-body is a self-contained structure (*ti*) that moves with regularity according to a triggering mechanism (*ji*). Like a living organism, but with inanimate mechanics, the Qing palace machine strove to achieve a dynamic balance between conflicting propensities as well as between the inside and the outside of the body.[4]

The materialist approach taken by the authors of this book reveals empire-making as an artisanal process of forging and tinkering; artefact-making *is* empire-making at its most elemental level. If each porcelain vase had to be fired from exacting formulas of the body and glaze, each tile gilded with an amalgamation of local and foreign methods, and if each object had to be counted and accounted for before, during, and after the making process, the same is true for the cobbling together of the multiethnic empire. As with all building projects, however, meticulous planning is no guarantee for a desirable outcome. As long as the empire was a co-production of people, materials, and nature, no amount of mastery and control on the parts of the Imperial Household Department could have anticipated all the possible contingencies. Lotus plants, medicinal herbs, and elephants were integral parts of the Qing palace machine no less than ceramic clay, jade boulders,

4 This concept of a machine resonates with that of the early modern German philosopher Christian Wolff (1679-1754) who wrote, 'A machine is an assembled work, whose movement is determined by the mode of its assemblage.' Wolff considered the world itself – including all its inanimate and animate constituents – as a 'Maschine'. To him the basic facets that defines a machine are *compositio* (assemblage) and *motus* (movement). Quote translated from the German as cited in Raunig, 'Abstrakte Maschinen', in Hans-Christian von Herrmann and W. Velminski, eds., *Maschinentheorien/Theoriemaschinen* (Bern: Peter Lang, 2012), 122.

gold metal, and the people who extracted and manipulated them. In the final analysis, the Qing palace machine was an organic machine-body that occupied definite time and space; therein lies the possibility of its history.

About the Authors

A native of Hong Kong, Dorothy Ko is Professor of History at Barnard College, Columbia University. She is a cultural historian of early modern China whose research focuses on gender, technology, and material culture. Her recent book, *The Social Life of Inkstones: Artisans and Scholars in Early Qing China* (Washington, 2017) is a finalist of the Morey Prize of the College Art Association. An earlier book, *Cinderella's Sisters: A Revisionist History of Footbinding* (California, 2005) won the Joan Kelly Memorial Prize of the American Historical Association.

Kai Jun Chen is Assistant Professor at Brown University. He specialises in the history of imperial institution, handicraft technology, and material culture studies. His monograph in progress examines the technocratic culture in the Imperial Ceramic Industry of the Qing dynasty. His other scholarly publications investigate the production of technological knowledge in the global circulation, collection, and replication of luxury artifacts, especially porcelains.

Martina Siebert works as subject librarian at the East Asia department of the Staatsbibliothek zu Berlin and as independent scholar. She researches and publishes on the history of Chinese exploration into nature and technology with a focus on the styles and agendas of presenting and organizing that knowledge in writing.

Contributors

ELIF AKCETIN is Visiting Assistant Professor at Haverford College. She is a cultural historian whose research focuses on the art of governance in Qing China. She is the co-editor of *Living the Good Life: Consumption in the Qing and Ottoman Empires of the Eighteenth Century* (2017). Her monograph, currently in preparation, takes a corruption trial and investigation as an entry point to explore the intricate link between knowledge and power.

KYOUNGJIN BAE is James P. Storer Assistant Professor of Asian history at Kenyon College. She is a historian of late imperial China, specializing in the history of craft, everyday technology, and Sino-European interaction.

KAI JUN CHEN is Assistant Professor at Brown University. He specialises in the history of imperial institution, handicraft technology, and material culture studies. His monograph in progress examines the technocratic culture in the Imperial Ceramic Industry of the Qing dynasty. His other scholarly publications investigate the production of technological knowledge in the global circulation, collection, and replication of luxury artifacts, especially porcelains.

XUELING GUAN 关雪玲 is a researcher (equivalent to curator) at the Palace Museum in Beijing. She specialises in the history of medicine of the Qing court on which she has published numerous articles and monographs such as *Qingdai gongting yixue yu yixue wenwu* 清代宮廷醫學與醫學文物 (The Court Medicine of the Qing Dynasty and Historical Medicinal Artifacts).

A native of Hong Kong, DOROTHY KO is Professor of History at Barnard College, Columbia University. She is a cultural historian of early modern China whose research focuses on gender, technology, and material culture. Her recent book, *The Social Life of Inkstones: Artisans and Scholars in Early Qing China* (Washington, 2017) is a finalist of the Morey Prize of the College Art Association. An earlier book, *Cinderella's Sisters: A Revisionist History of Footbinding* (California, 2005) won the Joan Kelly Memorial Prize of the American Historical Association.

HUI-MIN LAI 賴惠敏, Ph.D. (National Taiwan University, 1988) is a Research Fellow at the Institute of Modern History, Academia Sinica, in Taiwan. For many years, she has been studying the economic and cultural history of the Qing Dynasty. She is the author of a monograph on *Emperor Qianlong's Purse: Imperial Household Finance in Eighteenth Century China* (Qianlong huangdi de hebao 乾隆皇帝的荷包).

CHRISTINE MOLL-MURATA is Chair professor at the Department for History of China, Ruhr University Bochum, Germany, and honorary fellow of the International Institute of Social History, Amsterdam. Her research focuses on labour history of China, Taiwan, and Japan, the history of crafts and early industrialization in China and East Asia.

MARTINA SIEBERT works as subject librarian at the East Asia department of the Staatsbibliothek zu Berlin and as independent scholar. She researches and publishes on the history of Chinese exploration into nature and technology with a focus on the styles and agendas of presenting and organizing that knowledge in writing.

TE-CHENG SU 蘇德徵, Ph.D. (Imperial College London, 2019) is an Assistant Professor at the Department of Materials Science and Engineering, National Taiwan University. He specializes in manufacturing technologies and is the co-author of 'Brass consumption in the Qing Empire' in *Living the Good Life: Consumption in the Qing and Ottoman Empires of the Eighteenth Century.*

GUANGYAO WANG 王光堯 is researcher (equivalent to curator) at the Palace Museum in Beijing. He specialises in ceramic archaeology and the institutional history of imperial courts. He has published *Zhongguo gudai guanyao zhidu* 中國古代官窯制度 (The Systems of Official Ceramic Manufacture in Premodern China), book I and II.

YIJUN WANG is Assistant Professor of History at New York University. Her current project examines the transmissions of tin mining technology and changes in statecraft in China from 1700 to 1850s. She has recently contributed an article to *The Silver Age: Origins and Trade of Chinese Export Silver.*

YULIAN WU is Assistant Professor of History at Michigan State University. She specialises in material culture, gender relations, and Manchu studies in Qing China. She is the author of *Luxurious Networks: Salt Merchants, Status, and Statecraft in Eighteenth-Century China*. Her current project explores jade production and consumption in Qing China.

HUI-CHUN YU is Associate Professor, Department of Arts and Creative Industries, National Dong Hwa University, Taiwan. Her researches focus primarily on the history of institutional collectibles in early modern China, including imperial art collections, museum collections, animals and plants in parks and zoological gardens. By examining how politicians and institutions collected and displayed collectables, she explores the complicated politics of subject-making from the Qing Empire to early Republican China.

SHUXIAN ZHANG 張淑嫻 is Researcher at the Palace Museum in Beijing. She specialises in the architecture and interior decoration of Ming Qing palaces. Among her publications are the monographs *Ming Qing wenren yuanlin yishu* 明清文人園林藝術 (The Art of Literati Garden during the Ming and Qing Dynasties) and recently *Jinchuang xiuhu: qingdai huanggong neiyan zhuangxiu yanjiu* 金窗繡戶——清代皇宮內檐裝修研究 (Gilded Windows and Carved Doors: Study of Interior Decoration of the Qing Palaces).

QIONG ZHANG 張瓊 is a Researcher of the Palace Museum in Beijing. She is a specialist of pre-modern textiles and the history of dress and costumes of the Ming and Qing dynasties. Her publications on the subject include *Guxiu yuanliu bianxi* 顧繡源流辨析 (Investigation of the Origin and Development of the Gu Family Embroidery) and the edited volume *Qingdai gongting fushi* 清代宮廷服飾 (Costumes and Accessories of the Qing Court).

Bibliography

Published Sources

Ai Xingfu 艾性夫. *Sheng yu* 剩語 (Redundant Words). *Siku quanshu* 四庫全書 edition. Reprint. Taipei: Taiwan shangwu yinshuguan, 2008.

Akcetin, Elif and Suraiya Faroqhi, eds. *Living the Good Life: Consumption in the Qing and Ottoman Empires of the Eighteenth Century.* Leiden: Brill, 2017.

Akcetin, Elif. 'Consumption as Knowledge: Pawnbrokers in Qing China Appraise Furs'. In *Living the Good Life*, edited by eadem and Suraiya Faroqhi, 357-83. Leiden: Brill, 2017.

Anheuser, Kilian. 'Archaeotechnology: The Practice and Characterisation of Historic Fire Gilding Techniques'. *JOM-Journal of the Minerals Metals & Materials Society*, vol. 49 (1997): 58-62.

Anheuser, Kilian. 'Cold and Hot Mercury Gilding of Metalwork in Antiquity'. *The Bulletin of the Metals Museum*, vol. 26 (1996): 48-52.

Anon. 'Shangbu zuoqing bo guandi ban nongshi shiyanchang zhe 商部奏請撥官地興辦農事試驗場摺 (The Commerce Department's Memorial on Allocating Official Land for Agricultural Experiment Field'. *Dongfang zazhi* 東方雜誌, vol. 3, no. 6 (1906): 112-113.

Asimov, Isaac. *Understanding Physics*. New York: Barnes and Noble, 1966.

Balazs, Etienne, *Chinese Civilization and Bureaucracy*. New Haven: Yale University Press, 1966.

Bartlett, Beatrice S. *Monarchs and Ministers: The Grand Council in Mid-Ch'ing China, 1723-1820*. Berkeley: University of California Press, 1991.

Beijing Tongrentang shi 北京同仁堂史 (History of the Tongren House Drugstore), edited by Beijing Tongrentang Group Co., Ltd. 北京同仁堂集團公司 and the Editorial Committee of Beijing Tongrentang History 北京同仁堂史編委員. Beijing: Renmin ribao chubanshe, 1993.

Bello, David. *Across Forest, Steppe, and Mountain: Environment, Identity, and Empire in Qing China's Borderlands*. Cambridge: Cambridge University Press, 2015.

Bennett, Jane. *Vibrant Matter: A Political Ecology of Things.* Durham, NC: Duke University Press, 2010.

Berger, Patricia. *Empire of Emptiness: Buddhist Art and Political Authority in Qing China*. Honolulu: University of Hawai'i Press, 2003.

Bian, He. *Know Your Remedies: Pharmacy and Culture in Early Modern China.* Princeton, NJ: Princeton University Press, 2020.

Brook, Timothy. *The Confusions of Pleasure: Commerce and Culture in Ming China.* Berkeley, CA: University of California Press, 1998.

Brunnert, H.S. and V. V. Hagelstrom. *Present Day Political Organization of China*, translated by A. Beltchenko and E. E. Moran. Shanghai: Kelly and Walsh, 1912.

Bushell, Stephen. *Description of Chinese Pottery and Porcelain; Being a Translation of the T'ao Shuo*. Oxford: The Clarendon Press, 1910.

Butler, Judith. 'Performative Acts and Gender Constitution: An Essay in Phenomenology and Feminist Theory'. *Theatre Journal*, vol. 40, no. 4 (Dec. 1988): 519-31.

Cai Jiayi 蔡家藝. 'Qingdai Xinjiang yushi de kaicai yu shuchu 清代新疆玉石的開採與輸出 (Jade Mining and Output in Xinjiang at Qing Dynasty)'. *Zhongguo bianjiang shi di yanjiu* 中國邊疆史地研究, vol. 20, no. 3 (2010): 121-130.

Cao Yin 曹寅. *Lianting ji jianzhu* 楝亭集箋注 (Annotated Collection of Beadtree Pavilion). Beijing: Beijing tushuguan chubanshe, 2007.

Cellini, Benvenuto. *Due trattati, uno intorno alle otto principali arti dell'oreficeria, l'altro in materia dell'arte della scultura*. Fiorenza: Per Valente Panizzii & Marco Peri, 1568.

Cen Zhongmian 岑仲勉. *Hanshu Xiyu zhuan dili jiaoshi* (shang) 漢書西域傳地理校釋 (上) (Annotated Edition of the Chapters on Western Borderland's Geography in the Hanshu, A). Beijing: Zhonghua shuju, 1981.

Chang Te-Ch'ang. 'The Economic Role of the Imperial Household in the Ch'ing Dynasty'. *The Journal of Asian Studies*, vol. 31, no. 2 (1972): 243-273.

Chapman, Martin. 'Techniques of Mercury Gilding in the Eighteenth Century'. In *Ancient and Historic Metals: Conservation and Scientific Research*, edited by D. A. Scott, J. Podany, and B. B. Considine, 229-238. Marina del Rey, Ca.: Getty Conservation Institute, 1994. Accessed September 2020. http://www.getty.edu/publications/virtuallibrary/0892362316.html.

Chen Dazhang 陳大章. *Shi chuan mingwu jilan* 詩傳名物集覽 (Characteristics of Collections and Explanations of Natural Creatures in Classic Poetry). *Siku quanshu* 四庫全書 edition. Reprint. Taipei: Taiwan shangwu yinshuguan, 2008.

Chen Kuo-tung 陳國棟. 'Neiwufu guanyuan de waipai wairen – yu Qianlong gongting wenwu gongji zhijian de guanxi 內務府官員的外派外任—與乾隆宮廷文物供給之間的關係 (Imperial Bondservants' Provincial Positions and Their Contribution to the Acquisition of Objets d'Arts for Emperor Qianlong's Court)'. *Meishushi yanjiu jikan* 美術史研究集刊, no. 33 (Sep. 2012): 225-69.

Chen Kuo-tung 陳國棟. 'Wuyingdiang zongjianzao Heshihen – huoyue yu "Liyi zhi zheng" shijian zhong de yiwei Neiwufu renwu 武英殿總監造赫世亨—活躍於「禮儀之爭」事件中的一位內務府人物' (Heshihen, Superintendent of the Wuyindian Workshops: A "Rites Controversy" Figure of the Imperial Household Department)'. *Gugong xueshu jikan* 故宮學術季刊 = *The National Palace Museum Research Quarterly*, 30.1 (2012): 87-134.

Chen Qiyuan 陳其元 (1812-1881). *Yongxianzhai biji* 庸閒齋筆記 (Diary from the Yongxian Study). 1874 edition. Reproduced in *Xuxiu Siku quanshu* 續修四庫全書, vol. 1142. Shanghai, Shanghai guji chubanshe, 1995-2002.

Chen Wen-shih 陳文石 (Chen Wenshi). 'Qingdai de shiwei 清代的侍衛 (The Guards of the Qing Dynasty)'. In idem, *Ming Qing zhengzhi shehui shilun* 明清政治社會史論 (Historical Investigation of the Politics and Society during Ming and Qing), 623-650. Taipei: Taiwan xuesheng shuju, 1991.

Cheng Jufu 程鉅夫. *Xuelou ji* 雪樓集 (Anthology from the Snow Mansion). Reprint. Shanghai: Shanghai shuju, 1994.

Chi Jo-hsin 嵇若昕 (Ji Ruoxin). 'Cong 'huojidang' kan yongqian liangchao de neiting qiwu yishu guwen 從'活計檔'看雍乾兩朝的內廷器物藝術顧問 (The Advisors for Decorative Art to Emperor Yung-chen and Emperor Ch'ien-lung: Based on the Artisans of the Imperial Workshops)', *Dong Wu lishi xuebao* 東吳歷史學報 = *Soochow University Journal of History*, vol. 16 (2005): 53-105.

Chi Jo-hsin 嵇若昕 (Ji Ruoxin). 'Qing zhong houqi (1821-1911) Neiwufu Zaobanchu nanjiang jiqi xiangguan wenti 清中後期 (1821-1911) 內務府造辦處南匠及其相關問題 (Southern Craftsmen in the Workshops of the Imperial Household Department During the Middle to Late Qing Period (1821-1911) and Issues Related to Them)'. *Gugong xueshu jikan* 故宮學術季刊, vol. 32, no. 3 (2015): 63-89.

Chou, Ju-hsi. 'Tangdai: A Biographical Sketch'. In *Elegant Brush: Chinese Painting under the Qianlong Emperor*, edited by Chou Ju-hsi and Claudia Brown, 133-140. Phoenix, AZ: Phoenix Art Museum, 1985.

Chu, Raymond and William Saywell. *Career Patterns in the Ch'ing Dynasty: The Office of Governor-General*. Ann Arbor: Center for Chinese Studies, University of Michigan, 1984.

Chun Yuan 椿園, *Xiyu wen jian lu* 西域聞見錄 (Hearsay from the West Borderland). Preface 1777. Reproduced in *Zhongguo xibei wenxian congshu* 中國西北文獻叢書, 4th collection: *Xibei minsu wenxian* 西北民俗文獻, vol. 117. Lanzhou: Lanzhou guji shudian, 1990.

Clunas, Craig. *Fruitful Sites: Garden Culture in Ming Dynasty China*. London: Reaktion Books, 1996.

Corsi, Elisabetta (Yi Li 伊丽). 'Nian Xiyao de shengping jiqi dui yishu he kexue de gongxian 年希堯的生平及其對藝術和科學的貢獻 (A Biography of Nian Xiyao and His Contribution to Science and Art, published in Chinese)'. *Zhongguoshi yanjiu* 中國史研究 = *Journal of Chinese Historical Studies*, no. 3 (2000): 155-65.

Crossley, Pamela Kyle. *A Translucent Mirror: History and Identity in Qing Imperial Ideology*. Berkeley: University of California Press, 1999.

Crossley, Pamela Kyle. 'Manchu Education'. In *Education and Society in Late Imperial China, 1600-1900*, edited by Benjamin Elman and Alexander Woodside, 340-78. Berkeley: University of California Press, 1994.

Da Ming huidian (Wanli ban), see Shen Shixing et al., *Da Ming huidian* 228 *juan*.

Da Ming huidian (Zhengde ban), see Xu Pu et al., *Da Ming huidian* 180 *juan*.

Da Qing huidian (Yongzheng ban). See Yunlu et al., *Da Qing huidian* (Yongzheng ban).

Da Qing huidian shili (Guangxu ban). See Kungang et al., *Da Qing huidian shili* (Guangxu ban).

Da Qing huidian shili (Jiaqing ban). See *Qingdai wubu huidian*.

Da Qing huidian shili (Kangxi ban). See *Qingdai wubu huidian*.

Da Qing huidian shili (Yongzheng ban). See Yunlu et al., *Da Qing huidian shili* (Yongzheng ban).

Da Qing huidian zeli (Qianlong ban) 大清會典則例 (乾隆版) (Precedent Cases of Statutes of the Great Qing, Qianlong edition). Edited by Imperial Household Department. Beijing: Wuyingdian, 1764. Also see *Qingdai wubu huidian*.

Dai Hongci 戴鴻慈. *Chushi jiu guo riji* 出使九國日記 (Diary on the Diplomatic Visit to Nine Countries). Beijing: Diyi shuju, 1906.

Dai Nihon Bukkyō zensho 大日本佛教全書 (The Collected Works of Japanese Buddhism). Edited by Bukkyō Kankōkai 仏書刊行会. Tōkyō: Daihōrinkaku, 2007.

Daston, Lorraine and Peter Galison. *Objectivity*. Brooklyn: Zone Books, 2008.

De Boever, Arne, Alex Murray, Jon Roffe, and Ashley Woodward, eds. *Gilbert Simondon: Being and Technology*. Edinburgh: Edinburgh University Press, 2012.

Der Ling / Deling 德齡. *Qing gong ernian ji* 清宮二年記 (Two Years in the Forbidden City). Translated by Dongfang zazhishe 東方雜誌社. Shanghai: Shangwu yinshuguan, 1924.

Der Ling. *Two Years in the Forbidden City*. London: T. Fisher Unwin, 1912.

Ding Yizhuang 定宜莊 and Hu Hongbao 胡鴻保. 'Qingdai Neiwufu Gao-jia shijia de hunyin quan 清代內務府高佳世家的婚姻圈 (Marriage Circles within the Famous Lineages of the Imperial Household Department in Qing Period: A Case Study of the Gaojia Clan)'. *Qingshi yuanjiu* 清史研究, no. 3 (2005): 25-32.

Ding Yizhuang 定宜莊. 'Neiwufu Wanyan shijia kao 內務府完顏世家考 (A Study of Wanyan Aristocratic Family in the Imperial Household Affairs Administration)'. In *Qingshi luncong* 清史論叢, no. 14, edited by Qingshi luncong bian weihui 清史论丛编委会, 133-34. Shenyang: Liangning guji chubanshe, 1995.

Dodgen, Randall A. *Controlling the Dragon: Confucian Engineers and the Yellow River in Late Imperial China*. Honolulu: University of Hawai'i Press, 2001.

Du Halde, Jean-Baptiste. *Description of the Empire of China and Chinese-Tartary*. Translated from the French. London: J. Watts, 1741.

Du Hede 杜赫德 (J.-B. du Halde). *Yesuhui shi zhongguo shujian ji* 耶穌會士中國書簡集 (Collection of Chinese Writings by Jesuit Missionaries). Translated by Zheng Dedi 郑德弟, Lu Yimin 吕一民 and Shen Jian 沈坚. Zhengzhou: Daxiang chubanshe, 2005.

Du Zhengxian 杜正賢. 'Laohudong nansong guanyao zhi 老虎洞南宋官窑址 (Archaeological Site of the Official Kiln of the Southern Song at Laohudong)'. *Wenwu* 文物, no. 10 (2002): 4-31.

Duindam, Jeroen et al., eds. *Royal Courts in Dynastic States and Empires: A Global Perspective*. Leiden; Boston: Brill, 2011.

E'ertai 鄂爾泰 (Ortai) et al. *Baqi manzhou shizu tongpu* 八旗滿州氏族通譜 (Comprehensive Genealogy of the Eight Banner Manchu Clans). Beijing: Wuying dian, 1744.

E'ertai 鄂爾泰 (Ortai) et al. *Baqi tongzhi chuji* 八旗通志初集 (Comprehensive History of the Eight Banners, first collection). Changchun: Dongbei shifan daxue chubanshe, 1985.

Elias, Norbert. *Die höfische Gesellschaft – Untersuchungen zur Soziologie des Königtums und der höfischen Aristokratie*. Frankfurt: Suhrkamp Taschenbuch Wissenschaft, 1983.

Elliott, Mark C. *The Manchu Way*: *The Eight Banners and Ethnic Identity in Late Imperial China*. Stanford, CA: Stanford University Press, 2001.

Fan Liangshu 樊良樹. 'Beijing diyi diandeng shijian kao 北京第一盞電燈亮燈時間考 (Investigation on the Time of the First Electrical Light in Beijing)'. *Huabei dianli daxue xuebao* (Shehui kexue ban) 華北電力大學學報 (社會科學版), no. 6 (2016): 19-22.

Fang Yuqin 方裕瑾. 'Guangxu shiba nian zhi ershi nian Ningshou gong nei gaijian gongcheng shulüe 光緒十八年至二十年寧壽宮內改建工程述略 (Report on the Rebuilding of the interior of the Ningshou Palace between the 18th and the 20th year of the Guangxu reign'. In *Zhongguo zijincheng xuehui lunwen ji* 中國紫禁城學會論文集 (Proceedings of the Forbidden-city Research Society in China), 2nd collection, 260-264. Beijing: Zijincheng chubanshe, 1997.

Fengchenyuan xianxing zeli. In *Qinding Zongguan Neiwufu xianxing zeli er zhong*, vol. 307, *di er zhong* 第二種, chapter *Fengchenyuan* 奉宸苑 ('Garden Bureau').

Forêt, Philippe. *Mapping Chengde: The Qing Landscape Enterprise*. Honolulu: University of Hawai'i Press, 2000.

Fu Yuhong 傅育紅. 'Shaozao ciqi zhangcheng zhiding hou yuyao shaozao jingfei de hexiao 燒造瓷器章程制定後禦窯燒造經費的核銷 (The Accounting of Firing Expenses of the Imperial Kiln after the Establishment of Regulations Governing Ceramic Production)'. *Gugong bowuyuan yuankan* 故宮博物院院刊, no. 2 (2006): 73-80.

Fu Zhenlun 傅振倫 and Zhen Li 甄勵. 'Tang Ying ciwu nianpu changbian 唐英瓷務年譜長編 (Chronicle of Tang Ying and His Direction of Porcelain Production with Historical Data)'. *Jingdezhen taoci* 景德鎮陶瓷, no. 2 (1982): 18-67.

Fujun 富俊 et al. *Zongguan Neiwufu xianxing tiaoli: Guangchusi* 總管內務府現行條例: 廣儲司 (Current Regulations of the Imperial Household Department: The Grand Storage Office). In *Jindai Zhongguo shiliao congkan* 近代中國史料叢刊. Taipei: Wenhai chubanshe, 1972.

Gao Bin 高斌. 'Guzai caotang shiji 固哉草堂詩集' (Poetry from Guzai Thatch Hut). In *Qingdai shiwen ji huibian* 清代詩文集彙編 (Compilation of Poetry and Proses of Qing Dynasty), vol. 273. Shanghai: Shanghai guji chubanshe, 2010.

Gao Bin 高斌. *Zhili wudao chenggui* 直隶五道成規 (Precedent Rules Concerning Five Metropolitan Circuits in Zhili Province). 1743 edition (Starr Rarebook Collection, Columbia University).

George, Kenneth M. 'Objects on the Loose: Ethnographic Encounters with Unruly Artefacts – A Foreword'. *Ethnos: Journal of Anthropology* 64.2 (1999): 149-50.

Gongbuchabu 工布查布, trans. *Zaoxiang liangdu jing* 造像量度經 (Measurements of Buddhist Statues Sutra). Reprint. Taipei: Taiwan yinjingchu, 1956.

Gongzhong xianxing zeli (2), see *Qinding Gongzhong xianxing zeli di er zhong*.

Guo Runtao 郭潤濤. 'Mingdai de chuanpiao 明代的串票' (Revenue Tickets in the Ming Dynasty)'. In *Yang Guozhen jiaoshou zhishi wushinian jinian wenji* 楊國楨教授治史五十年紀念文集 (Collected Essays for the Fiftieth Anniversary of Professor Yang Guozhen), edited by Chen Chunsheng 陳春聲 and Chen Dongyou 陳東有, 32-55. Nanchang: Jiangxi jiaoyu chubanshe, 2009.

Haken, Hermann et al., eds. *The Machine as Metaphor and Tool*. Berlin: Springer, 1993.

Hangzhou shi wenwu kaogusuo 杭州市文物考古所 (The Institute of Cultural Relics and Archaeology in Hangzhou). *Nansong taimiao yizhi* 南宋太廟遺址 (The Archaeological Site of the Ancestral Shrine in the Southern Song). Beijing: Wenwu chubanshe, 2007.

Hayter-Menzies, Grant. *Imperial Masquerade: The Legend of Princess Der Ling*. Hong Kong: Hong Kong University Press, 2008.

He Xinhua 何新華. 'Qingdai dongnanya gong xiang maoyi yanjiu 清代東南亞貢象貿易研究 (Study on the Elephant Tribute Trade with Southeast Asia in the Qing Dynasty)'. *Dongnanya yanjiu* 東南亞研究 = *Journal of Southeast Asian Studies*, no. 1 (2011): 73-79.

He Xinhua 何新華. *Qingdai gongwu zhidu yanjiu* 清代貢物制度研究 (Study of Tributory System in the Qing Dynasty). Beijing: Shehui kexue wenxian chubanshe, 2012.

Hogge, David. 'The Empress Dowager and the Camera. Photographing Cixi, 1903-1904'. *Visualizing Cultures*. Massachusetts Institute of Technology, 2011. Accessed September 2020. http://vcarchive.com/mitvc/empress_dowager/index.html.

Hsieh, Bao Hua. *Concubinage and Servitude in Late Imperial China*. Lanham: Lexington Books, 2014.

Hsu Cho-Yun 許倬雲 et al., eds. *Di'erjie zhongguo shehui jingjishi yantaohun lunwenji* 第二屆中國社會經濟史研討會論文集 (Proceedings of the Second Conference on the Social Economic History of China). Taipei: Hanxue Yanjiu ziliao ji fuwu zhongxin 漢學研究資料及服務中心, 1983.

Huang Li-Chun 黃麗君 (Huang Lijun). 'Huangdi jiqi baoyi nucai: lun Qingdai huangquan yu Neiwufu guanliao tizhi 皇帝及其包衣奴才：論清代皇權與內務府官僚體制 (The Emperors and Their Bondservants: Interaction between Imperial Wills and the Bureaucratic Institution of the Imperial Household Department in the Qing Dynasty)'. PhD diss., National Taiwan University, 2014.

Huang Li-chun 黃麗君 (Huang Lijun). 'Qingchu manhanwen fanyijia Wanyan Ashitan shihuan shengping kao 清初滿漢文繙譯家完顏阿什坦仕宦生平考 (Study of the Official Career of Wanggiyan Ashitan (?-1683), An Early Qing Translator of Manchu and Chinese)'. In *Gongting yu haiyang de jiaohui* 宮廷與海洋的交匯, edited by Li Qilin 李其霖, 91-117. Taipei: Tamkang University Press, 2017.

Huang Li-chun 黃麗君 (Huang Lijun). 'Hesu jiqi yizuo *Qibentou* yanjiu 和素及其譯作"七本頭"研究 (A Study of Hesu and His Collection of Translations *Ci Ben Deo*)', *Taiwan Shida lishi xuebao* 臺灣師大歷史學報, no. 58 (2017): 91-150.

Huang Liuhong 黃六鴻. *Fuhui quanshu* 福惠全書 32卷. 1893 edition. Reproduced in *Siku weishou jikan* 四庫未收輯刊, *shibu* 史部, vol. 19.

Huang Wei 黃薇 and Huang Qinghua 黃清華. 'Yuan qinghua ciqi zaoqi leixing de xin faxian — cong shizheng jiaodu lun Yuan qinghua ciqi de qiyuan 元青花瓷器早期類型的新發現——從實證角度論元青花瓷器的起源 (New Discoveries about the Early Types of Blue and White Porcelain in the Yuan Dynasty – Empirical Examination of the Origin of Blue and White Porcelain)'. *Wenwu* 文物, no. 11 (2012): 79-88

Ingold, Tim. *Being Alive: Essays on Movement, Knowledge, and Description*. London; New York: Routledge, 2011.

Ji Huang 嵇璜 et al. *Qing chao Tongdian* 清朝通典 (Qing Dynasty Comprehensive Canon of Laws and Institutions). Reprint. Taipei: Taiwan shangwu yinshuguan, 1987.

Jia Yinghua 賈英華. *Modai taijian Sun Tingyao* 末代太監孫耀庭 (The Last Eunuch Sun Yaoting). Hong Kong: Zhonghua shuju, 2014.

Jiang Yasha 姜亞沙 et al., eds. *Qingdai gongyuan zeli huibian* 清代宮園則例彙編 (Collection of Qing Regulations and Precedents of Palaces and Gardens). Beijing: Quanguo tushu guan wenxian suowei fuzhi zhongxi, 2011.

Jianxi sheng kaogu yanjiu suo 江西省考古研究所 (The Archaeological Institute of Jiangxi Province) and Jingdezhen minyao bowuguan 景德鎮民窯博物館 (Museum of Popular Kilns in Jingdezhen), eds. *Jingdezhen Hutian yao zhi 1988-1999 nian kaogu fajue baogao* 景德鎮湖田窯址 1988-1999年考古發掘報告 (Report of Archaeological Excavation on the Hutian Kiln Site in Jingdezhen 1988-1999). Beijing: Wenwu chubanshe, 2007.

Jin Feng 金峯. 'Qingdai Xinjiang xilu taizhan (er) 清代新疆西路台站（二）(Stations in the Western Circuits of Xinjiang, 2)'. *Xinjiang daxue xuebao* 新疆大學學報, no. 2 (1980): 93-102.

Jin Feng 金峯. 'Qingdai Xinjiang xilu taizhan (yi) 清代新疆西路台站（一）(Stations in the Western Circuits of Xinjiang, 1)'. *Xinjiang daxue xuebao*新疆大學學報, no. 1 (1980): 60-73.

Jin Jian 金簡. *Wuying dian juzhen ban chengshi* 武英殿聚珍版程式 (Manual of Moveable Type Prints in the Wuying Palace). *Siku quanshu* 四庫全書 edition, *shibu* 史部 (History section), vol. 673. Reprint. Shanghai: Shanghai guji chubanshen, 2003.

Jin Yi 金易 and Shen Yiling 沈義羚. *Gongnü tangwang lu – Chuxiugong li suishi Cixi banian* 宮女談往錄宮女談往錄—儲秀宮裡隨侍慈禧八年 (Memoirs of a palace maid – service for Cixi during eight years at the Palace of Accumulated Beauty). Beijing: Zijincheng chubanshe, 1991.

Jin Yi. *Mémoires d'une dame de cour dans la Cité interdite* (He Rong'er 何榮兒). Translated and abridged by Dong Qiang. Arles: Philippe Picquier, 1996.

Jiuqing yiding wuliao jiazhi 九卿議定物料價值 (The Price of Materials Agreed Upon by Nine Ministers). Beijing, 1736. Reproduced in *Gugong zhenben congkan* 故宮珍本叢刊, vol. 317. Haikou: Hainan chubanshe, 2000.

Kerr, Rose and Nigel Wood. *Science and Civilization in China*, vol. V: *Chemistry and Chemical Technology*, part XII, *Ceramic Technology*. Cambridge: Cambridge University Press, 1980.

Kong Yingda 孔穎達. *Chunqiu zuozhuan zhengyi* 春秋左傳正義 (Rectified Meaning of Commentary of Zuo to Spring and Autumn Annals). Beijing: Wuying dian 武英殿 edition, 1739.

Kuhn, Philip. *Soulstealers: The Chinese Sorcery Scare of 1768*. Cambridge, MA: Harvard University Press, 1990.

Kungang 崑岡 et al. *Da Qing huidian shili* (Guangxu ban) 大清會典事例 (光緒版) (Precedent Cases of Statutes of the Great Qing, Guangxu edition), 1899. Reprint. Taipei: Chongwen shuju, 1963, or Beijing: Zhonghua shuju, 1991, or in *Xuxiu Siku quanshu* 續修四庫全書, vol. 798-814 (Shanghai: Shanghai guji chubanshe, 1995-2002), or in *Qingdai wubu huidian*.

Kungang 崑岡 et al. *Da Qing huidian tu* 大清會典圖 (Illustrations to the Statutes of the Great Qing), 1899. Manuscript version. Staatsbibliothek zu Berlin, Digitale Sammlungen. Accessed August 2020. http://resolver.staatsbibliothek-berlin.de/SBB000258D300000000.

Lai Hui-min 賴惠敏 (Lai Huimin). 'Cong Gao-pu an kan Qianlong chao de Neiwufu yu shangren 從高樸案看乾隆朝的內務府與商人 (Study on the Imperial Household Department and Merchants through Gao-pu's Criminal Case)'. *Xin shixue* 新史學 = *New History*, no. 13.1 (2002): 71-133.

Lai Hui-min 賴惠敏 (Lai Huimin). 'Tiegan zhuangjia: Qingmo Neiwufu xinzheku ren de jiahu yu shengji 鐵杆莊稼—清末內務府辛者庫人的家戶與生計 (An Iron Rice Bowl: The Households and Livelihoods of Sin Jeku Jetere Aha in the Imperial Household Department in Late Imperial China)'. *Zhongyang yanjiuyuan Jindaishi yanjiusuo jikan* 中央研究院近代史研究所集刊 = *Bulletin of the Institute of Modern History Academia Sinica*, no. 38 (Dec. 2002): 71-128.

Lai Hui-min 賴惠敏 (Lai Huimin). *Qianlong huangdi de hebao* 乾隆皇帝的荷包 (Emperor Qianlong's Purse). Taipei: Zhongyang yanjiuyuan jindaishi yanjiusuo, 2014.

Lai Hui-min 賴惠敏 (Lai Huimin). *Qingdai de huangquan yu shijia* 清代的皇權與世家 (Clans and Imperial Power of the Qing Dynasty). Beijing: Beijing daxue chubanshe, 2010.

Lam, Peter. 'Tang Ying (1682-1756): The Imperial Factory Superintendent at Jingdezhen'. *Transactions of the Oriental Ceramics Society*, vol. 63 (1998/1999): 65-82.

Lang Ying 郎瑛 (1487-1566). *Qixiu leigao* 七修類稿 (Compilation of Draft of Nine Kinds [of Knowledge]). Shanghai: shanghai shudian chubanshe, 2009.

Laohu 老虎. 'Canchikou jiaotang chaichu shijian kao 蚕池口教堂拆除时间考 (Investigation of the Date of the Demolition of the Church at Canchikou)'. Accessed September 2020. http://blog.sina.com.cn/s/blog_5ea5774b0100c1ea.html.

Latour, Bruno. *Reassembling the Social: An Introduction to Actor-Network-Theory.* Oxford; New York: Oxford University Press, 2005.

Lei Bingyan 雷炳炎. *Qingdai shehui baqi guizu zhijia shili yanjiu* 清代社會八旗貴族世家勢力研究 (Study of the Bannerman Aristocratic Clans in the Qing Society). Beijing: Zhongguo shehui kexue chubanshe, 2016.

Lettres édifiantes et curieuses écrites des missions étrangères. Toulouse: Sens et Gaude, 1811.

Li Dou 李斗. *Yangzhou huafang lu* 揚州畫舫錄 (The Pleasure Boats of Yangzhou). Annotated edition by Wang Beiping 汪北平 and Tu Yugong 涂雨公. Beijing: Zhonghua shuju, 1961.

Li Ji 李季, Wang Guangyao 王光尧, Dai Luoyuan 冀洛源, Dan Yingying 单莹莹, Wang Jin 王瑨, Zhao Jin 趙瑾 et al. 'Gugong bowuyuan Nan Daku cipian maicang fajue jianbao' 故宮博物院南大庫瓷片埋藏坑發掘簡報 (A Brief Excavation Report on the Porcelain Pit in the Southern Storehouse of the Palace Museum), *Gugong bowuyuan yuankan* 故宮博物院院刊, no. 5 (2016): 6-25.

Li Yuhang and Harriet Zurndorfer. 'Rethinking Empress Dowager Cixi through the Production of Art'. *Nan Nü*, 14/1 (2012): 1-20.

Liang Xudong 梁旭東. 'Zhongguo chuantong de liujin jishu 中國傳統的鎏金技術 (Traditional Fire Gilding in China)'. *Cailiao baohu* 材料保護 = *Material Protection*, vol. 23, no. 1/2 (1990): 83-86.

Lin Jun. 'Chinese Double-Entry Bookkeeping before the Nineteenth Century'. *The Accounting Historians Journal*, vol. 19, no. 2 (Dec. 1992): 103-122.

Lin Lu 林璐. 'Xiang ji 象記 (On Elephants)'. In *Yu chu xin zhi* 虞初新志 (New Anecdotes of Yu Chu), edited by Zhang Chao 張潮. 1700 edition. Reproduced in *Xuxiu Siku quanshu* 續修四庫全書, vol. 1783. Shanghai: Shanghai guji chubanshe, 1995-2002.

Ling Haicheng 凌海成, Yu Binhua 余斌華 et al. *Zui hou yige taijian* 最後一個太監 (The Last Eunuch). Changchun: Jinlin wenshi chubanshe, 1991.

Liu Duying 劉杜英. 'Qing di qingeng Fengzeyuan 清帝亲耕豐澤園 (Qing Emperor's [Ritual] Farming in the Fengze Garden'. *Zijincheng* 紫禁城, no. 133 (2005): 124-27.

Liu Jincao 劉錦藻, ed. *Huangchao xu wenxian tongkao* 皇朝續文獻通考 (Qing Dynasty's Continued Comprehensive Examination of Literature). Minguo edition. Reproduced in *Xuxiu Siku quanshu* 續修四庫全書, vol. 815-21. Shanghai: Shanghai guji chubanshe, 1995-2002.

Liu Wanhang 劉萬航. *Jinyin zhuangshi yishu* 金銀裝飾藝術. Taipei: Xingzhenyuan wenhua jianshe weiyuanhui, 1989.

Liu Wenpeng 劉文鵬. 'Lun Qingdai Xinjiang taizhan tixi de xingshuai' 論清代新疆台站體系的興衰 (The Rise and Fall of the Station System in Xinjiang during the Qing Dynasty). *Xiyu yanjiu* 西域研究, no. 4 (2001): 29-38.

Liu Wenpeng 劉文鵬. *Qing dai yichuan ji qi yu jiangyu xingcheng guanxi zhi yanjiu* 清代驛傳及其與疆域形成關係之研究 (Study on the Relation between the Formation of the Qing Territory and Its Official Relay System). Beijing: Zhongguo renmin daxue chubanshe, 2004.

Liu Xiaomeng 劉小萌. 'Guanyu qingdai neiwufu shijia 關於清代內務府世家 (On the Clans in the Imperial Household Department of the Qing Dynasty)'. In *Ming Qing shi luncong* 一 Sun Wenliang jiaoshou danchen qishi zhou nian jinian wenji 明清史論叢 – 孫文良教授誕辰七十周年紀念文集, 138-158. Shengyang: Liaoning daxue chubanshe 2004.

Liu Xiaomeng 劉小萌. *Manzu cong buluo dao guojia* 滿族從部落到國家 (From Tribe to State: An Early History of the Manchus). Beijing: Zhongguo shehui kexue chubanshe, 2007.

Liu Yan 劉燕. 'Qing mo Beijing xiyuan de yuyong tielu 清末北京西苑的御用鐵路 (The Imperial Railroad in the Westpark in Beijing in Late Qing)'. *Beijing dang'an* 北京檔案, no. 3 (2003): 56-57.

Liu Zhiwei 劉志偉. *Gongfu tizhi yu shichang: Ming-Qing shehui jingji shilun gao* 貢賦體制與市場: 明清社會經濟史論稿 (The Tribute-Tax System and the Market: Essays on Socio-Economic History of the Ming-Qing). Beijing: Zhonghua shuju, 2019.

Luo Wenhua 羅文華. *Longpao yu jiasha — Qing gong zangchuan fojiao wenhua kaocha* 龍袍與袈裟—清宮藏傳佛教文化考察 (Dragon Robes and Monks' Robes: A Study of Tibetan Buddhist Culture in the Qing Palace). Beijing: Zijincheng chubanshe, 2005.

Mao Xianmin 毛憲民. 'Qingdai Lunyiwei xunxiangsuo yang xiang 清代鑾儀衛馴象所養象 (Qing Dynasty's Imperial Guardsmen and Imperial Procession Guardsmen and Elephant Training)'. *Zijincheng* 紫禁城, no. 3 (1991): 16-21.

McElderry, Andrea. 'Frontier Commerce: An Incident of Smuggling'. *The American Asian Review*, no. 5 (1987): 47-82.

Meeker, Michael E. *A Nation of Empire. The Ottoman Legacy of Turkish Modernity*. Berkeley: University of California Press, 2001.

Milhauser, Kathy. 'Organizational Metaphors'. PPT presentation on YouTube. Last modified 24 Feb. 2014; accessed 28 Aug. 2020. https://www.youtube.com/watch?v=Xt_1qCHH_Qk.

Miller, Katherine. *Organizational Communication, Approaches and Processes*. Belmont, CA: Wadsworth, 2006.

Ming shilu 明實錄 (Veritable Record of the Ming). Reprint. Taipei: Zhongyang yanjiuyuan Lishi yuyan yanjiusuo, 1966.

Moll-Murata, Christine, Gijs Kessler, and Karin Hofmeester, eds. *Conquerors, Employers, and Arbiters: States and Shifts in Labour Relations 1500-2000* (*International Review of Social History*, vol. 61, Special issue no. 24). Cambridge: Cambridge University Press, 2016.

Moll-Murata, Christine. 'Tributary Labour Relations in China During the Ming-Qing Transition (Seventeenth to Eighteenth Centuries)'. *International Review of Social History*, vol. 61, Special issue no. 24 (2016): 27-48.

Moll-Murata, Christine. *State and Crafts in the Qing Dynasty*. Amsterdam: Amsterdam University Press, 2018.

Morse, Hosea Ballou. *The Chronicles of the East India Company, Trading to China 1635-1834*. Oxford: The Clarendon press, 1926-1929.

Nappi, Carla. *The Monkey and the Inkpot: Natural History and Its Transformations in Early Modern China*. Cambridge, MA: Harvard University Press, 2009.

Naquin, Susan. *Peking: Temples and City Life, 1400-1900*. Berkeley: University of California Press, 2000.

Needham, Joseph, Ho Ping-Yü, Lu Gwei-Djen, and Nathan Sivin. *Science and Civilization in China*, vol. V: Chemistry and Chemical Technology, part IV, *Spagyrical Discovery and Invention: Apparatus, Theories and Gifts*. Cambridge: Cambridge University Press, 1980.

Newby, L. J. 'The Begs of Xinjiang: between Two Worlds'. *Bulletin of the School of Oriental and African Studies*, vol. 61, no. 2 (1998): 278-297.

Nian Xiyao 年希堯. *Mianti bili bianlan* 面體比例便覽 (Handbook for Calculating Area and Volume). In *Gujin suanxue congshu* 古今算學叢書 (Book Series of Ancient and Modern Mathematics), Third collection. Edited by Liu Duo 劉鐸, vol. 429. Shanghai: Suanxue shuju, 1898.

Norman, Jerry. *A Concise Manchu-English Lexicon*. Seattle: University of Washington Press, 1978.

Ogawa, Kazuma. *Shinkoku Pekin kōjō shashinchō* 清國北京皇城寫眞帖 (Photography of the Imperial City of Beijing in the Qing Dynasty). Tōkyō, 1906. Accessed August 2020. http://pds.lib.harvard.edu/pds/view/46832640.

Pan Jie 潘潔. *Heishuicheng chutu qianliang wenshu zhuanti yanjiu* 黑水城出土錢糧文書專題研究 (Study of the Documents on Treasury Excavated from the Kharahoto Site). Yinchuan: Ningxia renmin chubanshe, 2013.

Park Ji-Won 朴趾源. *Rehe riji* 熱河日記 (Rehe Diary). Reprint. Shanghai: Shanghai shudian, 1997.

Perdue, Peter. *China Marches West: The Qing Conquest of Central Eurasia*. Cambridge, MA: Harvard University Press, 2005.

Qi Meiqin 祁美琴 and Cui Can 崔灿. 'Baoyi shenfen zaibian 包衣身份再辨 (The status of booi reconsidered)'. *Qingshi yanjiu* 清史研究, no. 1 (2013): 117-128.

Qi Meiqin 祁美琴. *Qingdai neiwufu* 清代內務府 (The Qing Imperial Household Department). Beijing: Zhongguo renmin daxue chubanshe, 1998.

Qian Shuoyou 潛說友. *Xianchun lin'an zhi* 咸淳臨安志 (The Gazetteer of Lin'an during the Xianchun Reign [1265-1272]). In *Song-Yuan fangzhi congkan* 宋元方志叢刊 (Series of Song-Yuan Gazetteer), vol. 4. Beijing: Zhonghua shuju, 1990.

Qianlong chao Neiwufu yinku jinxiang yuezhe dang 乾隆朝內務府銀庫進項月摺檔 (Monthly receipts reports of the Imperial Household Department's Bullion Vault from the Qianlong reign). Beijing: First Historical Archive, Microfilm, 2001.

Qianlong chao Neiwufu yinku yongxiang yuezhe dang 乾隆朝內務府銀庫用項月摺檔 (Monthly expenditure reports of the Imperial Household Department's Bullion Vault from the Qianlong reign). Beijing: First Historical Archive.

Qianlong chao Neiwufu zouxiao dang 乾隆朝內務府奏銷檔 (Accounting Reports of the Imperial Household Department during Qianlong Reign). The First Historcial Archives of China, Beijing. See also *Qing gong Neiwufu zouxiao dang*.

Qianlong chao zhengban tanwu dang'an xuanbian 乾隆朝懲辦貪污檔案選編 (A collection of archival materials on corruption cases from the Qianlong reign). Edited by Zhongguo diyi lishi dang'an guan 中國第一歷史檔案館. Beijing: Zhonghua shuju, 1994.

Qinding Gongzhong xianxing zeli di er zhong 欽定宮中現行則例第二種 (Current Regulations and Precedents of the Court, Imperially Commissioned, second Version). Compiled around 1883. Reproduced in *Gugong zhenben congkan* 故宮珍本叢刊, vol. 280. Haikou: Hainan chubanshe 2000.

Qinding Hubu zeli 欽定戶部則例 (Regulations and Precedents of the Board of Revenue). Reproduced in *Gugong zhenben congkan* 故宮珍本叢刊, vol. 284-286. Haikou: Hainan chubanshe 2000.

Qinding Zongguan Neiwufu xianxing zeli (yi) 欽定總管內務府現行則例 (一) (Current Regulations and Precedents for the Central Administration of the Imperial Household Department, first [version]). Reprint. Hongkong: Fuchi shuyuan chuban gongsi, 2014.

Qinding Zongguan Neiwufu xianxing zeli er zhong 欽定總管內務府現行則例二種 (Current Regulations and Precedents for the Central Administration of the Imperial Household Department, Imperially Commissioned, in two versions). Reproduced in *Gugong zhenben congkan* 故宮珍本叢刊, vol. 306-310. Haikou: Hainan chubanshe, 2000.

Qing shilu 清實錄 (Veritable Record of the Qing). Reprint. Beijing: Zhonghua shuju, 1986.

Qingdai Neifu keshu mulu jieti 清代內府刻書目錄解題目 (Annotated Bibliography of the Block-Print Books Issued by the Qing dynasty Imperial Household Department). Edited by Gugong bowuyuan tushuguan 故宮博物院圖書館 and Liaoning sheng tushuguan 遼寧省圖書館. Beijing: Zijincheng chubanshe, 1995.

Qingdai neige daku sanyi manwen dang'an xuanbian 清代內閣大庫散佚滿文檔案選編 (Selection of Loosen Manchu Archival Records from the Imperial Archive of the Qing Dynasty). Edited by Liaoning shehui kexueyuan lishi yanjiusuo 辽宁社会科学院历史研究所 et al. Tianjin: Tianjin guji chubanshe, 1991.

Qingdai wubu huidian 大清五部會典 (The Five Collected Statutes [and their Precedent Cases *Shili* 事例 and Illustrations *Tu* 圖] of the Qing [i.e. Kangxi, Yongzheng, Qianlong, Jiaqing, Guangxu reigns]). Wang Guangyue 王光越 et al., eds. Beijing:

Shutongwen Guji shujuku 書同文古籍數據庫, 2007 (https://guji.unihan.com.cn/). Accessed February 2020 via CrossAsia.org. http://erf.sbb.spk-berlin.de/han/qhd.

Qing gong Neiwufu Zaobanchu dang'an zonghui 清宮內務府造辦處檔案總匯 (Archives from the Imperial Workshops of the Imperial Household Department). Edited by First Historical Archive, Beijing, and Chinese University of Hong Kong Art Museum. Beijing: Renmin chubanshe, 2005.

Qing gong Neiwufu zouxiao dang 清宮內務府奏銷檔 (Accounting Reports of the Imperial Household Department during Qianlong Reign). Edited by Zhongguo diyi lishi dang'an guan 中國第一歷史檔案館. Beijing: Gugong chubanshe, 2014.

Qiu Fengjia 丘逢甲. *Lingyun hairilou shichao* 嶺雲海日樓詩鈔 (Poetry Anthology from the Pavilion of Clouds Behind Peaks and Sun on the Sea). 1937 edition. Reproduced in *Xuxiu Siku quanshu* 續修四庫全書, vol. 1576. Shanghai: Shanghai guji chubanshe, 1995-2002.

Raunig, Gerald. 'Abstrakte Maschinen'. In *Maschinentheorien/Theoriemaschinen*, edited by Hans-Christian von Herrmann und Wladimir Velminski, 121-132. Bern: Peter Lang, 2012.

Rawski, Evelyn S. *Education and Popular Literacy in Ch'ing China*. Ann Arbor, MI: University of Michigan Press, 1979.

Rawski, Evelyn S. *The Last Emperors: A Social History of Qing Institutions*. Berkeley: University of California Press, 2001.

Reuleaux, Franz. *Kinematics of Machinery: Outline of A Theory of Machines*. Translated and edited by Alexander B.W. Kennedy. New York: Dover Publications, 1963.

Roy, Kaushik and Peter Lorge, eds. *Chinese and Indian Warfare – From the Classical Age to 1870*. London: Routledge, 2014.

Ruitenbeek, Klaas. *Discarding the Brush: Gao Qipei 高其佩 (1660-1734) and the Art of Chinese Finger Painting.* Amsterdam: Rijksmuseum, 1992.

Schlesinger, Jonathan. *A World Trimmed with Fur: Wild Things, Pristine Places, and the Natural Fringes of Qing Rule*. Stanford, CA: Stanford University Press, 2017.

Schorkowitz, Dittmar and Ning Chia, eds., *Managing Frontiers in Qing China: The Lifanyuan and Libu Revisited*. Leiden: Brill, 2017.

Scott, James C. *Seeing Like a State: How Certain Schemes to Improve the Human Condition Have Failed*. New Haven, CT: Yale University Press, 1998.

Scripta Sinica Database: *Hanji quanwen ziliao ku* 漢籍全文資料庫, edited by Zhongyang yanjiu yuan lishi yanjiu suo 中央研究院歷史语言研究所. Taiwan, 1984-. Accessed May 2015. http://hanchi.ihp.sinica.edu.tw.

Shen Shixing 申時行 et al. *Da Ming huidian* 大明會典 (Collected statutes of the Great Ming). 15th year of Wanli 萬曆 reign (1587) edition in 228 *juan*. Reproduced in *Xuxiu Siku quanshu* 續修四庫全書, vol. 789-782. Shanghai: Shanghai guji chubanshe, 1995-2002.

Shi, Zhihong. *Central Government Silver Treasury: Revenue, Expenditure and Inventory Statistics, Ca. 1667-1899*. Leiden; Boston: Brill, 2016.

Shichahai yanjiuhui 什刹海研究会, ed. *Shichahai zhi* 什刹海志 (Shichahai Gazetteer). Beijing: Beijing chubanshe, 2003.

'Shouyong diaopi deng xiang 收用貂皮等項 (Collecting and Using Sable Furs and the Like)'. In *Zongguan Neiwufu xianxing tiaoli: Guangchusi* 總管內務府現行條例：廣儲司, 52-3. *Jindai Zhongguo shiliao congkan* 近代中國史料叢刊. Taipei: Wenhai chubanshe, 1972.

Simondon, Gilbert. *On the Mode of Existence of Technical Objects*. Minneapolis, MN: Univocal Publishing, 2017.

Sirén, Osvald. *Imperial Palaces of Peking*. Paris: Van Oest, 1926.

Siu, Victoria Cha-Tsu. *Gardens of a Chinese emperor: imperial creations of the Qianlong Era, 1736-1796*. Bethlehem, PA.: Lehigh University Press, 2013.

Song Jianze and Christine Moll-Murata. 'Notes on Qing Dynasty "Handicraft Regulations and Precedents" (*jiangzuo zeli*), with a Special Focus on Regulations on Materials, Working Time, Prices, and Wages'. *Late Imperial China*, vol. 23, no.2 (Dec. 2002): 117-18.

Song Lian 宋濂 (1310-1381). *Yuan shi* 元史 (History of the Yuan Dynasty). Beijing: Zhonghua shuju, 1976.

Song Minqiu 宋敏求 (1019-1079), ed. *Tang da zhaoling ji* 唐大詔令集 (Edicts and Decrees from the Tang Dynasty). Reprint. Beijing: Zhonghua shuju, 2008.

Spence, Jonathan. *Ts'ao Yin and the K'ang-hsi Emperor: Bondservant and Master*. New Haven: Yale University Press, 1966.

Stent, George Carter. 'Chinese Eunuchs'. *Journal of the North China Branch of the Royal Asiatic Society*, New Series, no. 11 (1877): 143-184.

Su Tianjue 蘇天爵 (1294-1352). *Guochao wenlei* 國朝文類 (Writings of the Yuan). Reproduced in *Sibu congkan chubian* 四部叢刊初編. Shanghai: Shanghai shudian, 1989.

Sun, Yaoting and Ling Haicheng. *Der letzte Eunuch: Das Leben Sun Yaotings, letzter Eunuch des Kaisers Puyi, erzählt von ihm selbst*. Translated by Uwe Frankenhauser. Leipzig: Gustav Kiepenheuer, 1993.

Sweeten, Alan Richard. *China's Old Churches: The History, Architecture, and Legacy of Catholic Sacred Structures in Beijing, Tianjin, and Hebei Province*. Leiden: Brill, 2019.

Tang Ying 唐英. *Tang Ying dutao wen dang* 唐英督陶文檔 (Archival Records of Tang Ying's Supervision of Porcelain Production), edited by Zhang Faying 張發穎. Beijing: Xueyuan chubanshe, 2012.

Tang Ying 唐英. *Tang Ying quan ji* 唐英全集 (Complete Works of Tang Ying), edited by Zhang Faying 張發穎. Beijing: Xueyuan chubanshe, 2008.

Tie Yuan 鐵源 et al., eds., *Qing gong ciqi dang'an quanji* 清宮瓷器檔案全集 (Complete Archives on Porcelains in the Qing Palace). Beijing: Zhongguo huabao chubanshe, 2008.

Tilly, Chris and Charles Tilly, *Work under Capitalism.* Boulder, CO: Westview Press, 1998.

Torbert, Preston M. *The Ch'ing Imperial Household Department: A Study of Its Organisation and Principal Functions, 1662-1796.* Cambridge, MA: Harvard University Press, 1977.

Tsai, Shih-shan Henry. *The Eunuchs in the Ming Dynasty.* Albany, NY: State University of New York Press, 1996.

Walker, Jill. 'Jade: A Special Gemstone'. In *Jade*, edited by Roger Keverne, 19-41. Boston: Springer, 1991.

Wall, Kate Dieffenbach. 'The Machine Metaphor in Modern Organizations'. Video presentation on YouTube. Uploaded 20 Apr. 2012, accessed 10 January 2020, https://www.youtube.com/watch?v=k6OXf_A99KU.

Wan, Maggie. 'Building an Immortal Land: The Ming Jiajing Emperor's West Park'. *Asia Major* 22, No. 2 (2009): 65-99.

Wang Guangyao 王光堯. *Mingdai gongting taoci shi* 明代宮廷陶瓷史 (History of Ming Court Ceramics). Beijing: Gugong chubanshe, 2010.

Wang Guangyao 王光堯. *Mingdai gudai guanyao zhidu*中國古代官窯制度 (The Systems of Official Ceramic Manufacture in Premodern China). Beijing: Zijincheng chubanshe, 2004.

Wang Guangyao 王光堯. *Zhongguo gudai guanyao zhidu* (er) 中國古代官窯制度(二) (The Systems of Official Ceramic Manufacture in Premodern China, vol. 2). Beijing: Gugong chubanshe, 2017.

Wang Guangyao 王光堯. 'Qingdai cizhi jiliqi luelun' 清代瓷質祭禮器略論 (A Brief Discussion of Porcelain Body Ritual Vessels of the Qing Dynasty)'. *Gugong bowuyuan yuankan* 故宮博物院院刊, no. 2 (2003): 70-79.

Wang Puzi 王璞子, an. *Gongcheng zuofa zhushi* 工程作法註釋 (Annotated Methods of Making Architectonic Projects). Beijing: Zhongguo jianzhu gongye chubanshe, 1995.

Wang Qiming 王啟明. *Tianshan langdao: Qing dai Tianshan daolu jiaotong yu yichuan yanjiu* 天山廊道：清代天山道路交通與驛傳研究 (Pathway across the Tianshan Mountains: Study on the Transport and Official Relay System on Roads across the Tianshan Mountains during the Qing Dynasty). Xi'an: Shanxi shifan daxue chubanshe, 2016.

Wang Shih-Ming 王士銘. 'Ji shi guanyuan yeshi nucai: qianlong chao changlu yanzheng 既是官員也是奴才: 乾隆朝長蘆鹽政 (Either A Government Official or A Private Representative of the Emperor – A Study of the Supervisory Officers to the Chang-Lu Salt Administration)'. MA thesis, National Chi Nan University, 2007.

Wang Shixiang 王世襄 ed. *Qingdai jiangzuo zeli* 清代匠作則例 (Handicraft Regulations of the Qing Dynasty). Zhengzhou: Daxiang chubanshe, 2000.

Wang Yao-t'ing (Wang Yaoting) 王耀庭. 'Yuci huazhuangyuan Tang Dai 御賜畫狀元唐岱 (Imperial Champion Painter Tang Dai)'. *Gugong wenwu yukan* 故宮文物月刊, no. 392 (Nov. 2015): 18-29.

Wang Zhonghan 王鐘翰. *Qingshi xukao* 清史續考 (More Investigation on the Qing History). Taipei: Huashi chubanshe, 1993.

Wang Zongmu 王宗沐 (1524-1592) et al. *Jiangxi sheng dazhi* 江西省大志 (Great Gazetteer of Jiangxi Province). 1597 edition. Reproduced in *Nanjing tushuguan cang xijian fangzhi congkan* 南京圖書館藏稀見方志叢刊. Beijing: Guojia tushuguan chubanshe 2012.

Wang, Ching-ling. *Praying for Myriad Virtues*. Berlin: Staatliche Museen zu Berlin & Kunsthistorisches Institut in Florenz – Max-Planck-Institute, 2017.

Wang, Ke. 'Between the "Ummah" and "China": The Qing Dynasty's Rule over Xinjiang Uyghur Society'. *Journal of Intercultural Studies* (Kobe University), no. 48 (2017): 183-219.

Wanyan Chonghou 完顏崇厚. *Hecha nianpu* 鶴槎年譜 (Biography of Crane Barge [Wanggiyan Chonghou). Reproduced in *Beijing Tushuguan cang zhenben nianpu congkan* 北京圖書館藏珍本年譜叢刊, vol. 169. Beijing: Beijing tushuguan chubanshe, 1999.

Wanyan Linqing (Wanggiyan Linkin) 完顏麟慶. *Hongxue yinyuan tuji* 鴻雪姻緣圖記 (Illustrated Records of A Wild Swan's Trail). Private print. Yangzhou, 1847 (Reproduced in *Riben cang Zhongguo dili wenxian zhenben huikan* 日本藏中國地理文獻珍本彙刊, vol. 46-48. Chengdu: Bashu shudian, 2018).

Wanyan Linqing (Wanggiyan Linkin) 完顏麟慶. *Yunyintang zougao* 雲蔭堂奏稿 (Memorials from the Yunyin Hall). Mikrofilm. Beijing: Quanguo tushuguan wenxian suowei fuzhi zhongxin, 2005 (also reproduced in print in *Guojia tushuguan lishi dang'an wenxian congkan* 國家圖書館藏歷史檔案文獻叢刊. Beijing: idem, 2005).

Wei Qingyuan 韋慶遠. 'Qingdai Neiwufu de jiangyi he yuyong shougongye 清代內務府的匠役和御用手工業 (Labour Force of the Imperial Household Department and the Imperial Handicraft Industry)'. In idem. *Ming Qing shi xuxi* 明清史續析 (Further Investigation of Ming and Qing History), 359-78. Guangzhou: Guangdong renmin chubanshe, 2006.

Will, Pierre-Étienne. *Bureaucracy and Famine in Eighteenth-Century China*. Translated by Elborg Forster. Stanford, CA: Stanford University Press, 1990.

Williams, Samuel Wells. *The Chinese Commercial Guide, Containing Treaties, Tariffs, Regulations, Tables, etc., Useful in the Trade to China & Eastern Asia: With an Appendix of Sailing Directions for Those Seas and Coasts*. Reprint. Taipei: Ch'eng-Wen Publishing, 1966.

Wolloch, Nathaniel. *Nature in the History of Economic Thought: How natural resources became an economic concept*. Abingdon and New York: Routledge, 2017.

Wu Meifeng 吳美鳳. '"Chongwen men bianjia" – Qingdai Neiwufu bianjia wujian zhi duoshao '交崇文門變價' 清代內務府變價物件知多少 (Price of Goods Sold from the Qing Imperial Household Department According to 'Sale on Discount at the Chongwen Gate')'. *Zijincheng* 紫禁城, no. 12 (2017): 106-123.

Wu Qian 吴千. 'Cixi zhuanyong de "youwan" tielu 慈禧專用的 "遊玩" 鐵路 (Railroad for Play, Reserved for Cixi)'. *Wenshi bolan* 文史博覽, no. 11 (2013): 21-22.

Wu Shan 吳山, ed. *Zhongguo gongyi meishu dacidian* 中國工藝美術大辭典 (Complete Dictionary of Chinese Crafts). Nanjing: Jiangsu meishu chuban she, 1989.

Wu Yuankang 吳元康 and Chu Rongbang 儲榮邦. 'Liudu – zhongguo gudai faming de yizhong zai cailiao biaomian shang du jinshu de jishu 鎏鍍—中國古代發明的一種在材料表面上鍍金屬的技術 (Amalgam Coating: One of Metal Coating Technologies on Surfaces of Materials that Invented by Ancient Chinese)'. *Tuzhuang yu diandu* 塗裝與電鍍, no. 1 (2011): 3-9.

Wu Yuankang 吳元康 and Chu Rongbang 儲榮邦. 'Liudu – zhongguo gudai faming de yizhong zai cailiao biaomian shang du jinshu de jishu (xu wan) 鎏鍍—中國古代發明的一種在材料表面上鍍金屬的技術 (續完) (Amalgam Coating: A Technique Invented in Pre-Modern China to Gild Metal)'. *Tuzhuang yu diandu* 塗裝與電鍍, no. 3 (2011): 25-28.

Wu Yulian. 'Chimes of Empire: The Construction of Jade Instruments and Territory in Eighteenth-Century China'. *Late Imperial China* 40 (2019): 43-86.

Wu Yulian. *Luxurious Networks: Salt Merchants, Status, and Statecraft in Eighteenth-Century China*. Stanford, CA.: Stanford University Press, 2017.

Wu Zhaoqing 吳兆清, 'Qingdai Zaobanchu de jigou he jiangyi 清代造辦處的機構和匠役 (The Institution and Personnel of the Imperial Workshops of the Qing Dynasty)', *Lishi dang'an* 歷史檔案, no. 4 (1991): 79-89.

Wu Zhenyu 吳振棫 (1792-1870). *Huayiguan shi chao* 花宜館詩鈔 (Wu Zhenyu's Huayiguan Poem Collection). 1865 edition. Reproduced in *Xuxiu Siku quanshu* 續修四庫全書, vol. 1521. Shanghai: Shanghai guji chubanshe, 2003.

Xing Yongfu 邢永福, Chengde shi wenwuju 承德市文物局, and First Historical Archive of China eds. *Qing gong Rehe dang'an* 清宮熱河檔案 (Archives of the Qing Palace on Rehe). Beijing: Zhongguo dang'an chubanshe, 2003.

Xu Ke 徐珂. *Qingbai leichao* 清稗類鈔 (Assorted Collection of Miscellaneous Historical Materials of the Qing Dynasty). Beijing: Zhonghua shuju, 1984 (3rd printing 2003).

Xu Pu 徐溥, Li Dongyang 李東陽 et al. *Da Ming huidian* 大明會典 (Collected Statutes of the Great Ming). Zhengde 正德 reign (1506-1522) edition in 180 *juan*. *Siku quanshu* 四庫全書, vol. 617-18. Reprint. Shanghai: Shanghai guji chubanshe, 2003.

Xu Shichang 徐世昌, ed. *Wanqingyi shi hui* 晚晴簃詩匯 (Collection of Poetry from the Hut of the Late Sun). 1929 edition. Reproduced in *Xuxiu Siku quanshu* 續修四庫全書, vol. 1629-33. Shanghai: Shanghai guji chubanshe, 1995-2002.

Xu Song 徐松. *Xiyu shuidao ji* 西域水道記 (1819) (Waterways in the West Borderland). Beijing: Zhonghua shuju, 2005.

Yan Jiesheng 嚴介生. 'Yiheyuan Cixi zhuanyong tielu 頤和園慈禧專用鐵路 (Sightseeing Railway for Empress Dowager in Summer Palace)'. *Wenhua jiaoliu* 文化交流 = *Cultural Dialogue*, no. 1 (2010): 72-74.

Yanagisawa Koji 柳澤光治 and Jo Ryoku Kinzo 定力金藏; translated by Xisheng 西生. 'Dianqi dujin gongye 電器鍍金工業 (The Industry of Gilding Electronic Products)'. *Hebei gongshang yuebao* 河北工商月報, vol. 1, no. 9 (1929): 181-190.

Yang Naiji 楊乃濟. 'Tanchun li jia zhi suo ben 探春理家之所本 (What is 'Tangchun's Domestic Management' Based on)'. *Hongloumeng xuekan* 紅樓夢月刊, no. 4 (1981): 39-40.

Yang Naiji 楊乃濟. 'Xiyuan tielu yu Guangxu chunian de xiulu dazhan 西苑鐵路與光緒初年的修路大論戰 (Debate over the Construction of Railroad in Early Guangxu Reign and the Westpark Railroad)'. *Gugong bowuyuan yuankan* 故宮博物院院刊, no. 4 (1982): 84-94.

Yang Xiaoyan 楊小燕. *Beijing dongwuyuan zhi* 北京動物園志 (History of Beijing Zoo). Beijing: Zhongguo linye chubanshe, 2002.

Yang Yifan 楊一凡, ed. *Huang-Ming Zhishu* 皇明制書 (The System of the August Ming). Beijing: Shehui kexue wenxian chubanshe, 2013.

Yao Yuanzhi 姚元之. *Zhuyeting zaji* 竹葉亭雜記 (Miscellaneous Notes at Pavilion of Bamboo Leaves). 1893 edition. Reproduced in *Xuxiu Siku quanshu* 續修四庫全書, vol. 1139. Shanghai: Shanghai guji chubanshe, 1995-2002; or rep. Beijing: Zhonghua shuju, 1982.

Yao Zhiyin 姚之駰. *Yuan Ming shi leichao* 元明事類鈔 (Records of Yuan and Ming). *Sikuquanshu* 四庫全書 edition. Reprint. Taipei: Taiwan shangwu yinshuguan, 2008.

Yeh, Kao-shu (Ye Gaoshu) 葉高樹. 'Qingchao de Qixue yu Qiren de Fanyi Jiaoyu 清朝的旗學與旗人的繙譯教育 (The School System and Translation Education of the Bannermen in the Ch'ing Dynasty)'. *Taiwan Shida lishi xuebao* 臺灣師大歷史學報, no. 48 (2012): 129-49.

Yu Hui-chun 余慧君. 'Cong huangjia lingyou dao wanshengyuan – Da Qing diguo de dongwu shoucang yu shanshi 從皇家靈囿到萬生園——大清帝國的動物收藏與展示 (From Imperial Menageries to Public Zoological Garden: Captive

Wild Animals at the Qing Court)'. *Xin shixue* 新史學 = *New History*, vol. 29, no. 1 (March 2018): 1-57.

Yu Minzhong 于敏中. *Qinding rixia jiuwen kao* 欽定日下舊聞 (Imperial Examination of Old Documents). *Siku quanshu* 四庫全書 edition. Reprint. Taipei: Taiwan shangwu yinshuguan, 2008.

Yuan Kaizheng 袁凱錚. 'Shilun cangchuan fojiao tongfoxiang waibu tezheng yu qi zhizuo gongyi 試論藏傳佛教銅佛像外部特徵與其製作工藝 (On the Exterior Features of Tibetan Buddhist Statues in Copper and Its Production Technique)'. *Xibei minzu daxue xuebao* (Zhexue shehui kexue ban) 西北民族大學學報(哲學社會科學版), vol. 5 (2009): 82-89.

Yuan Kaizheng 袁凱錚. 'Shixi cangzu liangzhong chuantong zhuzao gongyi de cunzai---you chuantong tong foxiang zhizuo yinfa de sikao 試析藏族兩種傳統鑄造工藝的存在—由傳統銅佛像製作引發的思考 (Discussion on Two Tibetan Traditional Forging Techniques: Reflection on the Making of Traditional Copper Buddhist Statues)'. *Zhongguo cangxue* 中國藏學 (March 2012): 175-185.

Yuan Quan 袁泉. 'Lue lun Song-Yuan shiqi shougongye de jiaoliu yu hudong xianxiang — yi qiqi wei zhongxin 略論宋元時期手工業的交流與互動現象——以漆器為中心 (A Brief Discussion of the Exchange and Interaction between Handicraft Workers During the Song and the Yuan Dynasties – with A Focus on Lacquerware)'. *Wenwu*文物, no. 11 (2013): 63-73.

Yunlu 允祿 et al. *Da Qing huidian* (Yongzheng ban) 大清會典事例 (雍正版) (Statutes of the Qing Dynasty, Yongzheng edition). In *Jindai Zhongguo shiliao congkan sanbian* 近代中國史料叢刊三編. Taipei: Wenhai chubanshe, 1985, or in *Qingdai wubu huidian*.

Yunlu 允祿 et al. *Da Qing huidian shili* (Yongzheng ban) 大清會典事例 (雍正版) (Precedent Cases of Statutes of the Qing Dynasty, Yongzheng edition). In *Jindai Zhongguo shiliao congkan sanbian* 近代中國史料叢刊三編. Taipei: Wenhai chubanshe, 1985, or in *Qingdai wubu huidian*.

Zhang Bao 張寶. *Xu Fancha tu san ji* 續泛槎圖三集 (Sequel to the Illustrated Journey, Third Collection). Shanghai: Dianshizhai, 1831.

Zhang Chao 張潮, ed. *Yu chu xin zhi* 虞初新志 (New Anecdotes of Yu Chu). 1700 edition. Reproduced in *Xuxiu Siku quanshu* 續修四庫全書, vol. 1783. Shanghai: Shanghai guji chubanshe, 1995-2002.

Zhang Jianhong 張劍虹. 'Cong Gao-pu an kan Qingdai yushi kuangchan zhengce 從高樸案看清代玉石礦產政策 (Study on the Policy Concerning Jadeite Mines in the Qing Dynasty through Gao-pu's Criminal Case)'. *Zhongguo kuangye* 中國礦業, vol. 25, no. 10 (2016): 168-71.

Zhang Tingyu 張廷玉 et al. *Mingshi* 明史 (History of the Ming Dynasty). Beijing: Wuying dian 武英殿, 1739. Reprint. Taipei: Dingwen shuju, 1980.

Zhang Wenhui 張文惠. *Zui hou yige taijian* 最後一個太監 (The Last Eunuch). Hangzhou: Zhejiang shaonian ertong chubanshe, 1984.

Zhang Yajing 張雅晶. 'Qingdai wanyan shi hesu shiji kao 清代完顏氏和素事蹟考 (Examination of Hesu of the Wanggiyan Clan from the Qing Dynasty)'. In *Manxue luncong diyi ji* 滿學論叢第一輯, edited by Zhao Zhiqiang 趙志強, 254-65. Shenyang: Liaoning minzu chubanshe, 2011.

Zhang Zhaoping 張兆平. 'Kangxi chao zhuming manwen fanyijia Hesu 康熙朝著名滿文翻譯家和素 (He Su, A Renowned Manchu Translator of the Kangxi Reign)'. *Gugong bowuyuan yuankan* 故宮博物院院刊, vol. 168, no.4 (2013): 143-150.

Zhao Bing 趙冰. 'Administration, Savoir Techniques et Reconnaissance Impériale: Tang Ying 唐英 (1682-1756)'. *Revue de Synthèse* 131, no. 3 (2010): 307-341.

Zhao Binwen 趙炳文. 'Shi lunshu Qingdai fanyi shijia Hesu 試論述清代繙譯世家和素 (Preliminary Discussion of Hesu from An Eminent Lineage of Translators of the Qing Dynasty)'. In *Qingdai dangwen yu qinggong wenhua – diyiqi Qinggongshi yantaohui lunwen ji* 清代檔案與清宮文化—第九屆清宮史研討會論文集 = *Proceeding of the 9th Conference of the Qing Court History: Archive and Culture*, 489-99. Beijing: Zhongguo dang'an chubanshe, 2010.

Zhao Erxun 趙爾巽 et al. *Qingshi gao* 清史稿 (Draft for History of the Qing Dynasty). Reprint. Beijing: Zhonghua shuju, 1977.

Zhao Shi'an 趙世安. *Renhe xianzhi* 仁和縣志 (Local Gazetteer of Renhe County). Printed Kangxi 26 (1687). Reproduced in *Zhongguo difangzhi jicheng: Zhejiang fuxianzhi ji* 中國地方志集成: 浙江府縣志輯, vol. 5. Nanjing: Jiangsu guji chubanshe, 1993 (also see the digital copy at Harvard-Yenching Library. Accessed February 2020. http://nrs.harvard.edu/urn-3:FHCL:14361176).

Zhongguo zijincheng xuehui lunwen ji 中國紫禁城學會論文集 (Proceedings on the Conference of the Zijincheng Society of China), 2nd collection. Edited by 于倬云, Zhu Chengru 朱誠如 and Zhongguo Zijincheng xuehui 中國紫禁城學會. Beijing: Zijincheng chubanshe, 2002.

Zhongyang yanjiuyuan lishi yuyan yanjiusuo 中央研究院歷史語言研究所, ed. *Ming-Qing shiliao*明清史料 (Collected Historical Materials of Ming and Qing Dynasties). Taipei: Zhongyang yanjiuyuan lishi yuyan yanjiusuo, 1957-1958.

Zhu Jinfu 朱金甫. 'Lun Kangxi shiqi de Nanshufang 論康熙時期的南書房 (On the Southern Study during the Kangxi Period)'. *Gugong bowuyuan yuankan* 故宮博物院院刊 = *Palace Museum Journal*, no. 2 (1990): 27-38.

Zito, Angela. *Of Body and Brush: Grand Sacrifice as Text/Performance in Eighteenth-Century China*. Chicago: University of Chicago Press, 1997.

Zongguan Neiwufu xianxing tiaoli: Guangchusi. See Fujun et al., *Zongguan Neiwufu xianxing tiaoli: Guangchusi*.

Archival Collections and Newspapers

Dagong bao 大公報, edited by Ying Lianzhi 英斂之 (1867-1926) et al. Tianjin, 1902-1937.

Dianshizhai huabao 點石齋畫報, 1898 edition, Shanghai: [Shenbao guan]. 'Shanghai's Lens on the New(s): Image Gallery'. *Visualizing Cultures*. Massachusetts Institute of Technology, 2015. Accessed September 2020, https://visualizingcultures.mit.edu/dianshizhai/dsz_gal.html.

Dongfang zazhi 東方雜誌. Shanghai, Chongqing, 1904-1948. Reproduced in *Zhongwen qikan quanwen shuju ku* 中文期刊全文数据库, 1911-1949, edited by Quanguo baokan suoyin, Shanghai.

Gongzhong gechu dang'an 宮中各處檔案 (Documents of the Palace Archives). The First Historical Archives of China (Zhongguo Diyi lishi dang'an guan 中國第一歷史檔案館), Beijing.

Guanhua zhuyin zimu bao 官話注音字母報 (Official Language Newspaper). Beijing, 1916-1920. Reproduced *in Zhongwen qikan quanwen shuju ku* 中文期刊全文数据库, 1911-1949 (Database of Chinese Periodicals, 1911-1949). Edited by Quanguo baokan suoyin 全國報刊索引, Shanghai.

Huke tiben 戶科題本 (Memorials of the Board of Revenue). The First Historical Archives of China (Zhongguo Diyi lishi dang'an guan 中國第一歷史檔案館), Beijing.

Junjichu dang'an 軍機處檔案 (Documents of the Grand Council). The First Historical Archives of China (Zhongguo Diyi lishi dang'an guan 中國第一歷史檔案館), Beijing.

Junjichudang zhejian 軍機處檔摺件 (Archives of the Grand Council). Taiwan National Palace Museum (Taiwan Guoli Gugong bowuyuan 台湾國立故宮博物院), Taipei.

Junqichu shangyudang (*fang ben*) 軍機處上諭檔 (方本) (Grand Council record book of ordinary matters). The First Historical Archive of China of China (Zhongguo Diyi lishi dang'an guan 中國第一歷史檔案館), Beijing.

Manwen yuezhe bao 滿文月摺包 (Manchu Monthly Archive). The First Historical Archives of China (Zhongguo Diyi lishi dang'an guan 中國第一歷史檔案館), Beijing.

NWF, NWF-FCY and NWF *xinzheng* 新整 (Neiwufu Fengchengyuan 內務府奉宸苑, Archives of the Garden Bureau of the Imperial Household Department). The First Historical Archives of China (Zhongguo Diyi lishi dang'an guan 中國第一歷史檔案館), Beijing.

Qianlong chao zouxiao dang 乾隆朝奏銷檔案. Microfilm. Taipei: Academia Sinica (Zhongyang yanjiuyuan lishi yuanjiusuo 中央研究院歷史研究所).

Shenbao (*Shun-Pao*) 申報, edited by Ernest Major (1841-1908). Shanghai, 1872-1949.

Shengjing shibao 盛京時報. Shenyang/Mukden, 1906-1944.

Zhongyang yanjiuyuan lishi yuyan yanjiusuo cang Ming-Qing shiliao 中央研究院歷史語言研究所 (Ming-Qing historical materials stored at the Institute of History and Philology). Academia Sinica, Taipei.

Zhupi zouzhe 朱批奏折 (Imperial Memorials with Red Rescripts). The First Historical Archives of China (Zhongguo Diyi lishi dang'an guan 中國第一歷史檔案館), Beijing.

Zou'an 奏案 (Memorials). The First Historical Archives of China (Zhongguo Diyi lishi dang'an guan 中國第一歷史檔案館), Beijing.

Index

Locators in italic refer to charts, maps, tables, and figures. Thus *11*map 1b refers to map 1b on page 11.

For Product Safety Concerns and Information please contact our EU representative GPSR@taylorandfrancis.com Taylor & Francis Verlag GmbH, Kaufingerstraße 24, 80331 München, Germany